W9-BCC-569
LIBRARY
DISCARD
621.3815
E25H

Heathkit/Zenith Educational Systems

A Step-by-Step Introduction

Training in Computers & Electronics Series

Prentice-Hall, Inc. Englewood Cliffs, New Jersey 07632

Library of Congress Cataloging in Publication Data
Main entry under title:

Electronic test equipment.

(Training in computers & electronics series)
"A Spectrum book."
"Adapted from a larger work entitled Electronic test equipment"—T.p. verso.
Includes index.
1. Electronic instruments. 2. Electronic measurements. I. Heathkit/Zenith Educational Systems (Group) II. Series.
TK7878.4.E575 1983 621.3815'48 83-3032
ISBN 0-13-252205-5
ISBN 0-13-252197-0 (pbk.)

This book is available at a special discount when ordered in bulk quantities. Contact Prentice-Hall, Inc., General Publishing Division, Special Sales, Englewood Cliffs, N. J. 07632.

A SPECTRUM BOOK

10 9 8 7 6 5 4 3 2 1

Printed in the United States of America
Manufacturing buyer Patrick Mahoney

ISBN 0-13-252197-0 {PBK.}

ISBN 0-13-252205-5

Prentice-Hall International, Inc., *London*
Prentice-Hall of Australia Pty. Limited, *Sydney*
Prentice-Hall of Canada Inc., *Toronto*
Prentice-Hall of India Private Limited, *New Delhi*
Prentice-Hall of Japan, Inc., *Tokyo*
Prentice-Hall of Southeast Asia Pte. Ltd., *Singapore*
Whitehall Books Limited, *Wellington, New Zealand*
Editora Prentice-Hall do Brasil Ltda., *Rio de Janeiro*

CONTENTS

Unit 1

ANALOG METERS

INTRODUCTION

In this unit, we will discuss analog meters, the types of movements used and how those movements are used to measure voltage, current, and resistance. You will learn the operation of the volt-ohm-milliammeter (VOM) and the electronic volt-ohm-milliammeter (EVOM). You will find out about some of the problems with these types of meters and learn how to use them effectively in making electrical measurements.

THE METER MOVEMENT

d'Arsonval Movement

The heart of any analog meter is the meter movement. The most popular type of meter movement is the permanent-magnet, moving coil called the d'Arsonval movement.

CONSTRUCTION

Figure 1 shows the construction of the d'Arsonval meter movement. Several important parts are listed. We will discuss these parts in detail, starting with the permanent magnet.

Figure 2 shows the permanent magnet system. A horseshoe magnet produces the stationary magnet field. To concentrate the magnetic field in the area of the moving coil, pole pieces are added to the magnet. These are made of soft iron and have a very low reluctance. Consequently, the lines of flux tend to concentrate in this area as shown. Also, a stationary soft-iron core is placed between the pole pieces. Enough space is left between the pole pieces and the core so that the moving coil can rotate freely in this space. As you can see, the pole pieces and core restrict most of the flux to the area of the moving coil.

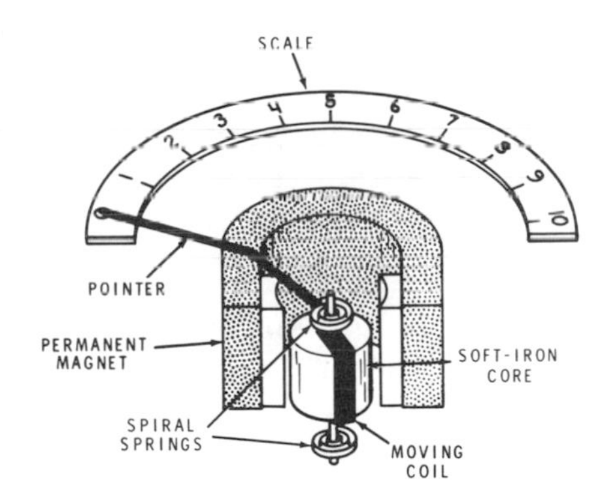

Figure 1

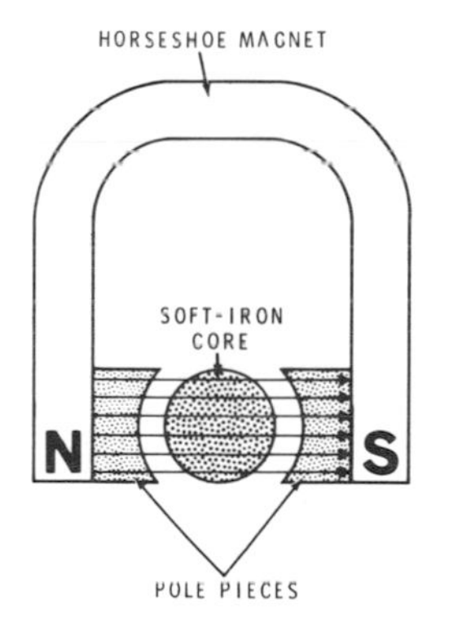

Figure 2

Figure 3 shows how the moving coil fits around the soft-iron core. The coil consists of many turns of extremely fine wire on an aluminum frame. The aluminum frame is very light so that little torque is needed to move it. The two ends of the coil connect to the leads of the ammeter, voltmeter, or ohmmeter.

Figure 4 shows the details of the pointer assembly. The pointer is attached to the moving coil so that it moves when the coil does. Counterweights are often attached to the pointer so that a perfect balance is achieved. This makes the pointer easier to move and helps the meter to read the same in all positions. A well balanced meter will read the same whether held vertically or horizontally. Retaining pins on either side of the movement limit the distance that the pointer and other rotating parts can move. Two spiral springs at opposite ends of the moving coil force the pointer back to the zero position when no current is flowing through the coil. In most movements, the spiral springs are also used to apply current to the moving coil. The two ends of the coil connect to the inner ends of the spiral spring. The outer end of the rear spring is fixed in place. However, the outer end of the front spring connects to a zero adjust screw. This allows you to set the pointer to exactly the zero point on the scale when no current is flowing through the coil.

The moving coil, pointer, and counterweight rotate around a pivot point. To hold the friction to an absolute minimum, jeweled bearings are used at this point just as they are in a fine watch.

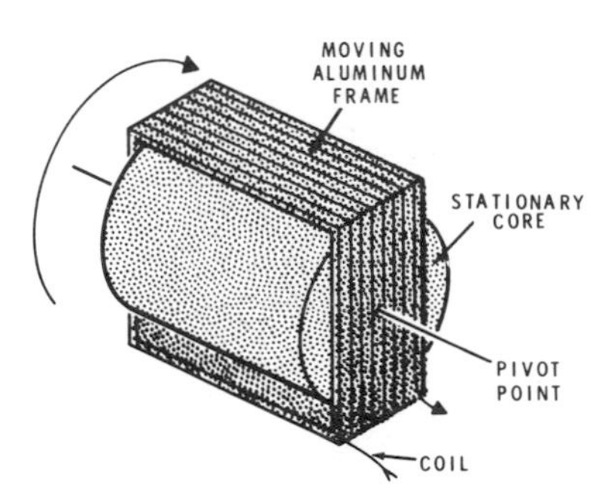

Figure 3

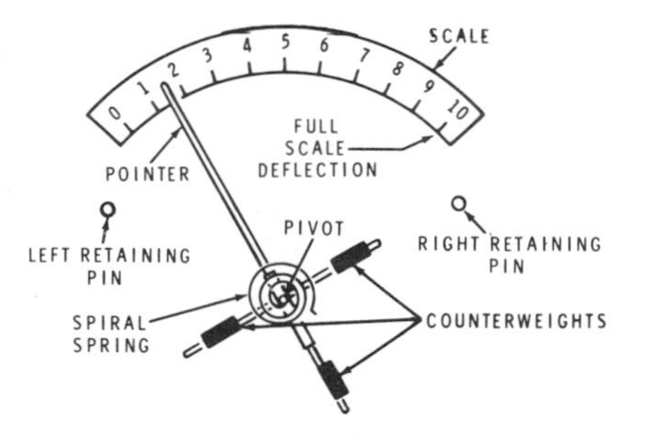

Figure 4

OPERATION

Now that you have an idea of how the meter is constructed, we will discuss how it operates. We know that a conductor is deflected at a right angle to a stationary magnetic field if current flows through the conductor. The right-hand motor rule describes this action. Figure 5 illustrates this rule and the motor action which causes the meter to deflect. An end view of one turn of the moving coil is shown. Current is forced to flow through the coil so that current flows "out of the page" on the left. Applying the right-hand rule to the coil at this point, we find that the coil is forced up on the left and down on the right. This forces the pointer to move up-scale or in the clockwise direction.

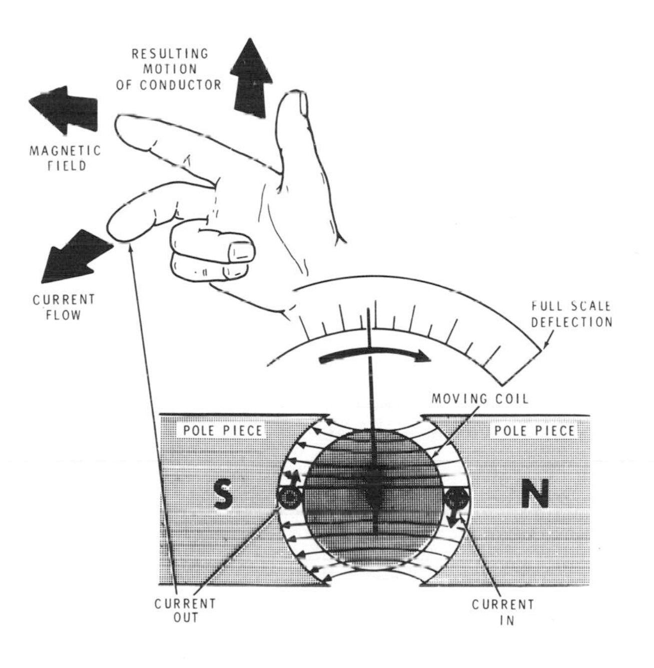

Figure 5

The amount of torque produced by this tiny "motor" is proportional to the magnitude of the current which flows through the moving coil. The more current, the greater the torque will be and the further the pointer will be deflected.

Meter movements are rated by the amount of current required to produce full-scale deflection. For example, a 50 μA (microampere) meter movement deflects full scale when only 50 μA of current flows through it. The 50 μA meter movement is one of the most commonly used d'Arsonval movements. Movements of 10 μA, 20 μA, 100 μA, and 200 μA are also popular.

If we apply an alternating current (AC) to the movement shown in Figure 5, each time the current reverses, the coil will attempt to reverse its direction of deflection. If the current changes direction more than a few times each second, the coil cannot follow the changes. Thus, AC must not be applied to this type of meter movement.

Taut-Band Movement

An important variation of the d'Arsonval movement is the taut-band meter movement. Figure 6 is a simplified diagram which shows the construction of this type of movement. The moving coil is suspended by two tiny stretched metal bands. One end of each band is connected to the moving coil while the other end is connected to a tension spring. The purpose of the springs is to keep the bands pulled tight. The bands replace the pivots, bearings, and spiral springs used in the conventional d'Arsonval movement. This not only simplifies the construction of the meter, it also reduces the friction to practically zero. Consequently, the taut-band movement can be made somewhat more sensitive than the movement discussed earlier. Taut-band instruments with 10 μA movements are available.

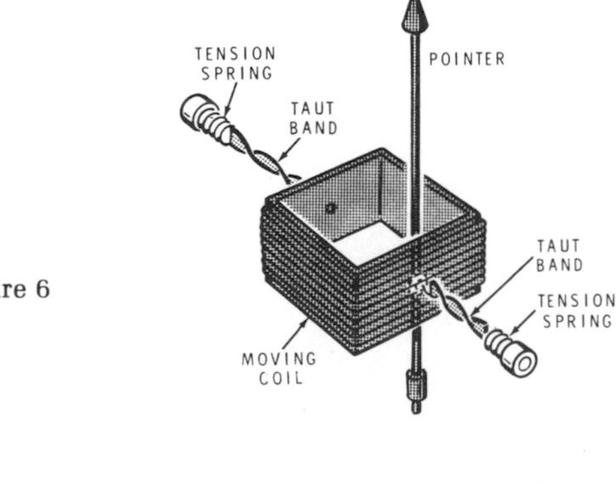

Figure 6

The taut-bands serve several purposes. First, they suspend the coil in such a way that the friction is nearly zero. When current is applied, the coil rotates and the bands are twisted. When current is removed, the bands untwist, returning the pointer to the zero position. The bands also serve as the current path to and from the coil.

The taut-band movement has several advantages over the original d'Arsonval type. As we have seen, it is generally more sensitive. It is also more rugged and durable. Mechanical shocks simply deflect the tension springs which can then bounce back to their original positions. The instrument remains more accurate for the same reason. Because of these advantages, the taut-band movement is becoming increasingly popular.

Electrodynamometer

A meter movement which is capable of measuring both AC and DC is the Dynamometer movement. It is capable of very accurate measurements at low frequencies. Figure 7 shows the construction of the electrodynamometer. It is similar to the d'Arsonval except that it has no permanent magnet and the magnetic field is produced by two field coils. When measuring current, all three coils are connected in series as shown in Figure 8. Since the current being measured must develop the magnetic

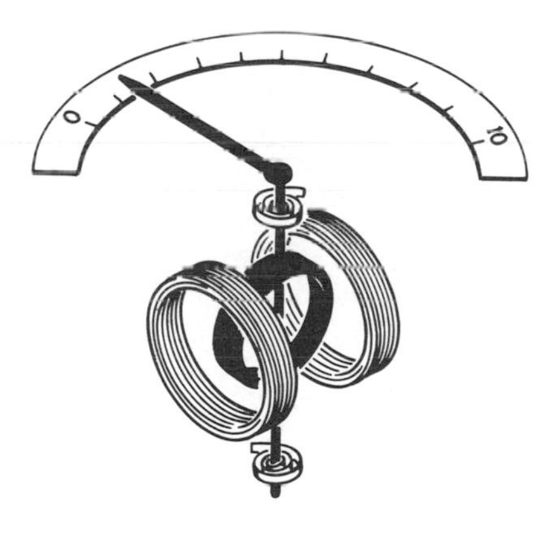

Figure 7

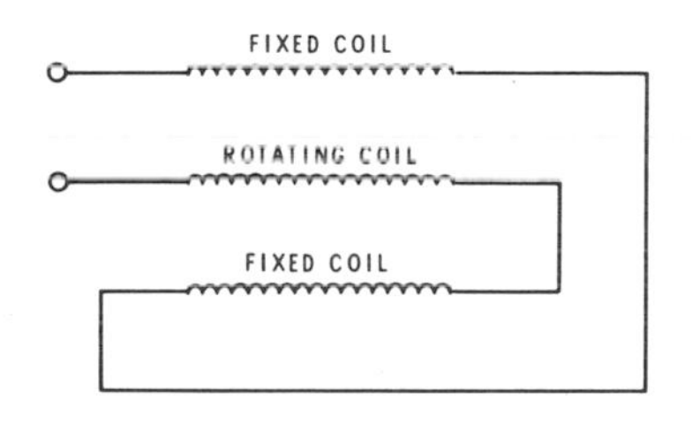

Figure 8

field, this movement is not as sensitive as the d'Arsonval. Figure 9A shows a cutaway of the three coils with the current flow shown by a ⊕ which means current flows "into the page" and a ⊙ which means current flows "out of the page." By applying the right-hand rule, we see that the moving coil would rotate in a clockwise direction. If we reverse the direction of current flow as shown in Figure 9B and apply the right-hand rule, we see that rotation is still clockwise.

We can see that an electrodynamometer movement responds properly to both AC and DC. The reading obtained will be a true rms current with any shape of input. This type of meter is most useful in audio and power circuits where adequate current is available without loading the circuit.

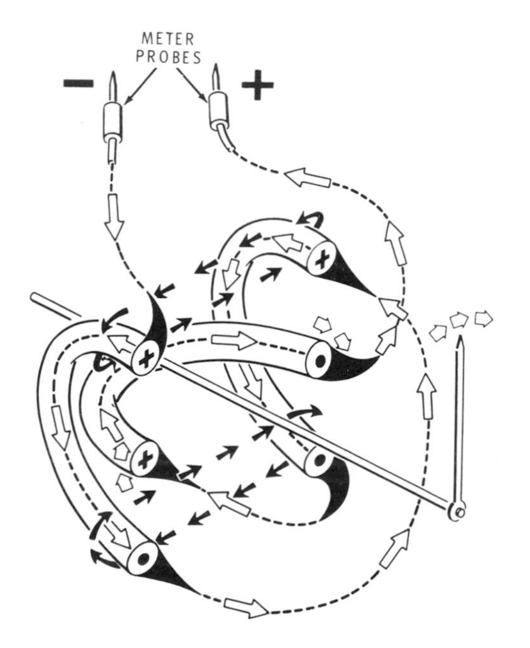

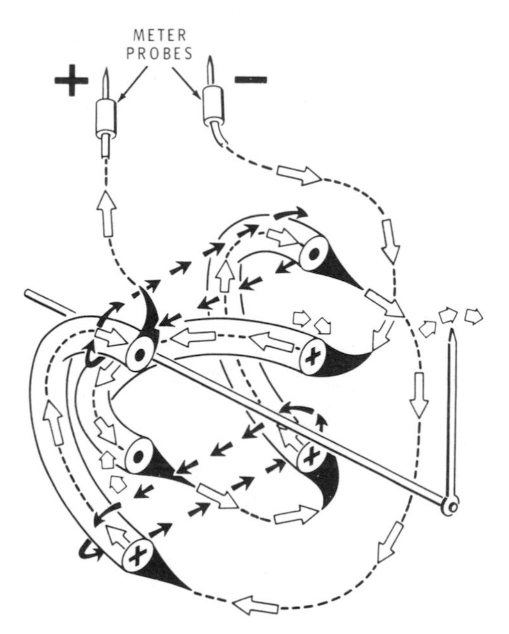

OPEN LARGE ARROW SHOWS DIRECTION OF CURRENT.

OPEN SMALL ARROW SHOWS DIRECTION OF ROTATION.

SOLID SMALL ARROW SHOWS DIRECTION OF MAGNETIC FIELD.

Figure 9A

Figure 9B

Iron Vane

Another meter movement which reads rms current is the moving iron or iron vane type. There are two methods of construction. Figure 10 shows the plunger type. The high permeability plunger will try to center itself in the coil when the current flows through the coil. The spring tries to hold the pointer at zero; therefore, the deflection is proportional to the current through the coil.

Radial-Vane

Another way this type of meter is built is shown in Figure 11. Here, two iron vanes are enclosed in a coil of wire. One of the vanes is stationary and the other, which has a pointer attached, is free to move. Current flowing through the coil magnetizes both vanes such that they repel each other; therefore, the amount of deflection is proportional to the current. This type of meter is called the radial-vane meter movement.

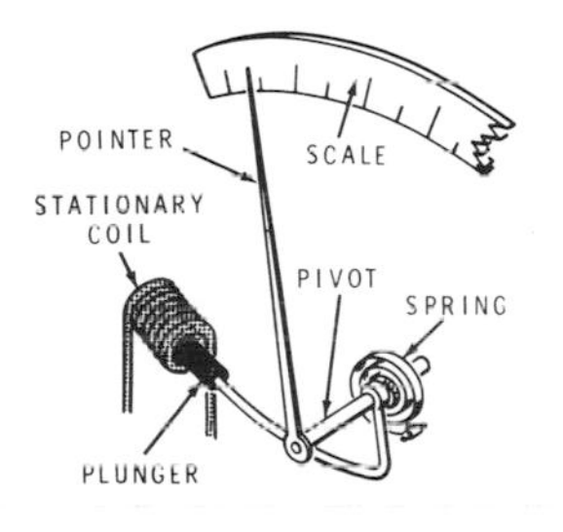

Figure 10

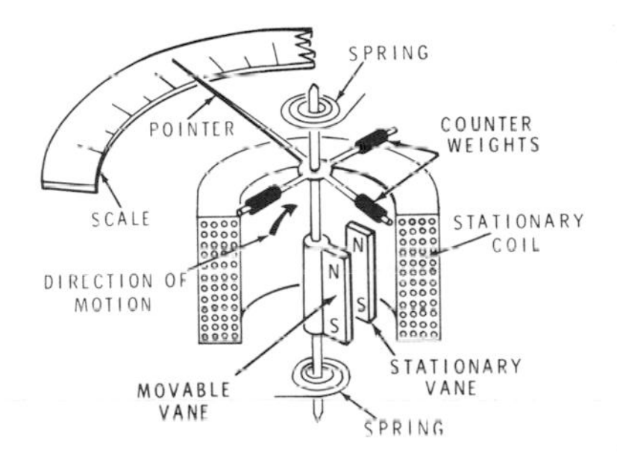

Figure 11

CONCENTRIC-VANE

Construction The basic concentric-vane meter movement is shown in Figure 12. Notice that it is similar to the radial-vane instrument but is stationary, and moving iron vanes are semicircular in shape. The moving vane is mounted inside the stationary vane and is attached to the pointer. The moving vane has square edges but the stationary vane is tapered along one edge.

When current flows through the coil, the magnetic field produced around the coil passes through the two vanes and causes them to become magnetized in the same direction. However, the magnetic lines of force are not distributed uniformly through the stationary vane because of its tapered edge. Fewer lines of force will pass through the narrow end than through the wide end because the narrow end offers higher resistance or opposition to the magnetic lines of force. The wider end provides an easier path for the lines of force, thus allowing more lines of force to be produced. Therefore, the wider end of the vane becomes more strongly magnetized than the narrow end. This means that the strongest repulsion will occur between the wide end of the stationary vane and moving vane. Therefore, the movable vane is forced to rotate toward the tapered end of the stationary vane and turn against the tension provided by the springs. This, in turn, causes the pointer to deflect up-scale. The higher the current through the meter movement, the greater the deflection of the pointer.

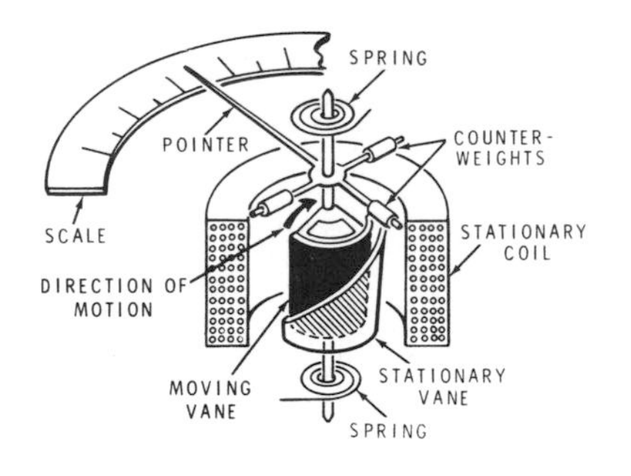

Figure 12

Electrical Characteristics Although moving-vane meters are primarily used to measure AC, they may also be used to measure DC if their scales are appropriately calibrated. This is possible because the moving and stationary vanes always repel each other no matter which direction the current is flowing.

In general, moving-vane meters require more current to produce full-scale deflection than d'Arsonval meters. Therefore, moving-vane meters are seldom used in low power circuits. They are more suitable for measuring the relatively high currents encountered in various types of AC power circuits.

Most moving-vane meters have a full-scale accuracy of approximately ±5% but they cannot provide accurate readings over a wide frequency range. Most of these meters cannot provide accurate readings at frequencies that are much above 100 Hertz and are used mostly in applications where 60 Hertz AC power is involved.

None of the moving-iron meters are very sensitive; however, all may be made very rugged. About the only use for these designs is in power applications where a rugged rms meter is needed.

Thermocouple

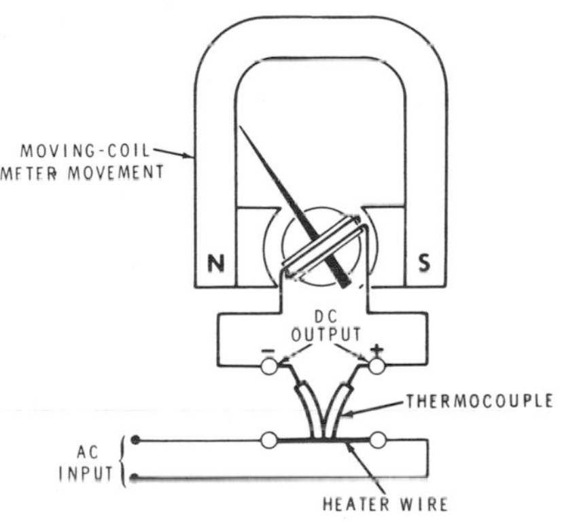

Figure 13

The thermocouple meter is, in reality, a d'Arsonval meter movement with a thermocouple input.

OPERATION

The thermocouple meter basically consists of a thermocouple and a moving-coil meter movement as shown in Figure 13. The thermocouple consists of two dissimilar metal strips or wires which are joined together at one end. When this junction is heated, the two metals react by producing a difference of potential or voltage across their opposite ends. The thermocouple is, therefore, used to convert heat into an electrical voltage.

CONSTRUCTION

A short heater wire is placed against the thermocouple's junction as shown in Figure 13. The AC input is applied directly to this heater wire and this wire heats up to a temperature which is determined by the amount of current flowing through the wire. The heat produced by the wire causes the thermocouple to produce a DC output voltage which, in turn, causes a current flow through the moving-coil meter movement. This current causes the pointer to deflect and indicate the value of the input AC.

A higher AC input current will produce more heat within the wire, a higher DC output voltage from the thermocouple, a higher current through the meter movement, and a greater pointer deflection. Therefore, higher or lower AC input currents result in more or less deflection of the meter's pointer, respectively.

ELECTRICAL CHARACTERISTICS

The thermocouple meter can measure alternating currents over an extremely wide frequency range. In fact, its upper frequency limit extends well up into the radio frequency (RF) range. These meters are often used at frequencies as high as several thousand megahertz. However, they may also be used to measure DC if their scales are appropriately calibrated. This is because these meters are completely insensitive to the rate at which the input current varies. They respond only to the amount of heat that the AC or DC input can produce.

When a thermocouple meter is used to measure an alternating current that has an extremely high frequency, it is necessary to calibrate the instrument at that particular frequency. In other words, the instrument must be adjusted so that it indicates the correct AC value (usually the effective value) at that frequency. This is because a phenomenon known as "skin effect" occurs at extremely high frequencies. Skin effect occurs because high frequency AC currents tend to flow near the outer surface of a wire and this phenomenon becomes even more pronounced as frequency increases. This means that most of the conductor's interior is not used to support current and its resistance is higher than it would normally be. Therefore, the effective resistance of the heater wire and the other wires within the meter vary with frequency, thus changing the internal resistance of the meter. This change in internal resistance affects the meter's response, thus making it necessary to calibrate the meter at the frequency at which it will be used.

Thermocouple meters provide quite accurate AC measurements, which are an indication of the true rms current. Typical meters will usually have a full-scale accuracy of ±2% or ±3%, and certain types of laboratory instruments may have an accuracy of ±1%.

Electrostatic

All of the meters discussed so far place a load on the circuit under test — that is — they draw some current from the circuit which can affect the operation of that circuit. The electrostatic meter overcomes this problem by using a capacitor as the meter movement. Construction is similar to a variable capacitor in that fixed plates are alternated with movable plates. The rotating plates have a pointer attached. When a voltage is applied, the plates become charged and attract each other, giving a pointer deflection proportional to the charge. The only current required by this meter is the initial charging current, which makes it the most sensitive of all meter movements.

Measuring AC

The d'Arsonval meter movement, which is the most sensitive and accurate meter movement, is capable of measuring DC only. However, this meter movement may be used to measure alternating current if the AC is converted to DC before it is applied to the meter movement. This is usually accomplished by using a group of rectifier diodes. These rectifiers are connected between the input AC and the meter, and they allow current to flow in only one direction through the meter.

RECTIFIER-TYPE

Operation The schematic diagram of a rectifier-type, d'Arsonval meter is shown in Figure 14A. Notice that four rectifiers are used in conjunction with one meter movement. The four rectifiers are identified as D_1, D_2, D_3, and D_4 and are arranged in what is called a *bridge rectifier* configuration. The two input terminals of the circuit are identified as A and B.

Assume that an AC generator is connected to input terminals A and B and that this generator is supplying a continuous alternating current which varies in a sinusoidal manner. Also assume that during each positive alternation of the AC sine wave, terminal A is positive with respect to terminal B. During each positive alternation, the circuit current would therefore be forced to flow along the path indicated by the solid arrows. It would flow from terminal B, through D_3, through the meter movement, through D_2, and back to terminal A. During each negative alternation, when B is positive with respect to A, current must flow along the path indicated by the open arrows. In other words, it would flow from terminal A, through D_1, through the meter movement, through D_4, and back to terminal B. Notice that even though the input current changes direction, the current through the meter movement is always in the same direction. The four rectifiers, therefore, convert the input AC into DC (actually a pulsating DC) as shown in Figure 14B.

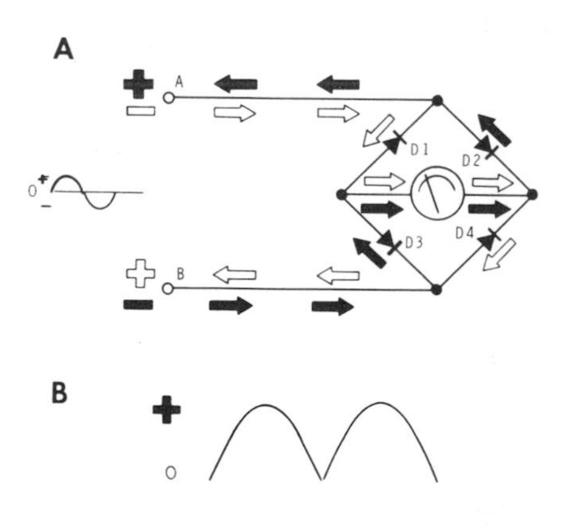

Figure 14

The process of converting AC to DC is referred to as rectification, and this is why the diodes used in this application are referred to simply as rectifiers. Furthermore, the rectifiers in this circuit convert both halves of the AC sine wave into a pulsating direct current and are therefore said to be providing full-wave rectification. Less complicated circuits, which use only one or two rectifiers, may also be used to provide half-wave rectification. In this last process, only one-half of the AC sine wave is allowed to flow through the meter movement.

The current through the meter movement flows in pulses, since each alternation rises from zero to a peak value and drops back to zero again. Unless the frequency of the AC input is extremely low, the meter movement will not be able to follow the variations in the pulsating current. Instead, the meter's pointer responds to the average value of the AC sine wave or, in other words, 0.637 times the peak value. However, the scale on the meter is usually calibrated in effective or rms values. In other words, the numbers on the meter scale represent effective values which are equal to 0.707 times the peak value. The effective value of a sine wave is much more important than the average value, since effective values are used in most AC calculations involving current and voltage.

Electrical Characteristics A variety of rectifier-type meters are available which are capable of measuring a wide range of alternating currents. Each meter is designed to measure up to a certain maximum current. For example, some meters may have calibrated scales which extend from 0 to 1 milliampere, 0 to 10 milliamperes, or 0 to 100 milliamperes.

Most rectifier-type meters have an accuracy less than that of DC meters, due to the rectifiers in the circuit. Accuracies of ±4% to 5% are common.

The rectifier-type meter is useful for measuring alternating currents over a specific frequency range. This type of meter is usually quite accurate over a frequency range that extends from approximately 10 Hertz to as much as 10,000 to 15,000 Hertz. Below a lower limit of approximately 10 Hertz, the meter's pointer tends to fluctuate in accordance with the changes in input current, thus making it difficult to read the meter. Above the upper frequency limits just mentioned, the meter readings are usually too inaccurate to be usable. In fact, the accuracy of the meter progressively gets worse as frequency increases beyond just a few hundred Hertz, because the rectifiers have a certain amount of internal capacitance and the moving-coil meter movement has a certain amount of internal inductance.

These two internal properties present a certain amount of opposition to the flow of alternating current through the meter, and this opposition varies with frequency.

A possible disadvantage of the d'Arsonval meter movement is its response to AC. This meter responds to the average value of current which is 0.637 of the peak value when the current or voltage is a pure sine wave. The rms value is more useful; therefore, most meter scales are calibrated to indicate rms. Figure 15 shows a sine wave with the various values marked and Figure 16 shows the relationship between the various values of a sine wave. We can see that if the waveform is such that the average value is not 0.901 rms, we will not get a proper indication.

Figure 17 shows some typical waveforms. We will assign a peak value of 10 to each wave and see what the meter would indicate. With the sine wave, the meter will read 7.07, which is a 1.11×average. With the square wave, the meter will still indicate 1.11× average, but in this case, average is equal to peak; therefore, the meter will read 11.1, which is wrong. With a pulse input, the meter will still respond to the average and indicate 1.11×average. Once more, this is wrong. The meter will respond to the average of the irregular waveform and indicate 1.11×average. We have no way of knowing if this is correct or not. Most likely, it is wrong. These limitations are of little importance in most applications; however, they must be considered when selecting a meter for a particular application.

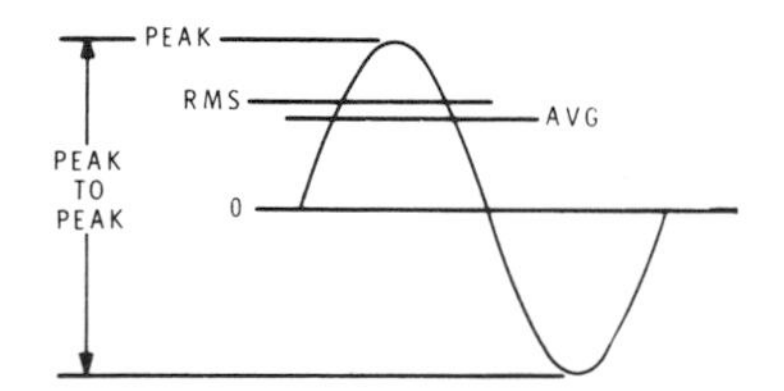

Figure 15

FROM \ TO	PEAK	RMS	AVERAGE	PEAK-TO-PEAK
PEAK	x 1.000	x 0.707	x 0.637	x 2.000
RMS	x 1.414	x 1.000	x 0.901	x 2.829
AVERAGE	x 1.570	x 1.110	x 1.000	x 3.140
PEAK-TO-PEAK	x 0.500	x 0.354	x 0.319	x 1.000

Figure 16

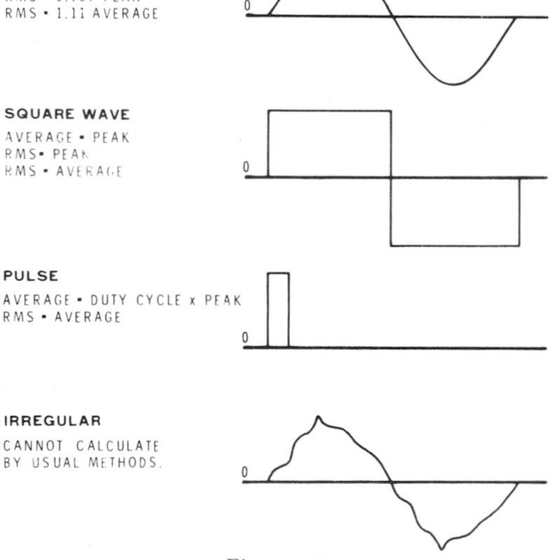

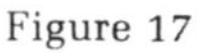

Figure 17

THE AMMETER

With the exception of the electrostatic meter, the meter movements discussed in the previous section are basically current meters. That is, they deflect when current flows through them. In each case, the moving coil consists of many turns of extremely fine wire. In most cases, current is carried to the coil via the fragile spiral springs or the taut-bands of the movement. Because of the delicate nature of the coil and the springs or bands, care must be taken not to feed excessive current through the movement. The current necessary for full-scale deflection will not harm the movement, but a 100% overload might. The coil may burn out; the spring may be damaged; or the aluminum needle may be bent if it is driven too hard against the right retaining pin.

Also, with the d'Arsonval meter, care must be taken to observe polarity when using the meter movement. A reverse current will cause the needle to deflect backwards. If the current is too great, the needle may be bent when it bangs against the left retaining pin.

Increasing the Range of the Ammeter

Each meter movement has a certain current rating. This is the current which will cause full-scale deflection. For example, an inexpensive meter movement may have a current rating of 1 mA (milliampere). To obtain a usable reading, the current through the movement cannot be more than 1 mA. By itself, the movement has a single usable range of 0 to 1 mA.

Obviously, the meter would be much more useful it it could measure currents greater than 1 mA as well as those less than 1 mA. Fortunately, there is an easy way to convert a sensitive meter movement to a less sensitive current meter. We do this by connecting a small value resistor in parallel with the meter movement. The resistor is called a shunt. Its purpose is to act as a low resistance path around the movement so that most of the current will flow through the shunt and only a small current will flow through the movement.

Figure 18A illustrates a 1 mA meter movement connected across a low resistance shunt to form a higher range ammeter. The range will depend on how much current flows through the shunt. In Figure 18B, the current applied to the ammeter is 10 mA. However, only 1 mA of this flows through the meter movement. The other 9 mA flows through the shunt. Thus, to convert the 1 mA movement to a 0-10 mA meter, the shunt must be chosen so that 9/10 of the applied current flows through the shunt. Once this is done, the full-scale position on the scale indicates 10 mA, since this is the amount of current that must be applied before full-scale deflection is reached.

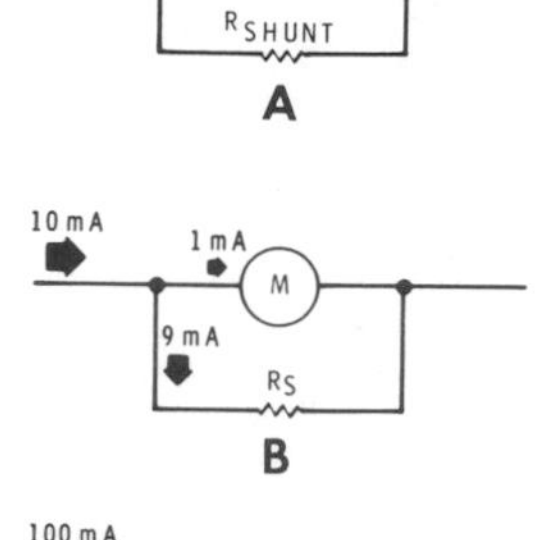

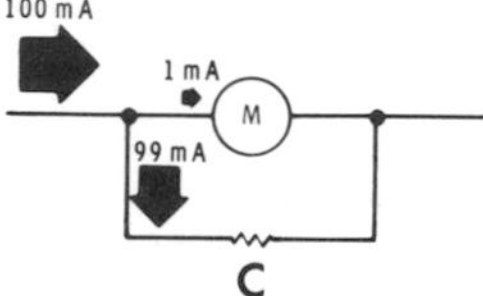

Figure 18

If the value of the shunt is made smaller, the meter can indicate even higher values of current. Figure 18C shows the requirements necessary to measure 100 mA. Here, 99 mA or 99 percent of the applied current must go through the shunt. Thus, the resistance of the shunt must be much smaller than the resistance of the meter movement.

Where relatively small values of current are to be measured, precision resistors are used as shunts; however, when measuring large currents, the shunt must have a very low resistance and be capable of carrying a large current. Resistive wire or even metal bars are frequently used for this.

COMPUTING SHUNT RESISTANCE

To determine the proper value of the shunt resistor, we must first know something of the characteristics of the meter movement. In the earlier example, we know that full-scale deflection requires 1 mA. However, we must also know either the resistance of the meter movement or the voltage dropped by the movement when the current is 1 mA. Of course, if we know one, we can compute the other.

The resistance value of the meter movement is given in the manufacturer's literature, catalog, or operating instructions. Often it is printed right on the meter movement itself, or you can find it by using the procedure in Experiment 1. Assume that the 0-1 mA movement has a resistance of 1000 Ω (ohms), or 1 kΩ (kilohm). In this case, 1 mA of current causes a voltage drop across the meter movement of 1 V (volt).

If you refer back to Figure 18, you will see that this is the voltage developed across the meter movement in each of the examples shown. Since the shunt resistance is connected in parallel with the meter movement, this same voltage must be developed across the shunt. This means that in the example shown in Figure 18B, the 9 mA current must develop 1 V across the shunt. Using Ohm's law, we can now compute the value of the shunt since we know the current and voltage. Thus, the value of the shunt should be 111 Ω.

This is the resistance necessary to shunt 9 mA around the meter when a current of 10 mA is flowing in the circuit.

DC ammeters such as those included in volt-ohm-milliammeters normally use a 50 μA meter movement. This means that the lowest range available is 0 to 50 μA. Additional ranges allowing the measurement of larger values of current up to as much as 10 A are not uncommon. Figure 19 shows such a meter. Since this meter uses a d'Arsonval movement, it can measure DC current only. Ranges of 0.05 mA, 1 mA, 10 mA, 100 mA, 500 mA and 10 A are available. If we assume that the meter resistance is 4000 Ω, what value of shunt resistance is required for the 10 mA range?

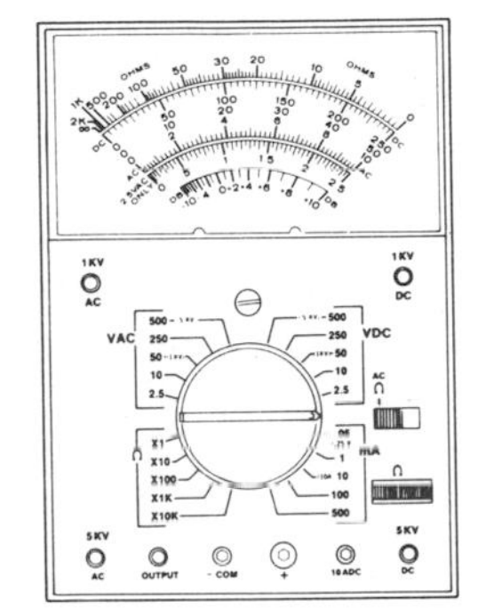

Figure 19

Figure 20 shows the basic circuit. Of the 10 mA input, 50 μA flows through the meter, leaving 9.95 mA to flow through the shunt resistor.

After finding the current through the shunt, we must find the voltage developed across the circuit. This can be computed by Ohm's law, and is equal to 0.2 V.

We can now use this voltage and shunt current to find shunt resistance, which is equal to 20.1005 ohms.

You will note that a very precise value of resistance is required for meter accuracy; however, an R_s of 20.1 ohms would be adequate. Using the same meter movement as in the last problem, what value of shunt is required for the 10 A range?

If you answered 0.0200001 ohms, you are right. Practically, 0.02 ohms would be the correct value.

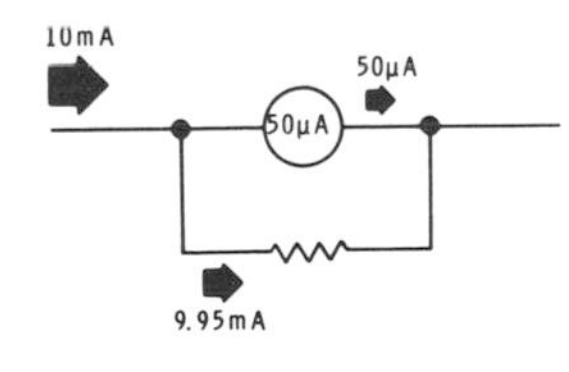

Figure 20

Scales

LINEAR

Figure 21

Ammeter scales will be calibrated for the type of current being measured and for proper full-scale deflection currents. A DC meter is usually straight-forward, having only one scale with different values assigned for each range. Figure 21 shows such a scale. The scale has 50 divisions so that on the .05 mA (50 μA) range, each division represents 1 μA of current. On the 500 mA scale, each division represents 10 mA, and on the 1 mA scale, each division represents 20 μA.

Assume that we wish to measure 265 μA. We would first select the proper range — 1 mA — , make sure the meter is connected properly in the circuit, and read the meter. The pointer should read as shown in Figure 22.

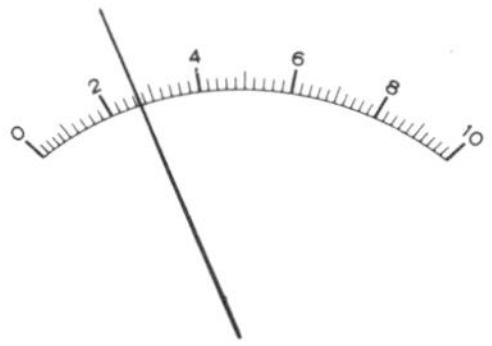

Figure 22

On meters that measure AC, scales might be provided for average, rms, and peak values. Since average = 0.638 peak, and rms = 0.707 peak, three separate scales would be required. The most common scale is the rms scale.

NONLINEAR

The scales we have just discussed are all linear, that is, the numbers or values on the scale are equally spaced. The pointer deflection is always directly proportional to the current flowing through the meter movement. The d'Arsonval meter movement has a linear current scale.

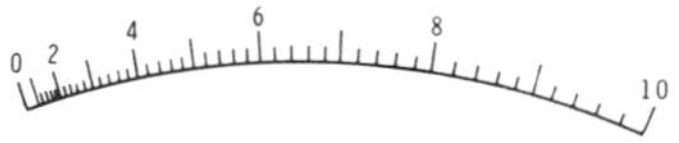

Figure 23

Other meter movements, such as the moving iron, the electrodynamometer, and the thermocouple, have nonlinear scales. The type of scale used with these meters is the "square law" scale. This means that pointer deflection increases with the square of the change in current. For example, suppose that 10 milliamperes of current causes the pointer to deflect a distance of 1 inch on the meter scale. If the current were increased to 20 mA, or doubled, the pointer would deflect four times as far, or 4 inches. If the current increased to three times its initial value, the pointer would deflect nine times as far ($3^2 = 9$). Figure 23 shows a typical square law scale. Notice that values near the zero end of the scale are closely spaced and difficult to read. Better resolution and accuracy is achieved at the high end of the scale.

Ammeter Accuracy

Every meter movement has a certain accuracy associated with it. The accuracy is specified as a *percentage of error at full-scale deflection.* Accuracies of ±2% or ±3% of full scale are common for good quality instruments. Figure 24 illustrates what is meant by ±3% of full scale. The scale shown is a 100 mA current scale. Remember, the meter accuracy refers to full-scale deflection. At full scale, ±3% equals ±3 mA. For this meter, a current of exactly 100 mA could cause the meter to read anywhere from 97 mA to 103 mA. Another way to look at it is that a meter reading of exactly 100 mA might be caused by an actual current of from 97 mA to 103 mA.

As you can see, a ±3% accuracy means that the reading may be off by as much as ±3 mA at full scale. More importantly, it means that the reading may be off by as much as ±3 mA at any point on the scale. For example, when the meter indicates 50 mA, the actual current may be anywhere from 47 mA to 53 mA. Thus, at half scale, the accuracy is no longer ±3%; it is now ±6%. For an indicated current of 10 mA, the actual current may be anywhere from 7 mA to 13 mA. Here, the accuracy is only ±30%.

Because meter accuracy is specified in this manner, the accuracy gets progressively worse as we move down the scale. For this reason, current measurements will be most accurate when a current range is selected that will cause near full-scale deflection of the meter. The nearer full scale, the more accurate the reading will be.

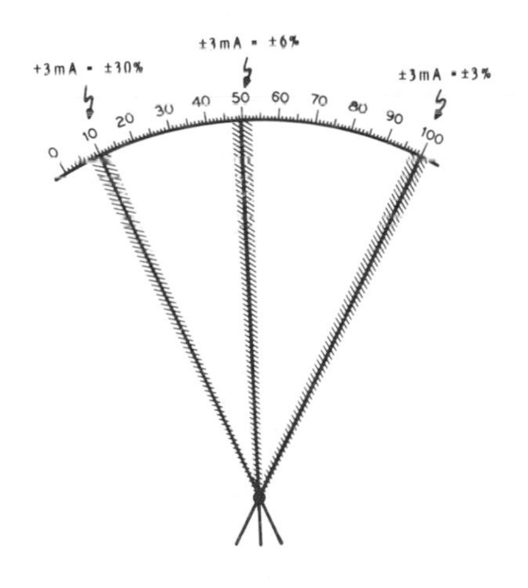

Figure 24

Meter accuracy and the accuracy of the reading are different things. One of the factors which can cause a relatively large error is PARALLAX. Parallax is caused by looking at the meter from an angle which will cause the pointer to appear left or right of the true position. Figure 25A shows an example of parallax error. With the eye directly in front of the pointer, the correct reading is obtained, but when the eye is moved to the left, the pointer appears to move right. Parallax is so common that some manufacturers place a mirror on the scale to help overcome the problem. To get maximum accuracy, close one eye and view the meter from a position which causes the pointer to appear directly over its reflection as shown in Figure 25B.

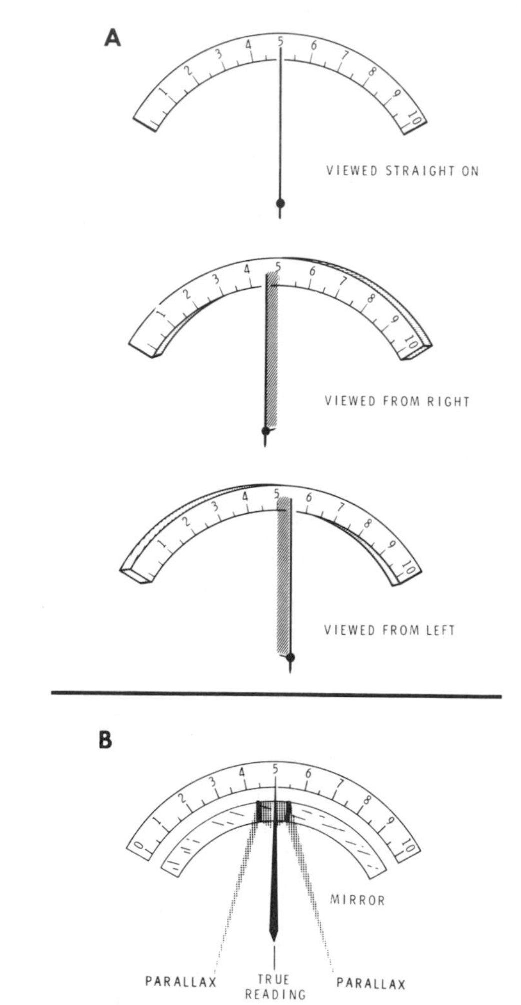

Figure 25

Circuit Connections

The ammeter must be connected so that the current to be measured flows through the meter. This means that the circuit must be broken so that the ammeter can be inserted in series. Figure 26A shows a basic transistor amplifier. Suppose we wish to measure total current through the transistor. The meter must be placed in the emitter circuit. One possible connection is shown in Figure 26B. Here, all of the emitter current will flow through the meter. Even this will not give a true reading of current because the meter changes the circuit. If we made our initial reading on the 10 mA range using the Heathkit IM-105 meter which we mentioned earlier, we would get a reading slightly less than 1 mA in this circuit. We would then switch to the 1 mA range in an attempt to make our measurement on the most accurate portion of the scale. The shunt resistance on the 1 mA range is 220 Ω, which is slightly more than 10% of the emitter resistance. By placing the meter in the circuit, we have changed the emitter resistor to 2.2 kΩ instead of 2 kΩ. That should not be noticed on a circuit of this type; however, there are some applications where the change is significant. You should constantly be aware that any time you take a measurement, you are changing the circuit somewhat, and you should take these changes into account.

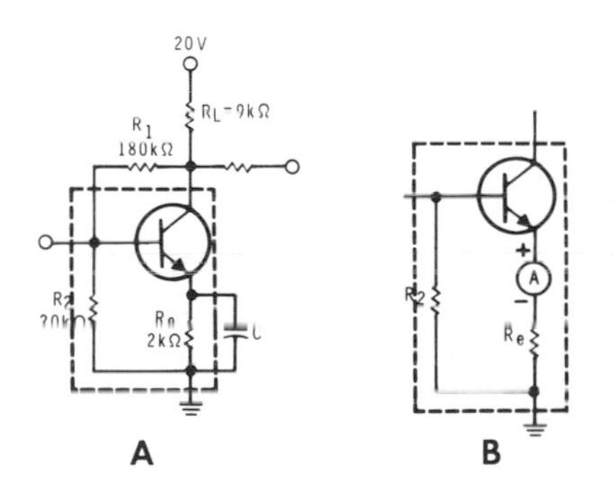

Figure 26

Clamp-on Meters

All of the meters previously described must be connected directly to an electronic circuit or device in order to obtain a current measurement. However, there is one type of measuring instrument which does not have to be physically connected to a circuit in order to provide a current measurement. This device can be simply clamped over a conductor and it will indicate the amount of current flowing through the conductor. These clamp-on type meters are also referred to as split-core meters, hook-on meters, or snap-around meters. We will now briefly examine one of these meters and consider its important electrical characteristics.

METER OPERATION

A clamp-on meter basically consists of a transformer (which has a split core) and a rectifier-type, moving-coil meter as shown in Figure 27. The instrument is usually mounted within a small plastic case so that it can be held in one hand.

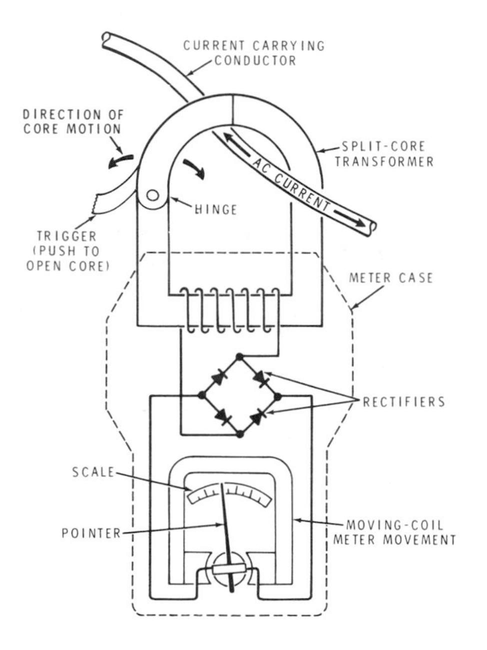

Figure 27

The meter uses a split-core transformer. One side of the transformer core is hinged and is movable. This section of the core may be opened by pressing a trigger which is attached to it. To measure the AC current in a conductor, the core (which is made of soft iron) must be opened so that the conductor can be inserted inside of the core. Then the core is closed, as shown, so that it completely surrounds the conductor. It is important to close the core completely so that no air gap is present.

The AC current flowing in the conductor produces a circular magnetic field which surrounds the conductor. The strength of this field is proportional to the current in the conductor. This magnetic field will expand and collapse as the AC increases and decreases in value and the direction of the field will change as the current changes direction. The iron core offers very little opposition to the magnetic field (much less than the surrounding air). This means that most of the lines of force will tend to flow through the core. However, in order for this to happen, the magnetic lines of force must cut across the coil of wire that is wound around the opposite side of the core. When this happens, a voltage is induced into the coil, which in turn causes an induced current to flow through the coil. The conductor, the core, and the coil, form a simple transformer with the conductor acting as the input or primary winding (which has only one turn) and the coil acting as the output or secondary winding.

The current induced in the secondary coil is an alternating current just like the current in the conductor. This AC is applied to the rectifiers which convert it to DC. The DC is then used to operate the moving-coil meter movement which causes its pointer to deflect. The meter is calibrated so that it will indicate the effective value of the AC flowing through the conductor.

ELECTRICAL CHARACTERISTICS

Since the clamp-on meter depends on transformer action for operation, it can be used only to measure AC. The moving magnetic field produced by the AC in the conductor is necessary to induce a voltage in the secondary coil of the transformer. The magnetic field produced by a direct current is constant and, therefore, cannot pass through the transformer.

In general, the clamp-on meter is useful for measuring only relatively high alternating currents. This is because the current in the conductor must be high in order to produce a magnetic field which is strong enough to induce a significant amount of current into the secondary coil. These meters are often used to measure currents as high as several hundred amperes.

Typical Ammeter

The typical VOM usually has provision for measuring current. In this case, we can only measure DC. Figure 28A shows a simplified circuit for the milliammeter set on the 1 mA range. The method of shunt selection is unique and the following paragraphs will explain how it works.

Consider the circuit as two parallel resistors as shown in Figure 28B. Rm represents the total resistance of the meter circuit and Rs represents the total resistance of the shunt resistors. With the meter in the 1 mA position, Rm equals 5000 ohms and Rs equals 263.1 ohms. Since the ratio of Rm: Rs is inversely proportional to the ratio of Im: I_S, we can calculate shunt current in the following manner.

$$\frac{Rm}{Rs} = \frac{Is}{Im}$$

Solving this equation, we see that current through the shunt is approximately 19 times as much as that through the meter. Therefore, if 50 μA flows through the meter, 950 μA will flow through the shunt for a total current of 1000 μA (1 mA).

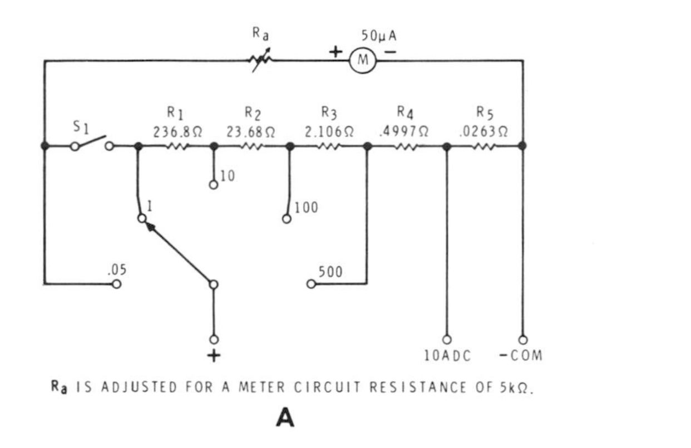

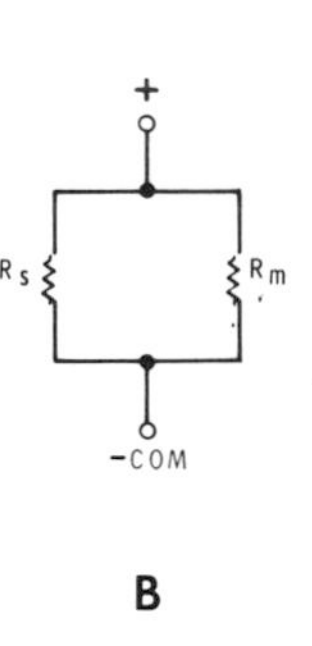

Figure 28

This operation is exactly like the operation of the basic ammeter which you studied earlier. However, when we switch to a higher range, things change. Notice in Figure 28A, all of the shunt resistors are connected in series. With this arrangement, the range switch is connected in series with the entire parallel network and never becomes a part of the shunt resistance. This eliminates any possibility of contact resistance affecting the accuracy of shunt resistance and thus the accuracy of the current reading. The circuit is called an Ayrton shunt and is used in several of the better VOM's.

To get a better idea of how this circuit works, assume the range switch is moved to the 10 mA position. Looking at Figure 28A, we see that R1 is now in series with the meter movement, making Rm in Figure 28B equal to 5236.8 ohms. Rs, which is now made up of R2 through R5, has decreased to 26.3 ohms. Applying the same ratio as before, we find that the Rm is 199 times Rs; therefore, I_s equals 199 times 50 μA or 9950 μA for a total current of 10,000 μA (10 mA).

Each time the switch is moved to a higher position, there is a slight increase in full-scale voltage drop across the meter and shunt to compensate for the increase in Rm.

To avoid overloading the switch, the 10 A position has a separate socket. When high current measurements are made, the test lead must be moved to the proper socket. However, because of the arrangement of the shunt resistors, the range switch must be set on the 10 A position.

Selection

Selecting a meter is a personal endeavor and depends on a number of factors. We will discuss some of those factors.

1. Accuracy: For most maintenance work, 6 to 10% mid-scale accuracy is sufficient. This will require a meter with 3 to 5% full-scale accuracy. You should be interested in how accurate the meter is at mid-scale, since this is the area where most readings will fall. To obtain accuracies in the 1% area, a mirrored scale is required to avoid parallax. Accuracies of better than 0.5% to 1% are beyond the capabilities of most analog meters.

2. Sensitivity: Get what you need, and no more. If you are never going to measure less than 100 mA of current, a 50 μA range is not necessary. The same thing applies to extra ranges. How often will you need a 50 amp range? If the meter you are selecting will also be used as a voltmeter, you may need all of the sensitivity you can get.

3. AC Measurements: AC current capability costs extra. The electrodynamometer is the most accurate and the thermocouple has the best frequency range. The electrodynamometer has the lowest impedance while the thermocouple is prone to burn-out, which can happen before the pointer even reaches full scale in some cases.

4. Resistance: Since an ammeter is connected in series with the circuit, it would appear that the lowest possible resistance would be desirable. However, this is not always the case. The coil of wire carries the current and makes up the resistance in the meter. To reduce the resistance, we reduce the turns on the coil. This, in turn, reduces sensitivity and accuracy. It is usually best to select the lowest resistance practical.

5. Physical Characteristics: This category includes many things from the size of the scale to how rugged the meter is built. Keep in mind that smaller scales are harder to read, and don't buy a smaller meter than you need. Controls should be convenient and easy to use. More sensitive meters are frequently more delicate as well as more costly. A meter for field use can be subjected to vibration, shock, heat, cold, moisture, and a number of other influences such as RF fields, etc. Here, again, the word is, "get what you need and no more."

6. Cost: It is often said that we get what we pay for. With test equipment, this may not be true. You can spend a lot of money for a meter that will not do your job. Conversely, the perfect ammeter for you may not cost much. Define your needs, then shop around.

Table 1 shows a comparison of some factors to consider when selecting a meter movement. The ratings are very general, and the individual units will vary considerably on a given factor. If practical, you should try to test several meters before making your selection.

TABLE 1

TYPE OF METER MOVEMENT	SENSITIVITY	ACCURACY	DURABILITY	AC MEASUREMENT	FREQUENCY RANGE	COST	REMARKS
d'Arsonval	Good	Good	Fair	Sine Wave*	Fair	Moderate	*Rectifier Required
Electro-Dynamometer	Poor	Good	Fair	RMS Any Waveform	Poor	High	
Taut Band	Very good	Good	Good	Sine Wave*	Fair	Moderate	*Rectifier Required
Moving Vane	Poor	Fair	Very good	RMS	Poor	Low	
Thermocouple	Fair	Very good	Fair	RMS	Excellent	High	
Electrostatic	Excellent	Good*	Poor	Average	None	High	Limited application *Poor at low voltage

THE VOLTMETER

The basic meter movement can be used to measure voltage as well as current. In fact, every meter movement has a certain voltage rating as well as current rating. This is the voltage which will cause full-scale deflection. Of course, the voltage rating is determined by the current rating and the meter resistance. For example, a 50 μA meter movement which has a resistance of 2000 Ω deflects full scale when connected across a voltage of 0.1 volt.

That is, the meter movement alone could be used to measure voltages up to 0.1 volt. Thus, the meter scale can be calibrated from 0 to 0.1 volt. However, if the meter movement is connected across a much higher voltage, such as 10 volts, it may be damaged. Obviously, to be practical, we must extend the voltage range of the basic meter movement.

Extending the Range

You have seen that a 50 μA, 2000 ohm meter movement can withstand a voltage of 0.1 volt without exceeding full scale. To extend the range, we must insure that the voltage across the meter does not exceed 0.1 volt when the meter movement is connected across a higher voltage. We do this by connecting a resistor in series with the meter movement as shown in Figure 29. The resistor is called a multiplier because it multiplies the range of the meter movement.

The purpose of the multiplier resistor is to limit the current which flows through the meter movement. For example, in the voltmeter shown in Figure 29, the current through the meter movement must be limited to 50 μA. Another way to look at it is that the multiplier must drop all the voltage applied to the voltmeter except the 0.1 volt allowed across the meter movement. For example, if the range is to be extended to 10 volts, then the multiplier must drop 10 volts $-$0.1 volt = 9.9 volts.

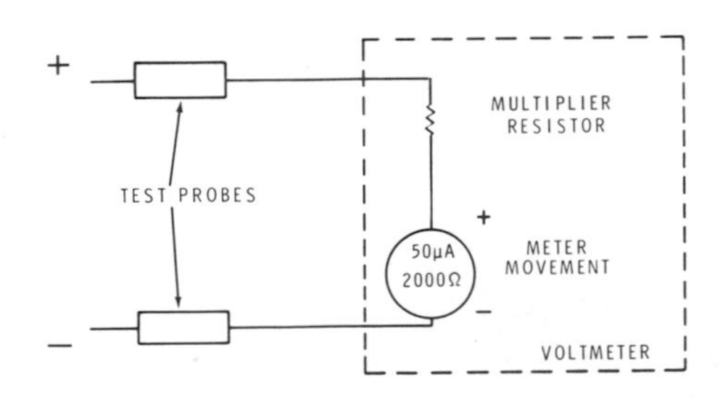

Figure 29

CALCULATING THE MULTIPLIER

You have seen that the value of the multiplier must be high enough to limit the current to the full-scale current rating of the meter movement for any applied voltage. If you keep this in mind, you can easily calculate the required value of the multiplier for any voltage range.

Assume you wish to convert the 50 μA, 2000 Ω meter movement to a 10-volt voltmeter by adding a multiplier in series. Obviously, a current of only 50 μA must flow when the voltmeter is connected across 10 volts. Thus, the total resistance of the voltmeter is equal to full-scale voltage divided by full-scale current or, in this case, 200,000 Ω.

However, the meter movement itself has a resistance of 2000 Ω. Thus, the multiplier must have a value of 200,000 Ω $-$2000 Ω = 198,000 Ω or 198 kΩ.

This means that the basic 50 μA, 2000 Ω meter movement can now measure 0 to 10 volts because 10 volts must be applied to cause full-scale deflection. From a voltage standpoint, the multiplier resistor drops 99% of the applied voltage, which in this case is 9.9 volts.

This leaves 0.1 volt across the meter. Because the total voltmeter resistance is 100 times larger than the meter resistance, the range of the meter is multiplied by 100. Of course, the meter scale should now be calibrated from 0 to 10 volts.

This procedure can be used to find the correct multiplier resistor for any meter movement and voltage as long as meter resistance is known.

MULTIPLE-RANGE VOLTMETERS

A practical voltmeter has several ranges. One arrangement for achieving multiple ranges is shown in Figure 30. Here, the voltmeter has four ranges which may be selected by the range switch. Again, the 50 μA, 2000 Ω meter movement is used. On the 0.1-volt range, no multiplier is required since this is the voltage rating of the meter movement itself.

On the 1-volt range, R_1 is switched in series with the meter movement. The value of R_1 is given as 18 kΩ.

Notice that on the 10-volt range, the 198 kΩ multiplier value computed earlier is switched in series with the meter movement. On the 100-volt range, a resistance of 1.998 MΩ was calculated; however, in practical application, a standard 2 MΩ resistor is used.

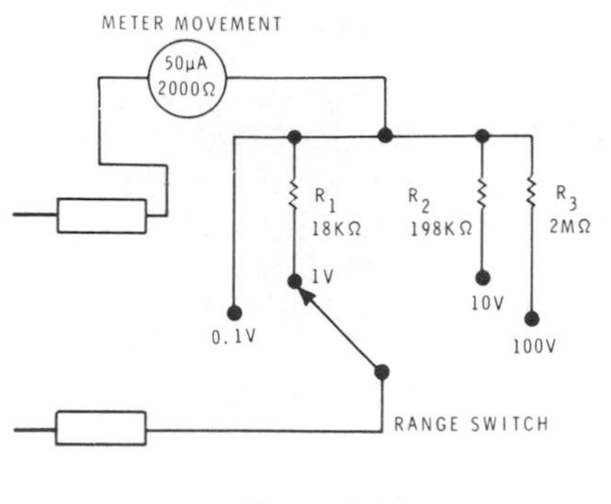

Figure 30

Figure 31 shows another arrangement sometimes used when several ranges are required. On the 0.1-volt range, no multiplier is required. On the 1-volt range, an 18 kΩ multiplier is switched in series with the meter movement. Up to this point, the arrangement is similar to that shown in Figure 30. However, here the similarity ends. On the 10-volt range, R_2 is switched in series with R_1. Thus, the total resistance in series with the meter movement is 18 kΩ + 180 kΩ = 198 kΩ. Notice that this is the same value of multiplier used on the 10-volt range in Figure 30. The only difference is that in Figure 30, a single 198 kΩ resistor is used, while in Figure 31, two resistors having a total resistance of 198 kΩ are used.

On the 100-volt range, R_3 is switched in series with R_1 and R_2. Thus, the total multiplier resistance is 18 kΩ + 180 kΩ + 1.8 MΩ = 1.998 MΩ. You will recall that this is the exact multiplier value computed earlier for the 100-volt range.

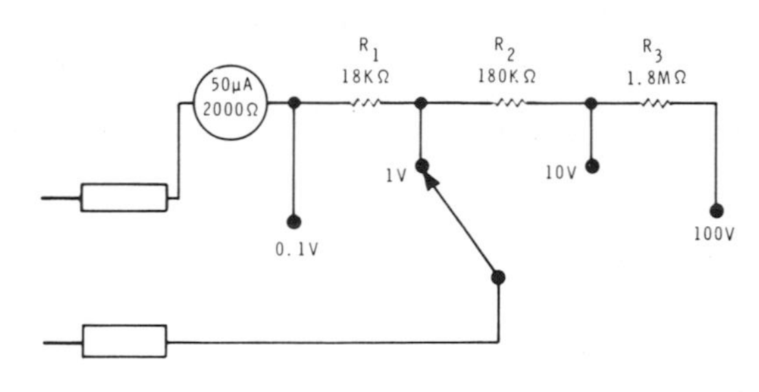

Figure 31

Sensitivity (Ohms per Volt)

An important characteristic of a voltmeter is its sensitivity. Sensitivity can be thought of as the amount of current required to produce full-scale deflection of the meter movement. For example, a 50 μA meter movement is more sensitive than a 1 mA meter movement because less current is required to produce full-scale deflection.

However, sensitivity is more often defined in another way. Sensitivity is normally expressed in ohms per volt (also written ohms/volt). The more sensitive the meter is, the higher the ohms-per-volt rating will be. The sensitivity in ohms per volt of any voltmeter can be determined simply by dividing the full-scale current rating of the meter movement into 1 volt.

$$\text{Sensitivity} = \frac{1 \text{ volt}}{\text{I full scale}}$$

Thus, the sensitivity of a voltmeter which uses a 50 μA meter movement is 20,000 ohms per volt.

This means that on the 1-volt range, the voltmeter has a total resistance of 20,000 ohms. You can prove this by referring back to Figure 30 or 31. On the 1-volt range, the multiplier has a value of 18 kΩ and the meter movement has a resistance of 2 kΩ. Consequently, the total resistance is 20,000 Ω.

The sensitivity is determined solely by the full-scale current rating of the meter movement. Thus, it has the same sensitivity regardless of the range used. Consequently, on any range, the voltmeters shown in Figure 30 and 31 have a resistance of 20,000 $\Omega \times$ V, where V is the selected full-scale voltage range. Thus, on the 10 volt range, the total voltmeter resistance is 20,000 $\Omega \times 10 = 200,000\ \Omega$

Try another example. What is the sensitivity of a voltmeter which uses a 1 mA meter movement? Remember, sensitivity is determined by dividing the full-scale current into 1 volt. If you answered 1000 ohms per volt, you are right.

What would be the total resistance of this voltmeter on the 5-volt range? The resistance would be 1000 $\Omega \times 5 = 5000\ \Omega$.

Loading Effect of Voltmeters

An unfortunate aspect of elecronics is that measuring an electrical quantity often changes the quantity that we are attempting to measure. When measuring voltage, we must connect a voltmeter across the circuit under test. Since some current must flow through the voltmeter, the circuit behavior is modified somewhat. Often, the effects of the voltmeter can be ignored, especially if the meter has a high ohms/volt rating. However, if the voltmeter has a low ohms/ rating or the circuit under test has a high resistance, the effects of the meter cannot be ignored.

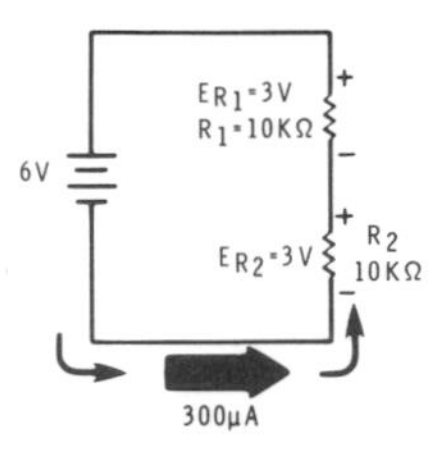

Figure 32

Consider the circuit shown in Figure 32. Here, two 10 kΩ resistors are connected in series across a 6-volt battery. Since the resistors are the same value, each drops one-half of the applied voltage or 3 volts. Thus, we would expect a voltmeter to read 3 volts if it were connected across either resistor. However, if the voltmeter has a low ohms/volt rating, the actual reading may be very inaccurate. Figure 33A shows the same circuit with a low-sensitivity voltmeter connected across R_2. The voltmeter has a sensitivity of 1000 ohms/volt. Since we expect the voltage across R_2 to be about 3 volts, the voltmeter is on the 0-10-volt range. Thus, its resistance (R_m) is $1000\ \Omega \times 10 = 10{,}000\ \Omega$. Because R_m is in parallel with R_2, the total resistance is reduced to one-half the original resistance or to 5,000 Ω.

Therefore, the circuit shown in Figure 33A reduces to the circuit shown in Figure 33B. Notice how this upsets the operation of the circuit. The total series resistance of R_1 and R_A is now only 15 kΩ instead of 20 kΩ.

Thus, the current increases from its previous value of 300 μA to a new value of 400 μA. The voltage distribution also changes since R_1 is now larger than R_A. The voltage dropped by R_A is now 2 volts and the voltage dropped by R_1 has increased to 4 volts.

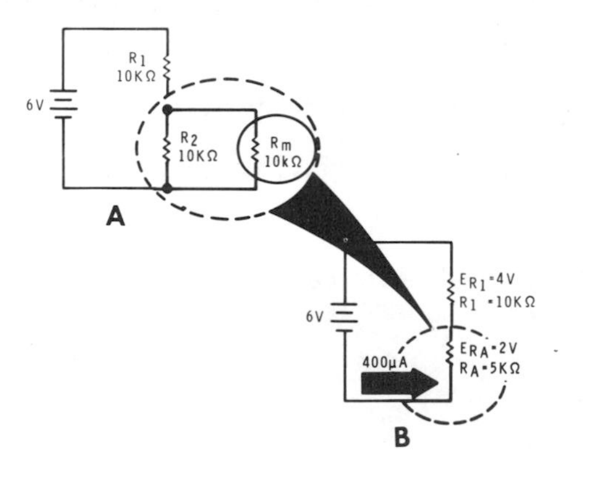

Figure 33

Thus, instead of reading 3 volts as we would expect, the meter measures only 2 volts across R_2. This is an inaccuracy of 33% . This effect is called **loading**. We say that the meter is loading down the circuit, causing the voltage across R_2 to decrease. The loading effect becomes noticeable when the resistance of the meter approaches that of the resistor across which the meter is connected. For example, if the resistance of the meter were made 10 times that of R_2, then the loading effect would be barely noticeable.

Figure 34A shows the same circuit with a 20,000 ohms/volt meter connected across R_2. On the 10-volt range the resistance of the meter is 20,000 $\Omega \times 10 = 200{,}000\ \Omega$ or 200 kΩ. Here, the equivalent resistance of R_m and R_2 in parallel is 9.52 kΩ as shown in Figure 34B.

Notice that R_A is very close to the value of R_2. Therefore, the circuit operation is only slightly upset. The current increases only slightly to about 307 μA while the voltage across R_1 rises to about 3.07 volts.

Meanwhile, the voltage across R_2 decreases slightly to about 2.93 volts.

Thus, instead of measuring 3 volts, the meter measures 2.93 volts. The inaccuracy is so small that it probably would never be noticed. The loading effect is minimized by using a voltmeter whose resistance is much higher than the resistance across which a voltage is to be measured.

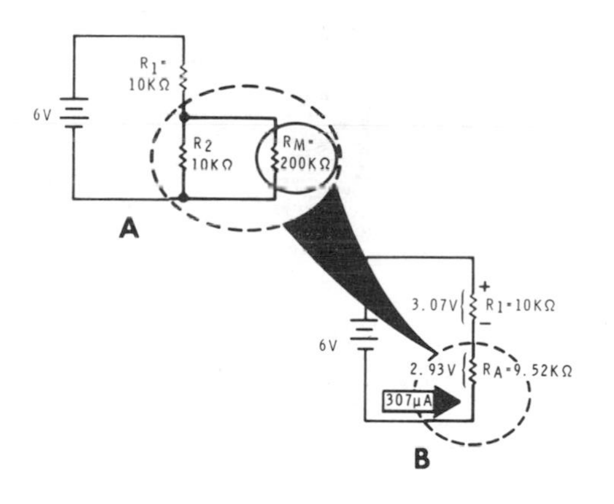

Figure 34

Accuracy

Voltmeter accuracy is determined by the accuracy of the movement and the tolerance of the components. A typical d'Arsonval movement has an accuracy of about 2% while the circuit components have a tolerance of ±1%. This will give an overall accuracy of approximately 3% of full scale. Because of the additional components required for rectification, the AC accuracy is usually less (4 to 5% of full scale).

Circuit Connections

All voltage measurements are made with the meter in parallel with the component or source across which the voltage is being measured. Figure 35 shows a simple DC circuit with the voltmeter properly connected. Notice that since this is a DC circuit, correct polarity is observed. If we were measuring AC, it would not matter which way the leads were connected.

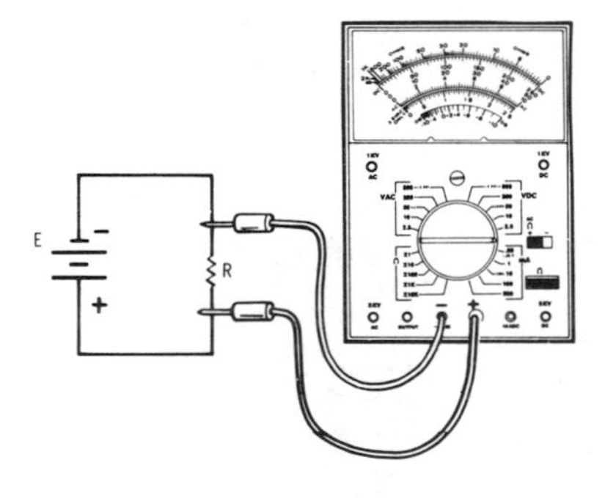

Figure 35

Typical Voltmeter (DC)

Since most voltmeters are part of a volt-ohm-milliammeter (VOM), we will look at the voltmeter section of a typical VOM. Figure 36 shows the circuit from such a meter. The meter movement has a full-scale deflection sensitivity of 50 μA and the multiplier resistors will limit the current to that amount for each range. Separate lead sockets are provided for 5,000 volts and 1,000 volts. Therefore, the lead will have to be moved for these measurements and the position of the range switch will have no effect.

The range switch connects the plus (+) socket to any one of the six taps on the DC voltage divider. Therefore, the probe need not be moved when changing between the lower six ranges.

R_A is the DC calibrate control and is adjusted to compensate for accumulative tolerances in the meter circuit. Adjustment can be made periodically to compensate for aging, etc.

R_T is a thermistor which compensates for meter resistance changes caused by ambient temperature change. Since R_T has a negative temperature coefficient and the meter has a positive temperature coefficient, any temperature change that would cause the meter resistance to increase would cause the resistance of R_T to decrease. Therefore, the total resistance would remain essentially the same.

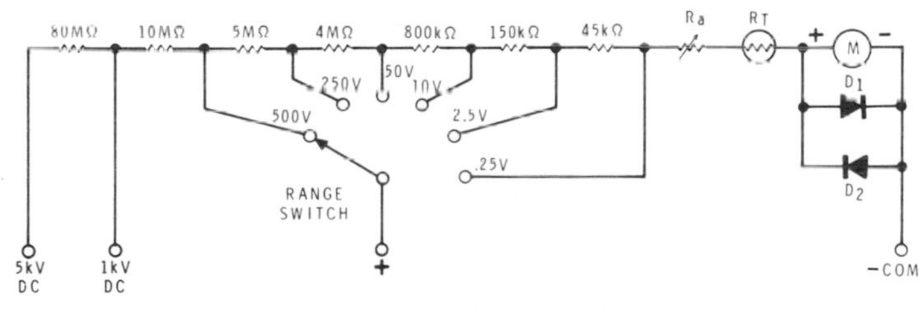

Ra IS ADJUSTED TO GIVE A METER CIRCUIT RESISTANCE OF 5kΩ.

Figure 36

Diodes D_1 and D_2 protect the meter movement from damage by limiting the voltage developed across it to about 0.2 volts. This should prevent burn-out of the movement due to excessive current; however, the sudden application of excessive voltage may bend the pointer as it hits the stop pins. The meter movement has an internal resistance of approximately 1600 Ω. With full-scale current of 50 μA, a voltage of 0.08 volts will be developed across the meter. If current through the meter increases to about 2-1/2 times full-scale current, one of the diodes will be forward biased, placing a low resistance in parallel with the meter and preventing any further increase in voltage. D_1 has its cathode connected to the negative terminal of the meter while D_2 has the cathode connected to the positive terminal. Therefore, no matter which polarity is connected to the meter it will be protected against burn-out.

Typical Voltmeter (AC)

In our typical meter, the AC section has many parts in common with the DC section and works much the same way. Figure 37 shows a simplified AC voltmeter circuit. Although not shown in Figure 37, the protective diodes, the thermistor, etc., will still be in the meter circuit. R_m represents the resistance of the multiplier resistors which are connected similar to Figure 36. Diodes D_1 and D_2 change the AC at the input to DC which can be measured by the meter. The diodes are "instrument diodes" and have a very low forward-bias voltage drop and low reverse-bias leakage current.

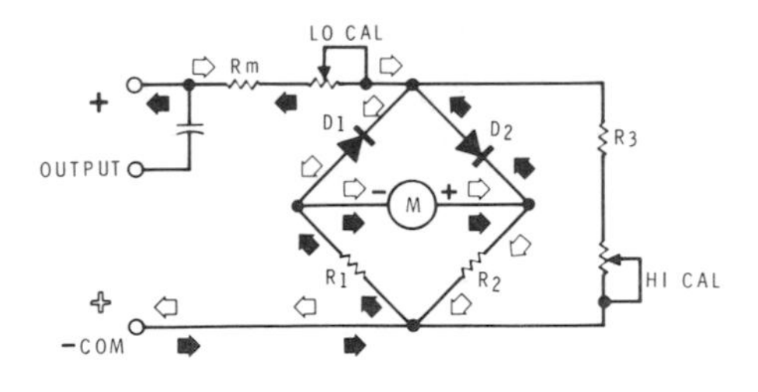

Figure 37

When the positive portion of the input is applied, D_2 conducts and current flows through R_1 and the meter, as well as through R_2. When the input goes negative, D_1 conducts through the meter and R_2, as well as R_1. With either polarity, current through the meter is in the same direction. R_1 and R_2 are used instead of diodes to help offset the effects of the resistance tolerance of the meter. The meter movement used here has a tolerance of ±10%. If the meter has a resistance of 1500 Ω, the variation could be ±150 Ω.

R_1 and R_2 are precision resistors with a value of several times the resistance of the meter, connected such that one of them is always in series with the meter movement. In this case, 5,000 Ω resistors with a ±1% tolerance are used.

To see how this affects the tolerance of the entire circuit, find the amount of deviation allowed by the meter movement.

10% of 1500 Ω = 150 Ω

Then find the deviation allowed by the resistor.

1% of 5,000 Ω = 50 Ω

Find the total deviation allowed.

150 Ω + 50 Ω = 200 Ω.

Determine what percentage this is of total resistance.

$$\frac{200}{6500} = 3.2\%$$

Therefore, by using precision resistors instead of diodes, we have improved the tolerance of the meter circuit from ±10% to +3%.

We have also added enough resistance in the circuit to allow us to use the following calibration method.

Low Calibrate (Lo Cal) and High Calibrate (Hi Cal) are adjusted to compensate for the tolerance of the AC multiplier resistors. Lo Cal will affect low voltage primarily while Hi Cal will have more effect on high voltages. If the range switch is in one of the low voltage positions, only a small portion of the voltage divider is connected in series with the meter movement and the applied voltage. Therefore, a change in the Lo Cal resistor will have a significant affect on the total resistance of the circuit. For instance, on the 2.5 VAC range, Rm in Figure 37 is about 7000 Ω and the Lo Cal resistor is 2200 Ω. You can see that a significant change is possible.

However, on the 500 VAC range, the input resistance (Rm) is approximately 2.5 MΩ. The Lo Cal resistor being in series would have very little effect on total resistance. Therefore, the Hi Cal resistor is connected in parallel with the meter and has a value which will allow approximately a 20% change in the resistance of R_3 and Hi Cal combined. Being in parallel with the meter movement, Hi Cal allows for adjustment of meter current at high voltage ranges.

An output circuit is included to allow the measurement of AC with the presence of DC. It is nothing but a blocking capacitor that keeps DC from the meter.

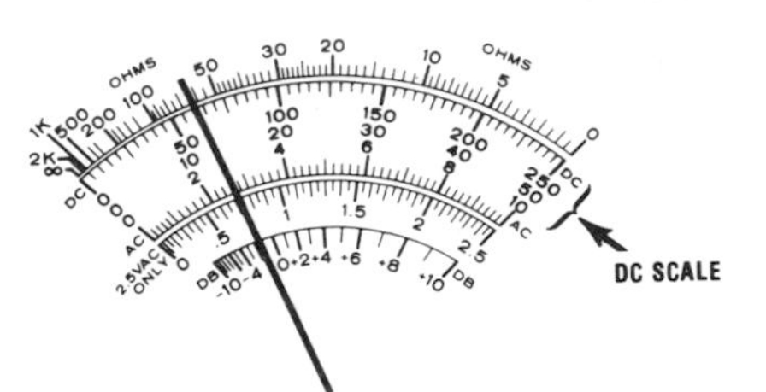

Figure 38

Scales

DC SCALE

Since the analog voltmeter uses the same type of meter movement as an ammeter, the scales will be similar. Whether the voltage scale is linear or non-linear depends on the type of movement used. Our typical meter uses a d'Arsonval meter movement; therefore, the current and voltage scales are linear. Figure 38 shows the scales of our typical meter. The DC voltage scale is shown by the arrow. Full-scale deflection can be 0.25 V, 2.5 V, 10 V, 250 V, 500 V, 1,000 V, or 5,000 V, depending on the range selected. Therefore, the pointer position in Figure 38 can have eight different interpretations. The pointer indicates 0.06 VDC on the 0.25 V range, 0.6 V on the 2.5 V range, 2.4 V on the 10 V range, 12 V on the 50 V range, 60 V on the 250 V range, 120 V on the 500 V range, 240 V on the 1,000 V range, and 1,200 V on the 5,000 V range.

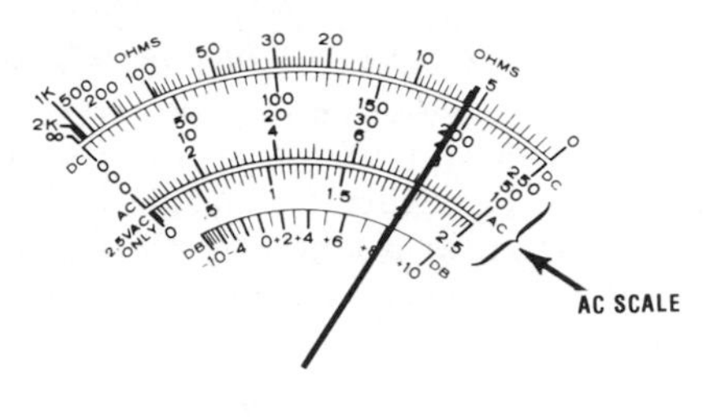

Figure 39

AC SCALE

The AC scales are shown in Figure 39. Notice that the 2.5 VAC scale is separate from the others. Here, 2.0 VAC is being measured on the 2.5 V range. All other AC ranges from 0.25 V to 500 V are measured on the next scale up. Reading this meter on the 2.5 VAC range, we get 2.0 VAC; on the 10 VAC range the reading is 7.9 VAC. Always be very careful when you read meters with multiple scales to insure that you are looking at the proper scale.

DECIBEL SCALE

Another scale found on many AC voltmeters is the decibel or dB scale. Figure 40 shows the dB scale of a typical voltmeter. The pointer is indicating 0 dB. Notice that this is not the zero voltage point on the scale.

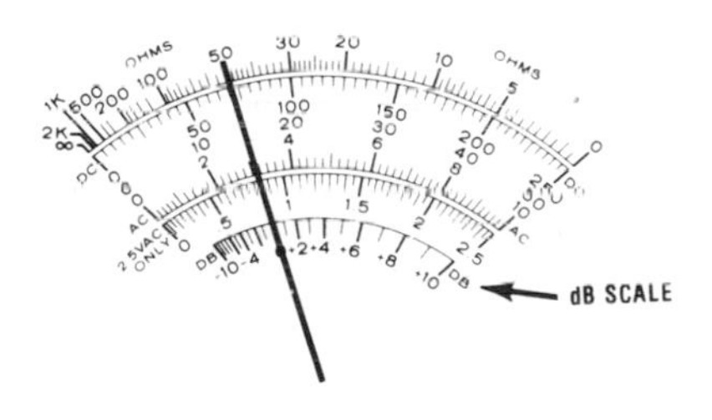

Figure 40

Also, this 0 dB point would only be 0 dB on one scale. On the meter we have been describing, the 2.5 VAC range is calibrated for 0 dB. Zero dB on this meter is the point where one milliwatt of power is produced by 0.775 volts rms across a 600 ohm load. Therefore, on the 2.5 VAC range, we can measure −10 to +10 dB on the dB scale. With this meter, we cannot measure less than −10 dB; however, we can measure more than +10 dB by going to a higher range. If we turn to the 10 VAC range, we must add 12 dB to the meter indication. The zero dB point on our meter becomes the 12 dB point, −10 dB is now +2 dB, and +10 dB is equal to +22 dB. This conversion applies to any point on the dB scale. Conversion factors for the meter are shown below:

RANGE	CORRECTION
10 VAC	Add 12 dB
50 VAC	Add 26 dB
250 VAC	Add 40 dB
500 VAC	Add 46 dB

Other meters will have different conversion factors. They can be found in the instruction manual and are frequently marked on the face of the meter.

Selection

Selecting a voltmeter involves many of the same decisions as selecting an ammeter. Will you be measuring DC or AC or both? Will the AC meter require a true rms measurement? What range of voltages will you measure? And, will you need a dB scale? Decibel measurement is very nice when working with audio. These are just some of the questions to be answered before you buy your meter. If you have — or plan to buy — an oscilloscope, the AC meter won't be as important. However, when adjusting for a null and other such alignment jobs, the meter is often better.

Range is a very important factor in selecting a meter. The lower limit will be determined mostly by the resistance of the meter movement. If you require very high voltage capability, consider a meter with a high voltage probe as an accessory, as it can be hazardous to measure 25,000 volts with ordinary probes.

Since the upper, or full-scale, portion of each range is more accurate, it is nice to have the different ranges spaced so most measurements can be made in the upper half of the scale. However, it all depends on how much accuracy you require. If you need more than two or three percent, it will cost more.

The sensitivity of a voltmeter can vary greatly from model to model. Sensitivities of 1,000 ohms/volt to 100,000 ohms/volt are readily available with 20,000 ohms/volt being the most common. While 20,000 ohms/volt may seem to be a lot, it is only 5,000 Ω of input resistance on the 0.25 V range. However, it is probably adequate on the higher ranges. When measuring voltages in solid-state circuits, voltages of a few millivolts are common, with circuit impedance of over 100,000 Ω not uncommon. In the extremes of this situation, the normal analog voltmeter will not be adequate, and an electronic voltmeter will be required (electronic voltmeters are discussed later). To keep circuit loading to a minimum, the total meter resistance on the range being used should be at least ten times the impedance of the circuit under test.

Remember, the more sensitive the meter, the more delicate and expensive it is. So once more, define your needs and compare.

THE OHMMETER

The basic meter movement can be used to measure resistance. The resulting circuit is called an ohmmeter. In its most basic form, the ohmmeter is nothing more than a meter movement, a battery, and a series resistance.

Basic Circuit

Figure 41 shows the basic circuit of the ohmmeter. The idea behind the ohmmeter is to force a current to flow through an unknown resistance, then measure the current. For a given voltage, the current will be determined by the unknown resistance. That is, the amount of current measured by the meter is an indication of the unknown resistance. Thus, the scale of the meter movement can be marked off in ohms.

The purpose of the battery is to force current through the unknown resistance. The meter movement is used to measure the resulting current. The test probes have long leads and they simplify the job of connecting the ohmmeter to the unknown resistor (R_x). Fixed resistor R_1 limits the current through the meter to a safe level. Variable resistor R_2 is called the ZERO OHMS adjustment. Its purpose is to compensate for battery aging.

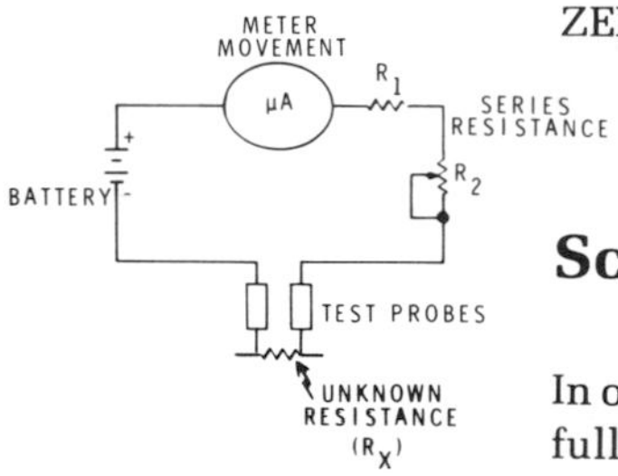

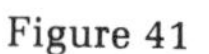

Figure 41

Scale Calibration

In ohmmeters of this type, 0 ohms appears on the right side of the scale (at full-scale deflection). The reason for this is shown in Figure 42A. Here, the two test probes are shorted together. Thus, the unknown resistance (R_x) between the probes is equal to 0 ohms. In this case, the meter should deflect full scale to the 0 ohms marking of the scale. Full-scale deflection for this meter is 50 μA. In order for the 1.5-volt battery to force 50 μA of current through the circuit, the total circuit resistance must be 30 kΩ.

The meter provides 2000 Ω while R_1 provides 22,000 Ω. Thus, R_2 must be set to exactly 6,000 Ω to insure a current of exactly 50 μA.

You may wonder why R_2 isn't a fixed 6,000 Ω resistor or why R_1 isn't a fixed 28,000 Ω resistor. The reason for this is that the battery voltage will change as the battery discharges. If the battery voltage drops to 1.45 volts, then to achieve full-scale deflection, the circuit resistance must be reset to 29 kΩ.

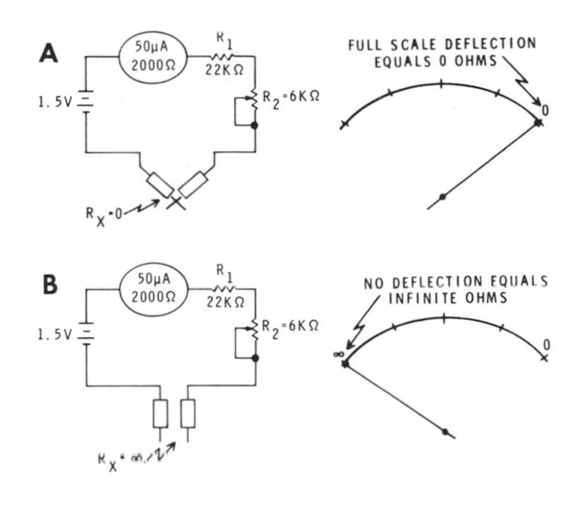

Figure 42

In this case, R_2 would have to be reset to 5000 Ω to compensate for the lower voltage. R_2 is called the ZERO OHMS adjust. Adjusting this resistance to "zero the ohmmeter" is the first step in every resistance measurement.

You have seen that full-scale deflection corresponds to an unknown resistance (R_x) of 0 ohms. Thus, the scale is marked 0 at this point. Now, what about the left side of the scale (no deflection)? Figure 42B illustrates this condition. Here, an open circuit exists between the two test probes. This corresponds to an infinite resistance. No current flows through the meter movement and the pointer rests at the left side of the scale. Consequently, this point on the scale is marked with the infinity symbol (∞). Thus, we have a scale with 0 ohms on the right and infinite ohms on the left.

Now we will discuss what resistance is represented by 1/2-scale deflection. The pointer will deflect to the center of the scale when the current is exactly 25 μA. This amount of current is caused by a total resistance of 60 kΩ.

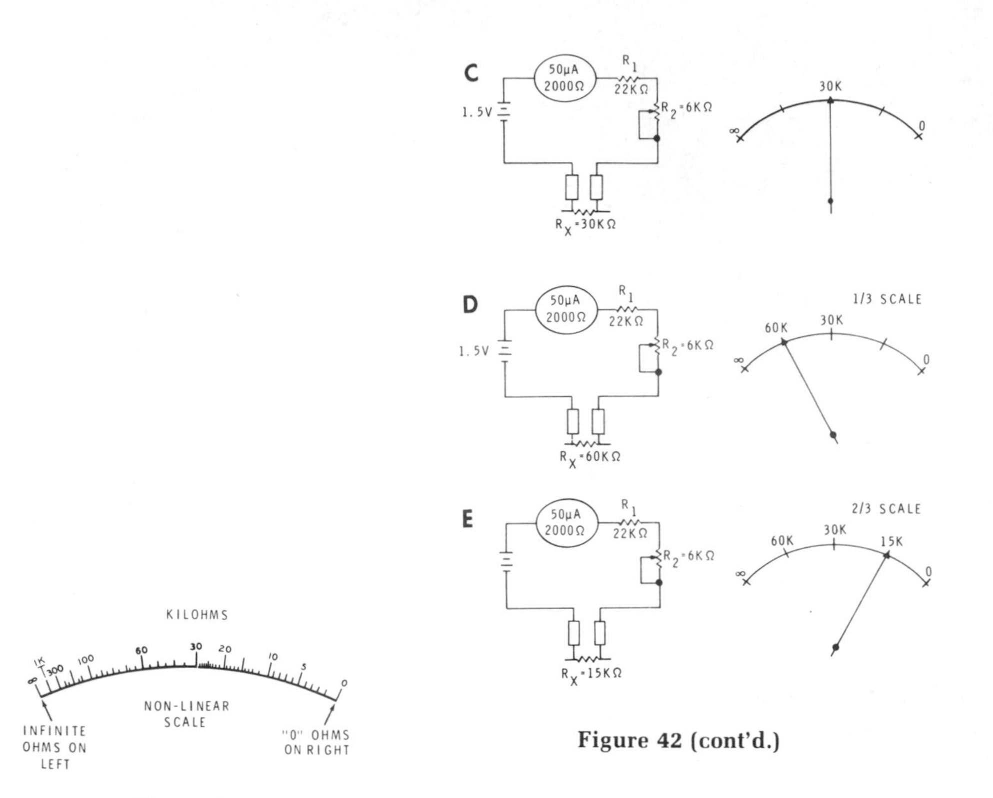

Figure 42 (cont'd.)

Figure 43

Since the meter, R_1, and R_2 have a combined resistance of only 30,000 Ω, the unknown resistance (R_x) supplies the other 30,000 Ω. That is, the meter deflects to half scale (25 μA) when the unknown resistance has a value of 30 kΩ. Consequently, the 1/2-scale point is marked 30 kΩ as shown in Figure 42C.

Using this same procedure, we can determine the amount of meter deflection for any value of R_x. Figure 42D illustrates that 1/3 full scale indicates an R_x of 60 kΩ, while 2/3 full scale shown in Figure 42E indicates an R_x of 15 kΩ. You can verify this by proving that 1/3 of 50 μA or 16.66 μA flows in the circuit shown in Figure 42D. Also, verify that 2/3 of 50 μA or 33.33 μA flows in the circuit shown in Figure 42E.

If enough points on the scale are found, the scale will take the form shown in Figure 43. There are two important differences between this scale and the ones used for current and voltage. First, the ohms scale is reversed with 0 on the right. Second, the scale is non-linear. For example, the entire right half of the scale is devoted to a range of only 30 kΩ; that is, from 0 to 30 kΩ. However, notice that the second 30 kΩ (from 30 kΩ to 60 kΩ) takes up less than 1/4 of the scale. The markings are squeezed closer together on the left side of the scale.

Creating Higher Ranges

A one-range ohmmeter would be of limited use. For this reason, multi-range ohmmeters have been developed. Two techniques have evolved for creating additional ranges. Both techniques are used in some ohmmeters.

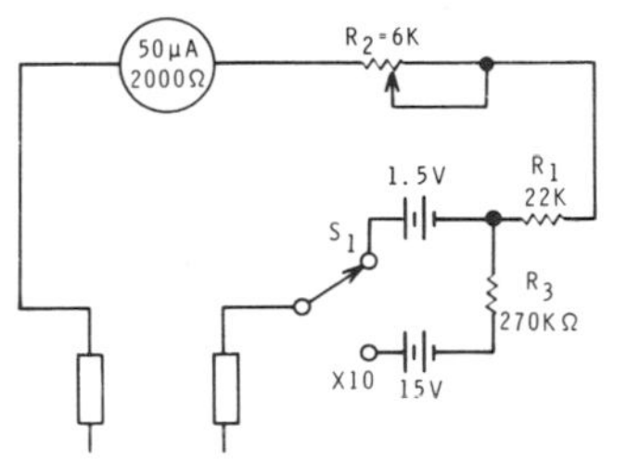

Figure 44

Figure 44 shows how a higher resistance range can be implemented. First, a switch is added to switch between the two ranges. Second, a higher voltage battery is added. Finally, a higher value series resistor is required. To increase the range by a factor of 10, both the voltage and the total series resistance must be increased by a factor of 10.

When S_1 is in the position shown, the meter operates exactly like the one shown earlier in Figure 42. However, when S_1 is switched to the × 10 position, the 15-volt battery is switched in series with R_3, R_1, R_2, and the meter. The higher voltage will not cause excessive current through the meter since the series resistance has been increased by the addition of R_3. Notice that the total resistance in the circuit is now 300 kΩ. Thus, when the leads are shorted together, the current is still 50 µA.

The right side of the scale still represents 0 ohms. However, 1/2 scale deflection (25µA) now occurs when the total resistance is 600 kΩ.

Of this 600 kΩ, the meter, R_1, R_2, and R_3 supply 300 kΩ. Therefore, the unknown resistance (R_x) must be 300 kΩ. This means that the center of the ohmmeter scale now represents 300 kΩ instead of 30 kΩ. The range has been increased by a factor of 10.

Of course, this technique cannot be carried much further because increasing the range by an additional factor of 10 would require a 150-volt battery. Fortunately, the range described above is sufficient for general-purpose use. It can measure resistance values up to several megohms. Resistors larger than this are rarely used in most electronic applications.

Creating Lower Ranges

The basic ohmmeter can also be modified to measure lower values of resistance. This is done by switching a small value shunt resistor in parallel with the meter and its series resistance.

Refer to Figure 45. With switch S_1 in the position shown, the ohmmeter operates exactly like the one shown earlier in Figure 42. However, when the position of S_1 is changed, R_3 is connected in parallel with the series combination of the meter, R_1, and R_2. The value of R_1 is 300 Ω or 1% of the combined series resistance of R_1, R_2, and the meter (30,000 Ω). Therefore, 99% of the current will flow through R_3 and only 1% will flow through the meter circuit. Recall that 25 μA of current is required for 1/2-scale deflection of the meter. What value of R_x will cause this amount of current to flow through the meter?

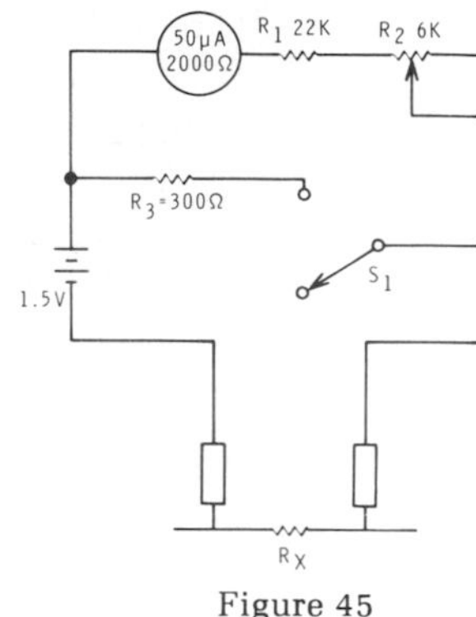

Figure 45

The resistance of the meter circuit is now about 300 Ω. If an unknown resistance R_x of 300 Ω is now connected between the probes, the current from the battery becomes 2.5 mA.

However, 99% of this current (2.475 mA) flows through R_3. Only 1% or 25 μA flows through the meter movement. Thus, 1/2-scale deflection now represents an unknown resistance of 300 ohms instead of 30 kΩ. Using this technique, lower ohmmeter ranges can be created.

Shunt Ohmmeter

The ohmmeters discussed up to this point are called series ohmmeters because the unknown resistance is always placed in series with the meter movement. A series ohmmeter can be recognized by its "backwards" scale. That is, 0 ohms is on the right while infinite ohms is on the left.

Another type of ohmmeter is called a shunt ohmmeter. Figure 46 illustrates the basic circuit of the shunt ohmmeter. This instrument gets its name from the fact that the unknown resistance is placed in parallel (shunt) with the meter movement. This completely changes the characteristics of the ohmmeter. For example, notice that when an open (infinite ohms) exists between the probes, 50 μA of current flows through the meter movement. This produces full-scale deflection. Consequently, infinite ohms is on the right side of the scale or at full-scale deflection. Notice that this is just the opposite of the series ohmmeter.

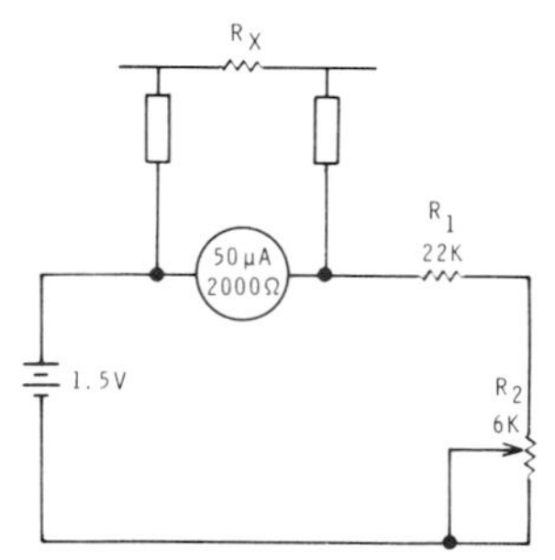

Figure 46

When the probes on the shunt ohmmeter are shorted together (representing an R_x of 0 ohms), the meter movement is shorted out. This produces no deflection. Thus, 0 ohms is on the left.

Recall that with the series ohmmeter, the 1/2-scale reading was 30,000 Ω for the 50 μA, 2000 Ω meter movement. However, with the shunt ohmmeter, this too is different. Here 25 μA of current flows through the meter movement when R_x is the same resistance as the meter movement. Thus, 1/2-scale deflection on the shunt ohmmeter is marked 2000 Ω.

The shunt ohmmeter has some disadvantages. For one thing, the battery discharges any time the ohmmeter is turned on. This is not the case with the series ohmmeter. It draws current from the battery only when a resistance is being measured.

Also, the meter movement in the shunt ohmmeter is more easily damaged if the meter is inadvertently connected across a voltage source. In the series meter, the 28,000 Ω in series with the meter movement tends to limit the current. Even so, we should never connect either type of ohmmeter to a live circuit.

Finally, because the 1/2-scale reading of the shunt ohmmeter is much less than that of the series ohmmeter, it is more difficult to accurately measure high resistance values on the shunt meter. However, it is easier to read low resistance values on the shunt meter for the same reason.

Accuracy

The accuracy of an ohmmeter is usually given in degrees of arc since it is difficult to express as a percentage figure on a non-linear scale. A typical meter has a movement accuracy of ±2% and resistor tolerances of ±1%. We can see that the accuracy of our ohmmeter is similar to that of our other meters.

Measurements

The ohmmeter is at times, one of the simplest instruments to use. At other times, it is one of the most difficult instruments to use. Checking resistance is very easy if the component is out of a circuit; connect a lead to each end of the resistor under test, select the proper resistance range and read the value of resistance from the scale. Multiply the reading times the range factor. For instance; if on the R × 100 range, you read 15 on the scale, you would multiply 15 × 100 for a resistance of 1500 Ω.

Components other than resistors can be tested with the ohmmeter. Capacitors, for instance, can be checked for leakage. All we are really doing is measuring a very high resistance. A good capacitor will have a resistance of several hundred megohms; therefore, the highest resistance range should be used. The capacitor will first indicate a low value of resistance which will increase to near infinity as the capacitor charges. With smaller values of capacitance, the increase may be so rapid that the initial reading appears to be infinity.

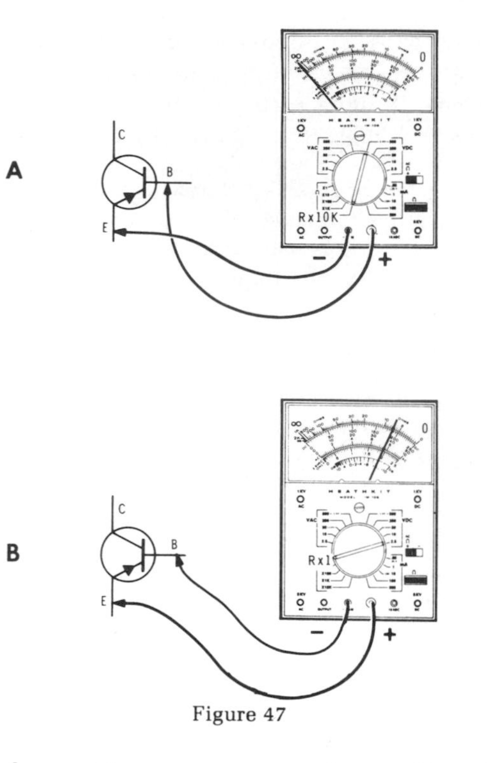

Figure 47

Some special precautions are necessary when working with electrolytic capacitors. First, correct polarity must be observed. This means you must know which is the positive and which is the negative lead on your ohmmeter. The handbook that comes with your meter can tell you this or you can measure the output with a voltmeter. You must also know the voltage at the lead tips of your ohmmeter so as not to exceed the voltage rating of the capacitor under test. With high values of capacitance, it may be best to start on the R × 1 range in order to keep the resistance low and reduce charge time, then increase the ranges as the capacitor charges.

Diodes are very easy to check with an ohmmeter. What we are interested in is the front-to-back resistance ratio of the diode. Remove the diode from the circuit, measure the resistance, then reverse the leads and measure the resistance again. One value should be considerably more than the other. If the diode is forward biased by the meter, the resistance will be low; if it is reverse biased, the resistance will be high. The actual ratio may vary from a minimum of 100 to 1 to several thousand to one, depending on the type of diode being tested.

Transistors may also be tested with an ohmmeter. There are three tests to be conducted. Figure 47A shows an ohmmeter connected to the base-emitter junction of a PNP transistor. The positive lead is connected to the base (N type material) and the negative lead to the emitter (P type material). This reverse biases the junction and gives a high resistance reading. Figure 47B shows the leads reversed, which forward biases the junction and gives a low resistance reading. The ratio should be similar to that achieved with a diode.

A similar check is made from base to collector. About the same results should be expected.

If both of these checks are good, an emitter-to-collector check should be made. In this check, both readings should be high.

Use caution in making these checks. On the R × 1 scale, a typical ohmmeter will allow a current of about 75 mA, which is enough to damage some diodes and transistors. On the R × 10 k range, a 15 V battery is used for power. The actual battery voltage may vary with different brands of meter. This is enough to cause junction breakdown in many diodes and transistors. Make sure you choose a range that will not destroy the component under test. Each meter will vary in the ranges that can be safely used.

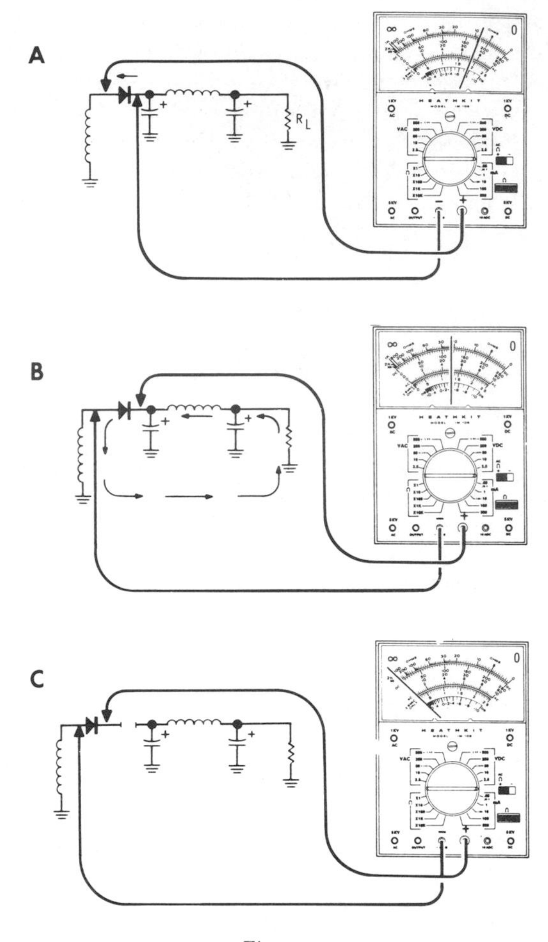

Figure 48

When making tests within a circuit, these and additional precautions apply. You want to test the diode in the power supply shown in Figure 48A. You might connect the ohmmeter as shown. You would get a low resistance reading. If you reverse the leads as in Figure 48B, you find that this also gives a relatively low reading due to the parallel path through the transformer secondary, R_L, and the choke, all of which have a low resistance. This same type of problem applies to any other circuit component being tested.

In addition to erroneous readings, it is possible to damage circuit components. Low voltage electrolytic capacitors are common in solid-state circuits and with a connection as shown in Figure 48B, we could exceed the working voltage of the capacitor. In Figure 48A, the electrolytic capacitors have the wrong polarity of voltage applied. This too can damage the capacitors.

About the only sure way to test a component is to isolate it from the circuit, which means disconnecting one lead as shown in Figure 48C. Here, we will read only the resistance of the diode. Disconnecting one lead of a component can be difficult in printed circuits. Therefore, this is usually done only to confirm a suspected failure.

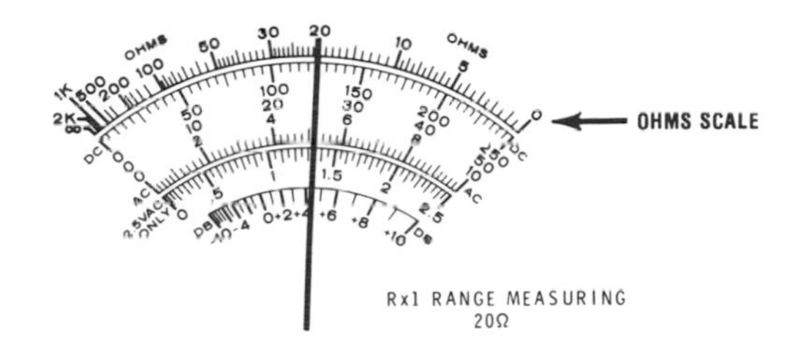

Figure 49

Scales

The ohmmeter has a non-linear scale which we discussed earlier. A typical ohmmeter scale is shown by the arrow in Figure 49. With the pointer positioned as shown, the meter indicates 20 Ω on the R × 1 range, 200 Ω on the R × 10 range, and so on. To find the resistance, read the meter directly and multiply the reading times the range factor. If the pointer goes higher than 50, it is usually best to go to the next higher range as the scale gets too crowded for accurate reading on the upper end. Also, notice that the scale divisions change as we go up the scale. From 0 to 10, we have divisions of .5; from 10 to 30, divisions of 1; from 30 to 50 divisions of 2, etc. As we move to the left of the scale, the divisions grow larger in numerical value and smaller in physical size. This is another reason for taking the reading at the lower end of the scale.

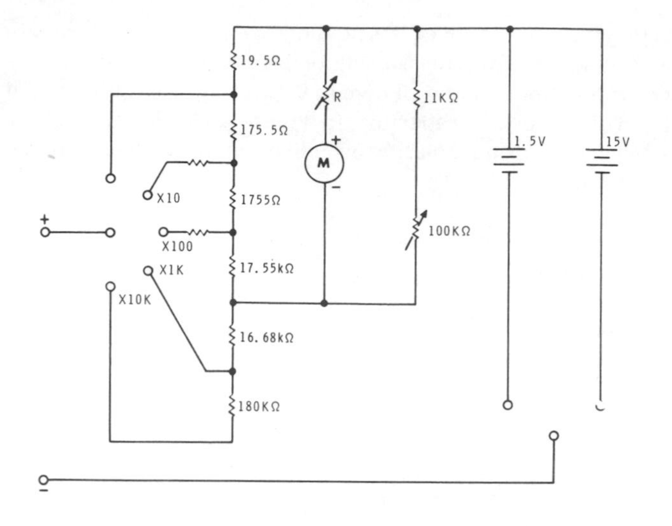

Figure 50

Typical Ohmmeter

Figure 50 shows a typical ohmmeter circuit. This circuit varies somewhat from the circuits we have been discussing. Here, the unknown resistance is connected in series with a battery and a precision internal resistor. The meter movement is connected in parallel with the precision standard resistor and acts as a voltmeter. To help you better understand this circuit, we will discuss the meter on the R × 1 range. Figure 51 is a simplified diagram of this range.

In Figure 51, the 19.5 Ω resistor is connected in a parallel with the rest of the meter circuitry. The 175.5 Ω resistor, the 1755 Ω resistor, and the 17.55 kΩ resistor in Figure 50 are added together to make the 19.5 kΩ resistor which is in series with the meter circuit. The resistance is actually 19.48 kΩ but is so much larger than the 19.5 Ω standard resistor that the small difference will not be noticed.

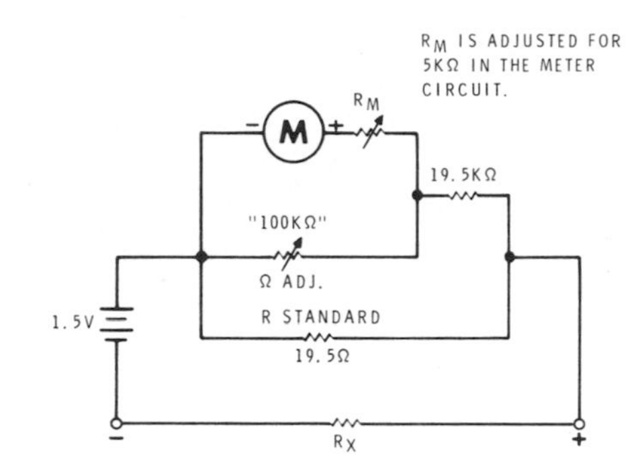

Figure 51

In Figure 51, R_x represents the unknown value of resistance to be measured. If the ohmmeter terminals are not connected — R_X is infinite, then no current will flow through the meter and the meter will indicate infinity. With the leads shorted together, 1.5 V is developed across $R_{standard}$ and across the meter circuits. $R_{standard}$ will carry a current of 76.923 mA. The meter movement in this case is a 50 μA; therefore, with the leads shorted, it should carry 50 μA. Since R_m is adjusted for a meter circuit resistance of 5 kΩ, the voltage drop across the meter circuit is 0.25 volts, leaving 1.25 volts to be dropped by the 19.5 kΩ resistor. In order to drop this voltage, a current of 64.103 μA must flow through the resistor. If 50 μA flows through the meter, then 14.103 μA must flow through the "ohms adjust" resistor. This resistor must be adjusted for 17.727 kΩ to "zero" the meter. Thus, the meter can be adjusted to compensate for changes in battery voltage or aging of certain components.

You have seen that the total current available on the R × 1 range is approximately 77 mA, which is enough to destroy some components.

Now look at the R × 1 k range shown in Figure 52. Once more, the meter has a 50 μA movement and R_m is adjusted for a meter circuit resistance of 5 kΩ. $R_{standard}$ is now 19.5 kΩ, the sum of the 19.5 Ω and the 17.55 kΩ resistors. The Ω ADJ will be adjusted for full scale current (0 ohms) with the leads shorted together. This will provide a voltage of 0.25 volts across $R_{standard}$ and the meter circuit leaving 1.25 volts to be dropped by the 16.68 kΩ resistor.

Current through the 16.68 kΩ resistor is 74.94 μA which is total current. This is much less than then 77 mA on the R × 1 range. Of this current, the meter carries 50 μA, $R_{standard}$ carries 12.8 μA, and the ADJ carries 12.12 μA. Therefore, the Ω ADJ must have a resistance of 20.63 kΩ.

Refer back to Figure 50 to see what happens on the R × 10 k range. The only change is to add 180 kΩ in series with the 16.68 kΩ resistor. You can see that total circuit resistance will be approximately 200 kΩ. With the 1.5 V battery, current will be about 7.5 μA, not enough to give full-scale deflection with the leads shorted. Therefore, the 15 V battery would be switched in, which provides enough voltage for 75 μA.

Selection

Choosing an ohmmeter is much the same as choosing an ammeter or voltmeter. Since the meter will most likely be used with solid-state circuits, make sure the voltage and current outputs are low enough that they will not damage circuit components. The range switch should go from ×1 to ×10 to ×100, etc. with no skips on a good meter. However, if the meter meets all other requirements, skips may be acceptable in the interest of keeping the price down.

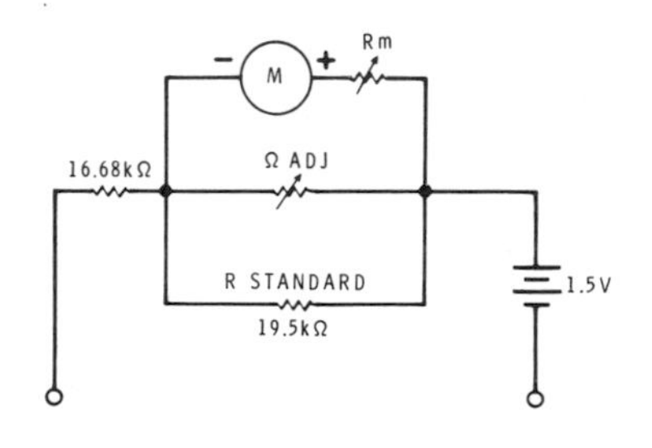

Figure 52

VOLT-OHM-MILLIAMMETER

The Volt-Ohm-Milliammeter (VOM), or multimeter, is nothing more or less than the three basic instruments — Ammeter, Voltmeter and Ohmmeter — combined into one case. Cost and convenience are the primary reasons for the combination. Using a single movement and single case can allow a reduction in cost over the three instruments. Carrying a single instrument is much easier and usually more convenient than carrying three. The major drawback of the VOM is that only one instrument can be used at a given time. This makes it difficult to monitor current while adjusting voltage, etc. This type of measurement is required in some alignment procedures.

Figure 53 is a schematic diagram of a very basic multimeter. This meter is very limited and not at all representative of what is available. It has four current ranges and four voltage ranges with only two resistance ranges, a high and a low. The basic meter movement makes up the 0.1 V range and the 50 μA range. Both are measured with the function switch in the mA position.

The function switch (S2) selects the parameter to be measured while the range switch (S1) selects the maximum amount.

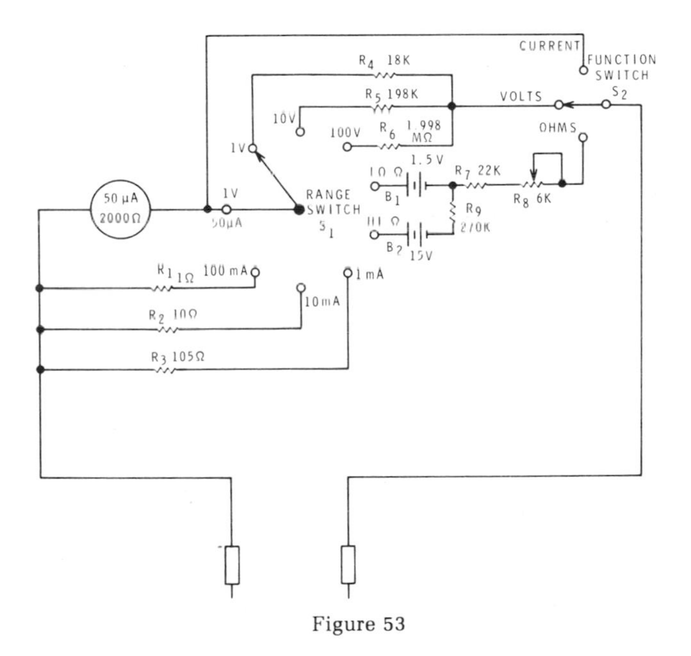

Figure 53

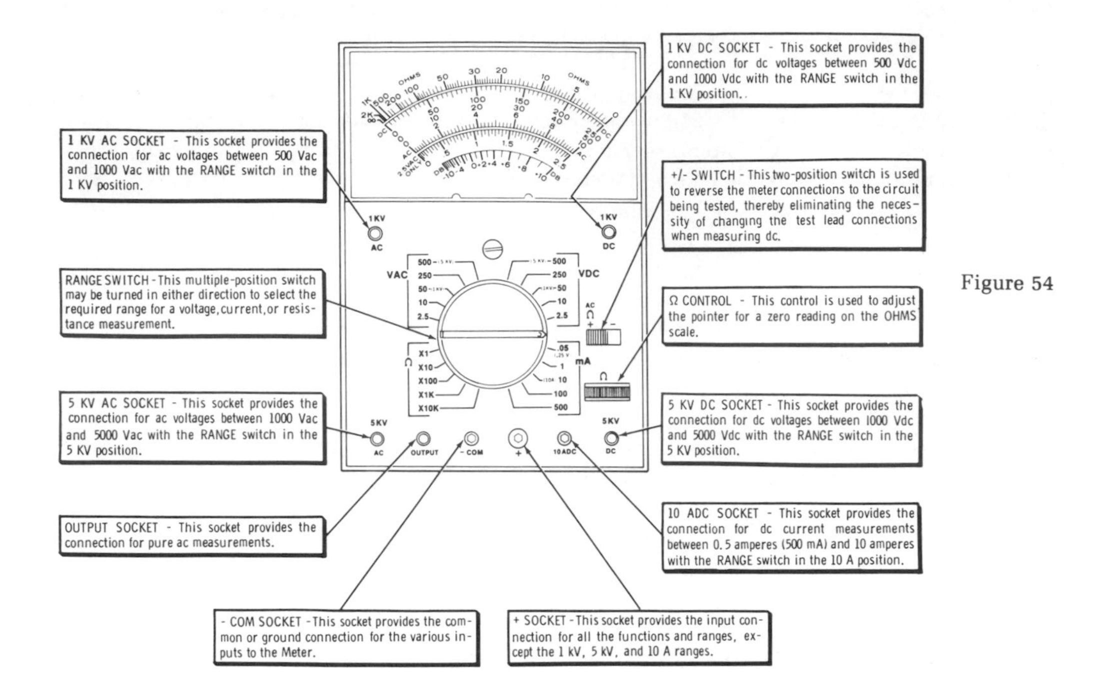

Figure 54

Typical VOM

A typical VOM is shown in Figure 54. The function and range switches are combined. The purpose of each control and socket is explained. We have already discussed many of the scales and markings on this meter, therefore, we will only discuss some special features here.

The markings on the voltage and current positions of the range switch are the full-scale deflection readings of the meter. Be sure that the quantity to be measured does not exceed the range you have selected. Also, be sure to select the proper function. Trying to measure voltage on the mA or Ω ranges can damage the meter circuits.

The probes must be inserted in the proper socket. Attempting to measure 1 kV or 5 kV without moving the probe to the proper socket can burn out the switch. The polarity switch must be in the proper position or the pointer can deflect in the wrong direction and perhaps bend the pointer.

Notice that some of the ranges are shared. The .05 mA and the .25 V DC range is one. It uses the basic meter with no multipliers or shunts and may be used for either measurement.

The 10 mA and 10 A ranges are shared by the switch but the probe must be moved from the "+" socket to the 10 A DC socket to measure the larger current. The 50 V/1 kV and 500 V/5 kV ranges are shared for both AC and DC; the probe must be moved to the proper socket on these ranges also.

DC VOLTMETER

A complete schematic of the meter which we have been discussing is shown in Figure 55. All of the switches in this drawing are shown in the 10 V DC position. Next, the current flow will be traced through this meter.

Start in the lower left-hand corner of the meter at the large minus (−) sign. Follow the arrows to pin 16 of switch S2 rear. Notice that the movable contact of the switch is making contact between pin 16 and pin 18. From pin 18, follow the arrows to pin 1 of the sliding switch (S3). This is the positive/negative switch. It is shown in the positive position. Notice that pins 1 and 2 are making contact in this position. Go in pin 1 and out pin 2 to the negative terminal of the meter.

Current flows through the meter from left to right, through the variable resistor and the thermistor, to pin 5 of switch S3. Since this switch is in the positive position, pins 4 and 5 are making contact.

Follow the arrows from pin 4, through R27 and R26 to pin 10 of S1 rear. Notice that the tab on the movable contact of the switch is making contact with pin 10. This tab also goes to a wiper arm which makes contact continuously in all switch positions. The wiper arm is connected to pin 18. From pin 18 of switch S1 rear, current flows through fuse F1 to the positive terminal of the meter. Thus, we have traced current flow from the negative terminal of the meter, through the meter movement to the positive terminal. A similar path exists on all DC voltage positions. The only major difference is the number of multiplier resistors. In this particular case, we went through only two, R26 and R27. Going to a higher range means going through more multiplier resistors.

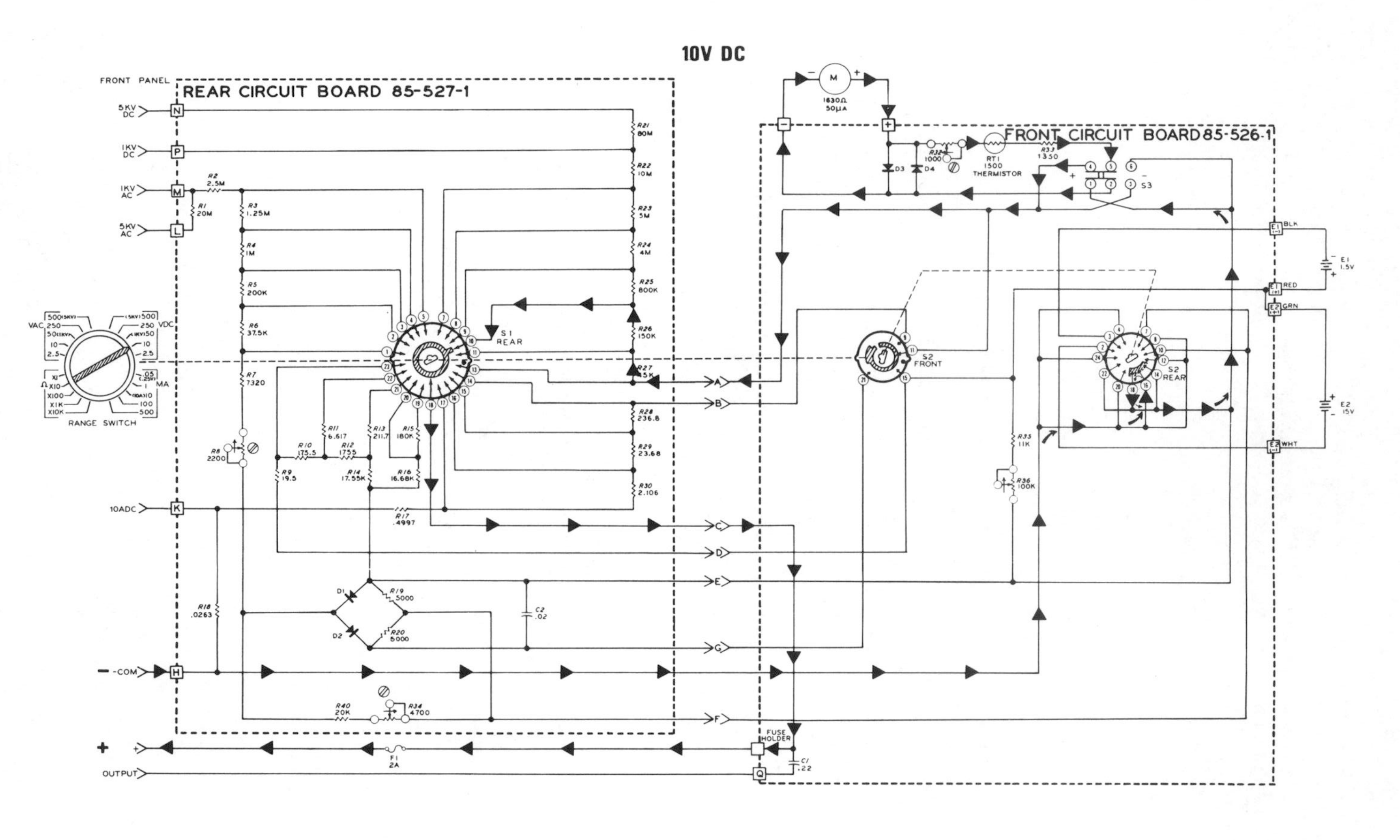

10V DC
FRONT PANEL
REAR CIRCUIT BOARD 85-527-1
FRONT CIRCUIT BOARD 85-526-1
5KV DC
1KV DC
1KV AC
5KV AC
10ADC
-COM
+
OUTPUT
RANGE SWITCH
VAC
VDC
MA
R1 20M
R2 2.5M
R3 1.25M
R4 1M
R5 200K
R6 37.5K
R7 7320
R8 2200
R9 19.5
R10 175.5
R11 6.617
R12 1755
R13 211.7
R14 17.55K
R15 180K
R16 16.68K
R17 .4997
R18 .0263
R19 5000
R20 5000
R21 80M
R22 10M
R23 5M
R24 4M
R25 800K
R26 150K
R27 15K
R28 236.8
R29 23.68
R30 2.106
R33 1350
R34 4700
R35 11K
R36 100K
R40 20K
C1 .22
C2 .02
F1 2A
FUSE HOLDER
S1 REAR
S2 FRONT
S2 REAR
S3
D1
D2
D3
D4
RT1 1500 THERMISTOR
1630Ω 50μA
E1 1.5V
E2 15V
BLK
RED
GRN
WHT

Figure 55

MILLIAMMETER SECTION

Now we will discuss the switches in one of the current positions. Figure 56 shows the switches in the 10 milliamp position. Trace the current through the meter. Start with the large negative sign by the common terminal. Coming in through the common terminal, you will notice that there are two types of arrowheads. This indicates two current paths. Recall that in an ammeter, a portion of the current flows through the meter movement and the rest flows through a shunt resistor which is in parallel with the meter movement. The shunt path is indicated by the solid arrows while the meter path is indicated by the open arrows. Trace the path for current through the meter movement first.

Follow the open arrows to pin 20 of switch S2 rear. Current flows out pin 18 of switch S2 rear and into pin 1 of the sliding switch S3. The switch is in the positive position, so pin 1 makes contact with pin 2. Trace from pin 2 to the negative terminal of the meter. Notice that the current flow in this position is still from the left to the right through the meter movement. Current flow through the meter movement will always be in the same direction.

From the meter movement, current flows through the resistor network and into pin 5 of the sliding switch. Follow the arrows to contact 11 of switch S2 front. The switch position is such that pin 11 is making contact with pin 9. Trace the current from pin 9 of switch S2 to switch S1 rear. Notice that the current which went through the shunt has rejoined the meter movement current.

We will stop at this point, go back, and pick up the shunt current which flows in parallel with the meter movement. Back at the common terminal again, both currents come in through pin H. They split at the first tie point with meter current going to the right. This time, follow the shunt current through R17, R30, and R29 to join the meter current.

The total current now flows into pin 15 of switch S1 rear. The switch connects pin 15 to pin 18. Follow the arrows to fuse F1 and out the positive terminal of the meter. Thus, there is a complete path of the current flow through the ammeter section of the meter. One of the paths goes through the meter, the other path goes through the shunt resistance. The other current ranges operate approximately the same and should be easy to trace following a similar procedure.

10 mA

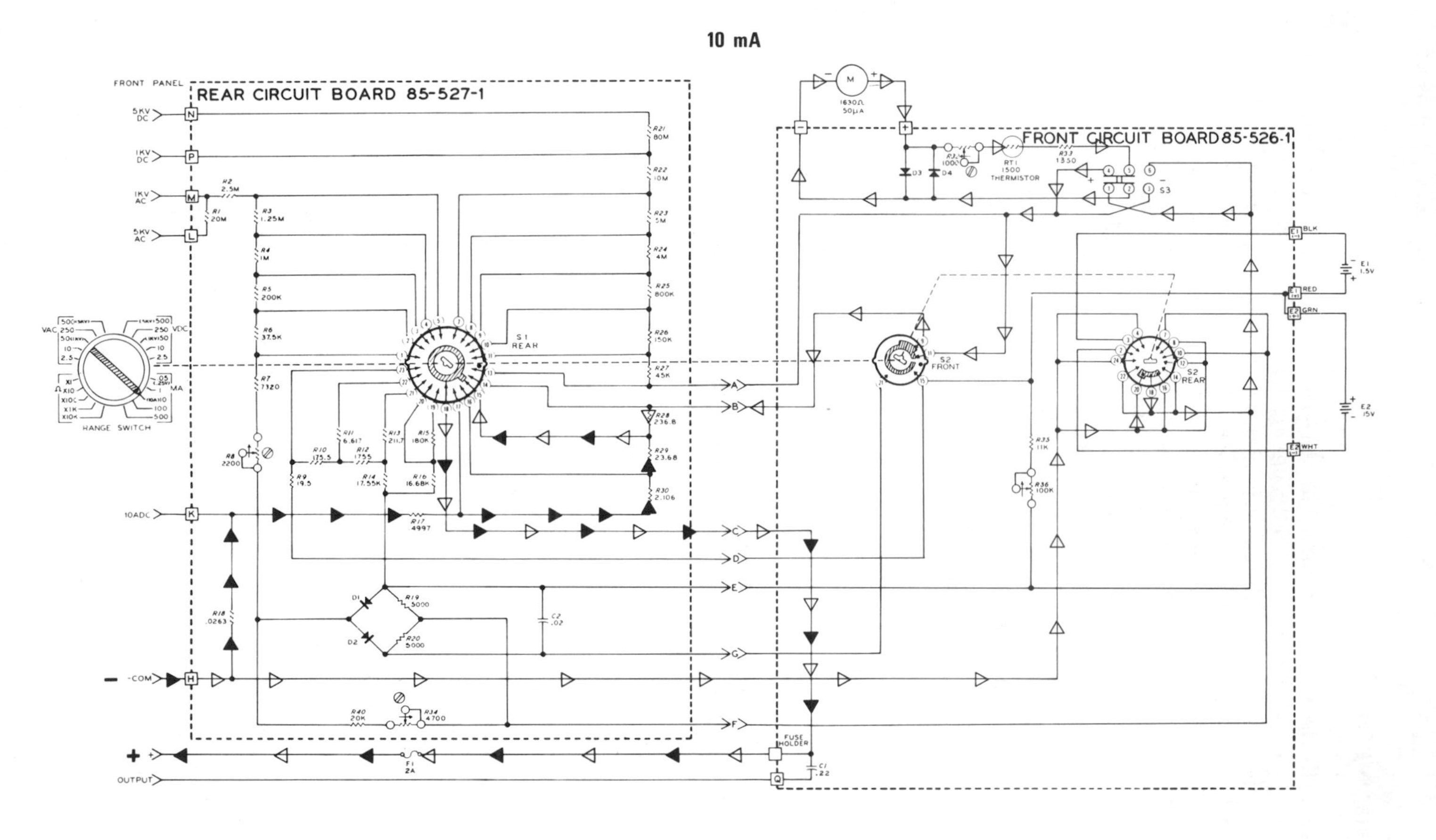

Figure 56

AC VOLTAGE

Figure 57 shows the meter in the 2.5-volt AC position. When measuring AC, we can connect the leads to the common and the positive terminal or to the common and the output terminal. The choice depends on whether the AC we are measuring has a DC component. If there is no DC component, we generally use the positive terminal of the meter. If there is a significant DC component in the voltage being measured, it could affect the outcome of our reading. Therefore, we would use the output terminal for making this measurement. In this example, we will use the positive terminal.

You know from the study of AC that the voltage changes polarity at the input, being positive on one terminal and negative on the other, then reversing to positive and negative in the opposite direction. We will start the discussion with a negative at the common terminal. This polarity of current will be indicated by solid arrows. Current flow is from the negative common terminal to pin 8 of switch S2 rear. The contact positions are such that pin 8 is now making with pin 7. Trace the current from pin 7 to the tie point between R19 and R20 of the bridge circuit which is used for a rectifier. The primary path path for current flow at this time is through R19. Follow the solid arrows to pin 1 of switch S3. Pins 1 and 2 are making contact. Thus, current flows from pin 2 to the negative terminal of the meter. Once again, current flow is from left to right through the meter movement.

Current continues from the meter movement through the adjustable resistor and the thermistor to pin 5 of switch S3. Pin 5 is making contact with pin 4. Trace the current from pin 4 of S3 to pin 11 of switch S2 front. Pin 11 is making contact with pin 21. Follow the arrows to the tie point at the junction of D2 and R20. With this polarity of input, current flow will be through D2, adjustable resistor R8, and resistor R7, to pin 1 of switch S1 rear. The tab of S1 makes contact with pin 1. Therefore, you can go from pin 1 to pin 18. From here the current flows through F1 and out the positive terminal of the meter. You have thus completed the path for current flow through the meter in the 2.5-volt AC range with a negative at the common terminal and a positive at the positive terminal.

When the AC signal at the input terminals reverses, current will flow through the meter as indicated by the open arrows. For the most part, the path is exactly the reverse of that which you just traced. Therefore, you should have no problem tracing the current through the meter.

You have now studied the switching arrangements for three functions of a typical volt-ohm-milliammeter. Current paths for other functions and ranges can be found using similar methods. Of course, switching arrangements will vary from one meter to another.

2.5V AC

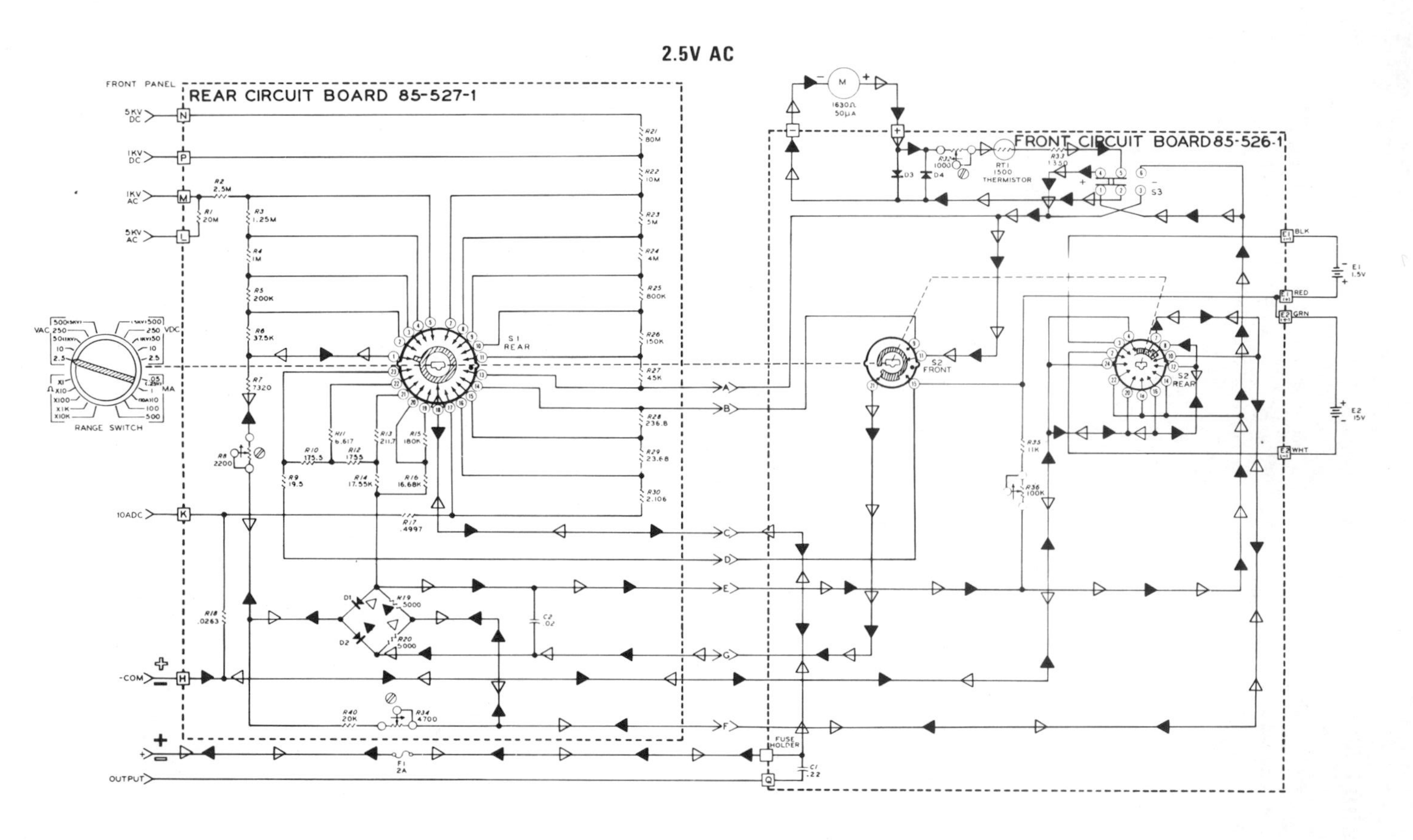

Figure 57

Selection

You have seen that the VOM offers a versatile and convenient way to make a variety of measurements. It frequently goes with the technician from bench to field.

Portability is, therefore, a big factor in selecting a VOM.

To meet this requirement, the VOM must not depend on any external source of power, nor should it be too large. However, smaller meters tend to be more delicate. A VOM should be rugged as it will often see rough use. Unfortunately, there are a number of units on the market which will not stand up to hard use. It is often impossible to tell from the specifications how sturdy a meter is. A good method is to check with some local servicemen to see how particular brands hold up.

AC current measuring is often absent from the VOM. Also missing is accuracy better than 2 to 3% . True RMS readings of voltage and current are not normally available. What the VOM does provide is a general-purpose instrument capable of making from 80 to 95% of a technician's measurements.

When selecting a VOM, you will be restricted in the types of meter movements available. Almost all VOM's are built around the d'Arsonval movement. Since the voltmeter section is usually the most used, compromises are often made in other areas to keep the cost down.

ELECTRONIC METERS

The electronic meter (EVOM) has several advantages over the passive type of VOM. By electronically processing the incoming voltage or current, much greater sensitivity with less circuit loading can be obtained. Accuracy in an analog unit meter is limited primarily by the accuracy of the meter movement; however, for a given movement, we can expect somewhat better accuracy.

Typical Meter

OPERATION

Figure 58 shows the front of a typical EVOM with an explanation of all the controls. You may want to refer to this diagram from time to time as you study the operation of this meter. The following, are some items that are different from the passive VOM.

1. The ZERO CONTROL R4 is used to electrically zero the meter on all ranges and functions. There is also an OHMS ADJUST which is used to adjust for full-scale deflection with the leads not touching when in the OHMS position. Note that the OHMS scale has infinity (∞) at the full-scale deflection side of the meter. This is typical of EVOMs; however, it is reversed on some units.

2. This meter has an "auto-polarity" function. With this function, the test leads may be connected either way and the meter will indicate the relative polarity of voltage at the "+" input, which is the red input probe. The indication is accomplished by two, light-emitting diodes (LED's) located in the low corners of the meter face. If the voltage at the "+" input, which is more positive than that at the "−" input, the DC LED will light. If the "+" voltage is negative with respect to the "−" input, the DC− LED will light. With this type of circuit, there is no danger of damaging the meter by connecting it backward. With an AC input, the polarity is constantly changing, therefore, both LED's will light.

3. Another useful feature of this meter is the Ω LV range. LV stands for low voltage. The voltage provided by this range is too low to forward bias a PN junction in a semiconductor. Therefore, this range is very useful for measuring resistance values in semiconductor circuits. This range cannot be used to test transistors or diodes as the voltage is too low to give a proper indication. The DC/Ω range must be used for this purpose.

Other than the features just mentioned, the operation of our typical EVOM is very similar to that of a VOM.

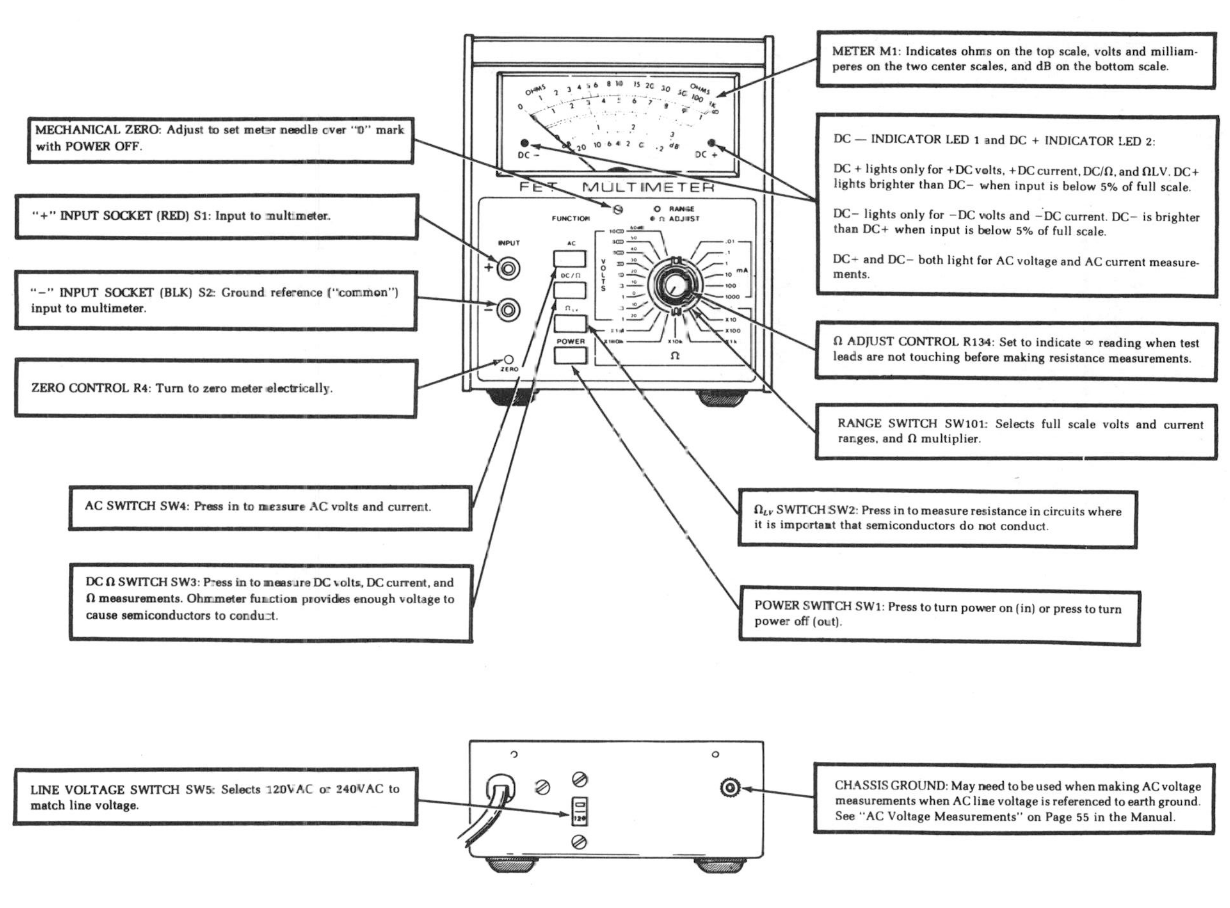
METER M1: Indicates ohms on the top scale, volts and milliamperes on the two center scales, and dB on the bottom scale.
DC — INDICATOR LED 1 and DC + INDICATOR LED 2:
DC + lights only for +DC volts, +DC current, DC/Ω, and ΩLV. DC+ lights brighter than DC− when input is below 5% of full scale.
DC− lights only for −DC volts and −DC current. DC− is brighter than DC+ when input is below 5% of full scale.
DC+ and DC− both light for AC voltage and AC current measurements.
Ω ADJUST CONTROL R134: Set to indicate ∞ reading when test leads are not touching before making resistance measurements.
RANGE SWITCH SW101: Selects full scale volts and current ranges, and Ω multiplier.
Ω_{LV} SWITCH SW2: Press in to measure resistance in circuits where it is important that semiconductors do not conduct.
POWER SWITCH SW1: Press to turn power on (in) or press to turn power off (out).
MECHANICAL ZERO: Adjust to set meter needle over "0" mark with POWER OFF.
"+" INPUT SOCKET (RED) S1: Input to multimeter.
"−" INPUT SOCKET (BLK) S2: Ground reference ("common") input to multimeter.
ZERO CONTROL R4: Turn to zero meter electrically.
AC SWITCH SW4: Press in to measure AC volts and current.
DC Ω SWITCH SW3: Press in to measure DC volts, DC current, and Ω measurements. Ohmmeter function provides enough voltage to cause semiconductors to conduct.
LINE VOLTAGE SWITCH SW5: Selects 120VAC or 240VAC to match line voltage.
CHASSIS GROUND: May need to be used when making AC voltage measurements when AC line voltage is referenced to earth ground. See "AC Voltage Measurements" on Page 55 in the Manual.
FET MULTIMETER
FUNCTION
RANGE
Ω ADJUST
INPUT
AC
DC/Ω
POWER
ZERO
DC −
DC +
OHMS
VOLTS
mA
Ω

Figure 58

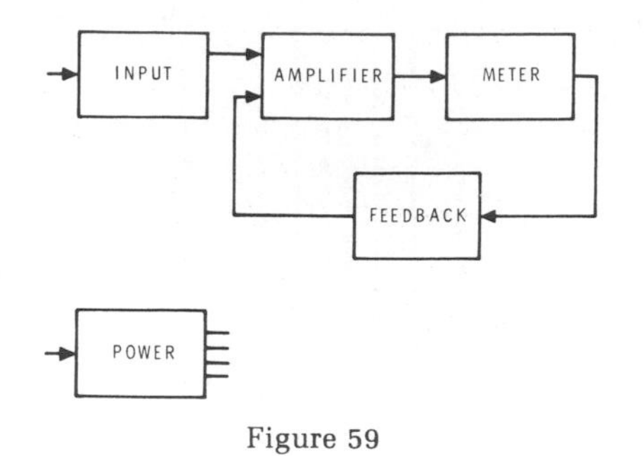

Figure 59

BLOCK DIAGRAM

Figure 59 is a simplified block diagram of an EVOM. The parameter being measured is changed by the input circuit to a voltage which biases the amplifier. This bias voltage is amplified and causes a current to flow through the meter. The meter current develops a feedback voltage which is returned to the amplifier to limit the gain and stabilize the output. The power supply provides all of the necessary voltages for the circuit.

THE OPERATIONAL AMPLIFIER

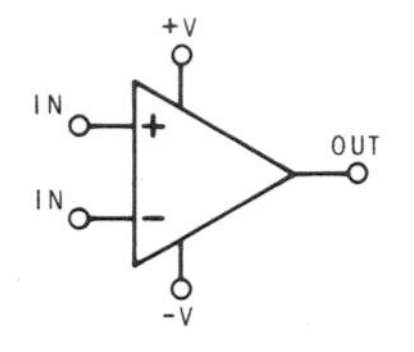

Figure 60

Figure 60 shows the basic operational amplifier (op-amp). +V and −V are the power supply voltages which are required by the op-amp. There are two inputs. The "+" represents the non-inverting input and the "−" represents the inverting input. That is, a change in voltage at the "+" input will cause the output to change in the same direction while a change in voltage at the "−" input will cause the output to change in the opposite direction. Since the gain of an op-amp is very high (100,000 is not uncommon), a very small change in the input causes a very large change in the output. In order for the op-amp to be useful in a meter circuit, this gain must be limited.

Figure 61 shows how this is accomplished in our meter circuit. Assume that R1 and R2 have a ratio of nine to one, with R1 being the larger resistor. This means that one tenth of the output voltage will be developed across R2 and returned to the inverting input of the op-amp. A change in voltage applied to the non-inverting input causes the output to change in the same direction as the input. Assume a positive going change at "+". There would be an increase in current through R1 and R2, which is one tenth of the output voltage. This voltage is applied to the "−" input where it decreases the effective size of the input signal. When the feedback voltage at the "−" input equals the voltage at the "+" input, the output will stabilize at ten times the "+" input. The entire circuit is now a non-inverting amplifier with a gain of ten. The gain of the circuit can be changed at any time by changing the ratio between R1 and R2. The gain of the op-amp is constant over a wide range of input voltage and frequency. The circuit is also relatively immune to variations in supply voltage and temperature.

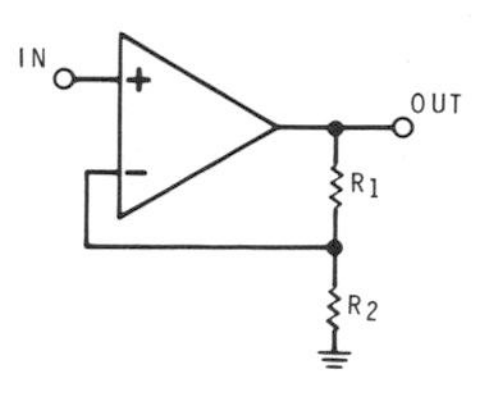

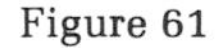
Figure 61

To make our op-amp work even better as the active component of an electronic VOM, we need to increase its input impedance. This can be accomplished by the addition of field-effect transistors (FET's) as shown in Figure 62. The FET's have a very high input impedance and a low output impedance. The high input impedance minimizes loading while the low output impedance provides a more stable driving point for the op-amp. Actually, the FET's are a part of the op-amp chip and will not be shown in any of the schematic drawings.

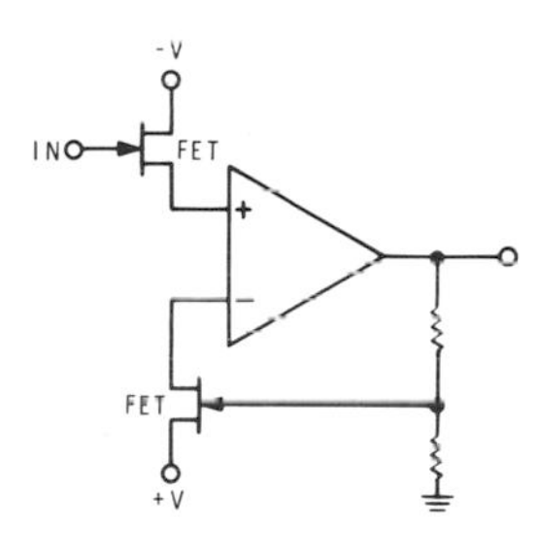

Figure 62

METER CONNECTIONS

Figure 63 is a simplified drawing of how a meter might be connected in the circuits. With the feedback configuration shown, the voltage across R will always be equal to the input voltage at "+"; therefore, the current through R and through M will be proportional to the input voltage. This type of circuit is known as a voltage-to-current converter. By adjusting the value of R, current through the meter can be adjusted for full-scale current to correspond with any given value of input voltage. If 0.1 volt is chosen, the meter could be calibrated linearly from 0 to a full scale of 0.1 V.

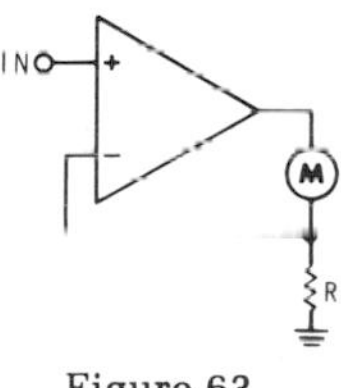

Figure 63

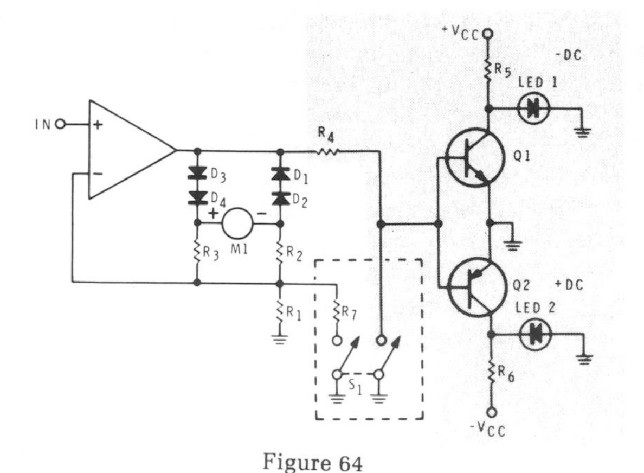

Figure 64

AUTO-POLARITY

In order to measure AC and have auto polarity, the meter must be part of a rectifier circuit. The circuit used in our typical meter is a full-wave rectifier as shown in Figure 64. We will use a positive input first. With a positive input, a positive voltage will be developed at the output of the op-amp. This will cause current to flow from ground through R1, R2, M1, D4, and D3 to the op-amp. Some current also flows through R3 which is in parallel with R2 and M1. Current through the meter is from right to left. If the input goes negative, the output of the op-amp will be negative. This causes current to flow thorugh D1, D2, M1, and R3 (also R2), then through R1 to ground. Current through the meter is still from right to left.

Since M1 always deflects in the same direction, regardless of voltage polarity, some method is needed to indicate the polarity of the input. In Figure 64, the circuitry shown in the shaded area performs that function. If the input is negative, LED1 will light; if it is positive, LED2 will light.

With an input voltage of 0 volts, the output from the op-amp to the bases of Q1 and Q2 will be 0 volts. Since the emitters of both transistors are grounded, the transistors will not be biased on and will have a high resistance. Therefore, current will flow from −Vcc through LED2 to ground and through LED1 to +Vcc, causing both LED's to light.

As the input voltage goes in a positive direction, the op-amp output goes positive, applying a positive voltage to the bases of Q1 and Q2. Q1 will be biased on, dropping the voltage on its collector below the conduction voltage of LED1. LED1 will turn off. With a positive voltage on the base, Q2 will remain cut off and LED2 will stay lit, indicating a positive voltage input.

With a negative input, Q1 will be biased off and Q2 biased on, turning LED1 on and LED2 off to indicate a negative voltage.

MEASURING AC

The circuit we have just discussed is capable of measuring AC with no further modifications. However, since the meter uses a d'Arsonval movement, it will respond to the average value of current. A more useful value is the RMS value of voltage or current. Since average is equal to 0.637 times the peak of a sine wave and RMS is 0.707 times the peak, some form of conversion is required.

For a given AC input, we would like the meter to read about 11% higher than the average value. We know that the amount of current through the meter controls its response, and the feedback voltage stabilizes the gain of the op-amp. If we change the feedback ratio, we can require a greater current in order to achieve the desired feedback.

In Figure 64, the circuit is changed to AC by closing switch S1. This places R7 in parallel with R1, reducing the resistance and requiring more current for a given voltage. The values are selected such that the new current is 111% of the old current.

At the same time, we place a ground at the bases of Q1 and Q2, holding both cut off and lighting both LED1 and LED2. LED's 1 and 2 will be lit with either a near zero DC input or when one of the AC ranges is selected.

MEASURING RESISTANCE

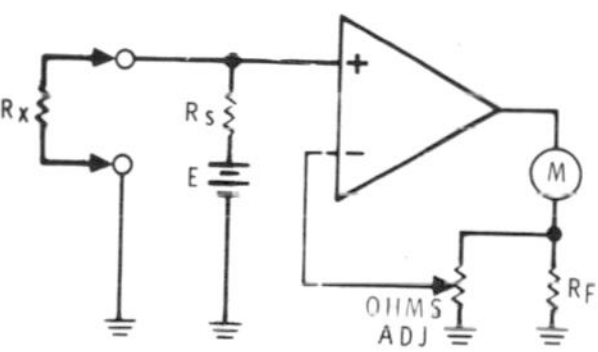

Figure 65

The method used for measuring resistance is shown in Figure 65. A precision standard resistor is placed in series with a voltage source in the input circuit. With Rx equal to an infinite value (leads open), there is no current through the circuit, and the full battery voltage is felt at the input of the op-amp.

The "ohms adjust" is adjusted for full-scale deflection of the pointer in this position. If the leads are shorted, all of the battery voltage is dropped across R_S and the input is at zero or ground potential and no current flows through the meter. The "zero adjust" (not shown) is adjusted for zero under these conditions. If the value of Rx is equal to R_S, the meter will deflect to half scale. We can thus calculate the voltage for any value of Rx and the meter scale can be calibrated.

Typical EVOM

Having studied the basics of how the electronic voltmeter works, you will now study some actual circuits from our typical meter.

DC SECTION

The circuitry for the DC voltage portion of our typical EVOM is shown in Figure 66. The input is fused with a fast-blow fuse to protect the meter from excessive input. After being switched and selected, the incoming voltage is applied to a voltage divider consisting of a string of precision resistors. The total resistance of this string is 10 megohms. This is the input resistance to our meter. The input impedance of 10 megohms will not change as the meter is switched from range to range; therefore, this meter will have very little loading effect on most circuits.

The resistors are connected in a voltage divider network such that each step is a ten-to-one ratio. The capacitors which are shown by dashed lines are actually in the circuit, but are used on the AC ranges only and have no effect on DC measurement.

If the 1-volt range is selected as shown in Figure 66, 1 volt of input will cause 0.1 volt to be developed between R104 and R105. This voltage is fed through SW101B to pin 3 of the op-amp. R2, C2, and R136 compose a low-pass filter to eliminate any AC component of the DC voltage. The transistors Q105 and Q106 are connected such that they function as clamping diodes to prevent any excess voltage from reaching the op-amp. Connected collector-to-collector as shown, they will function with either a positive or a negative voltage.

The output of the op-amp will cause current to flow through the diode network, the meter, and resistor network R145 through R149. The voltage at the wiper arm of R146 is fed through SW101E to the inverting input, pin 2, of the op-amp. When this voltage equals the voltage on pin 3, the current through the meter will stabilize.

When the range switch is changed to the 3-volt position, the voltage to the op-amp is still one-tenth of the voltage being measured. Therefore, if nothing else was changed, full-scale deflection would still be 1 volt. The change is in SW101E, which changes from position 1 to position 2, connecting the op-amp to the top of R145 instead of the wiper arm of R146. In this position, it takes less current to develop the necessary feedback. The values are such that a three-to-one division occurs. Now, a 1-volt input will only produce one-third deflection of the meter; thus, 3 volts are required for full-scale deflection. The remainder of the circuit works exactly as described in the simplified section.

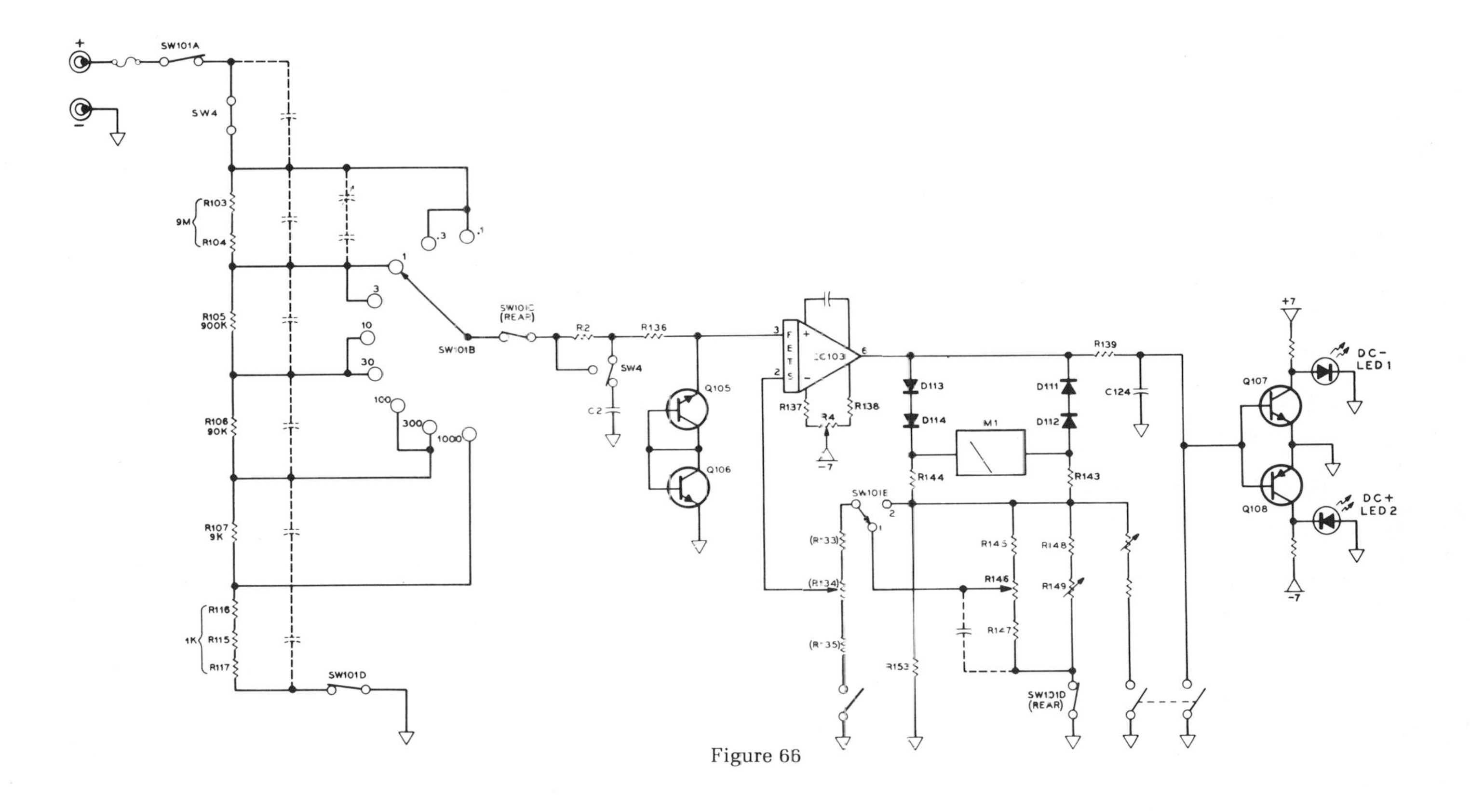
SW101A
SW4
R103
R104
9M
R105
900K
R106
90K
R107
9K
R116
R115
R117
1K
SW101D
.1
.3
1
3
10
30
100
300
1000
SW101B
SW101C
(REAR)
R2
R136
C2
Q105
Q106
FETS
IC103
R137
R138
R4
-7
D113
D114
D111
D112
M1
R144
R143
R139
C124
SW101E
R145
R146
R147
R148
R149
R153
SW101D
(REAR)
Q107
Q108
+7
DC-
LED1
DC+
LED2

Figure 66

AC SECTION

When measuring AC, capacitors C102 through C109 are an active part of the input voltage divider as shown in Figure 67. However, the filter capacitor, C2, must be disconnected from the input or the signal will be shorted to ground. Since the meter must give an accurate indication over a wide frequency range, (10 Hz to 100 kHz) values of capacitance must be selected such that the reactance is linear with a change in frequency, keeping the voltage division ratio relatively constant over the entire frequency range. To compensate for reduced op-amp and meter response, C121 and C126 adjust feedback to the op-amp by providing more non-inverting feedback and a reduced feedback to the inverting input.

Meter scale correction is provided by SW101D, as discussed earlier, which also disables the polarity indicator.

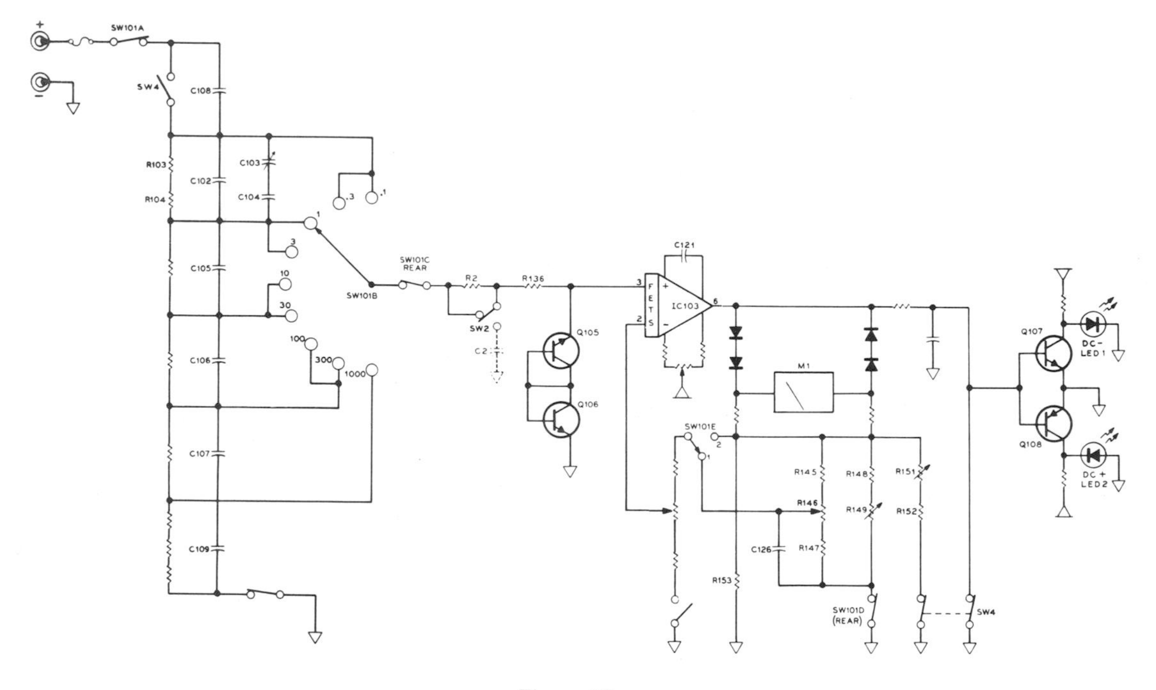

Figure 67

CURRENT SECTION

The circuit for measuring current is shown in Figure 68. Circuit current flows through F101, SW101A, SW101C, and the selected portion of the resistor network of R108 through R114. If the current is DC, the voltage thus developed will be coupled through SW4 to pin 3 of the op-amp. We have already discussed the operation of the rest of the circuit. If we are measuring AC, there is only one difference. C111 is in the circuit to block any DC component and R3 is connected to ground to provide a charge and discharge path from C111.

Over-current protection is provided by the diode network D101 through D104. If current increases to 1.5 times the rated amount, the diodes will conduct, decreasing the voltage developed across the resistor network. If current exceeds 2 amperes, F101 will blow, opening the input circuit.

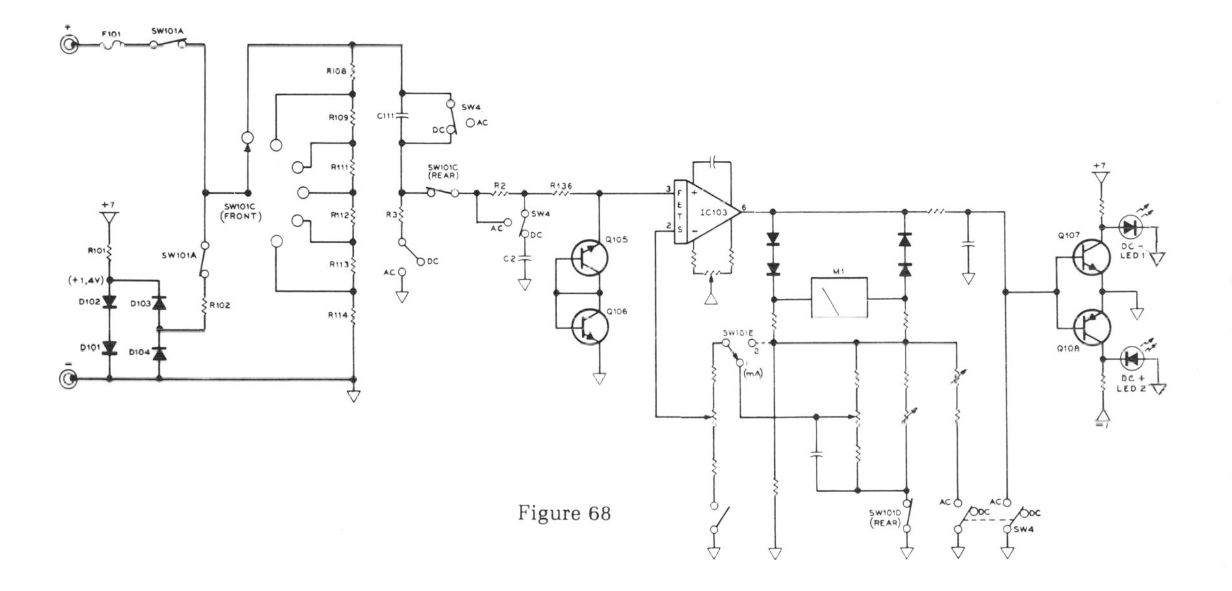

Figure 68

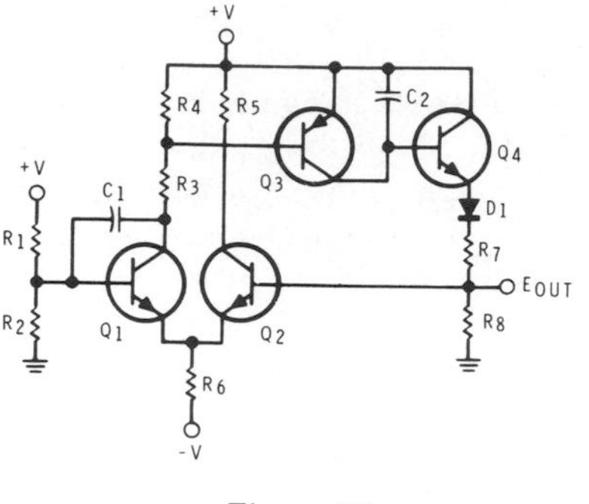

Figure 69

RESISTANCE SECTION

Reference Voltage To measure resistance accurately, a stable source of voltage is required. You learned in the section on ohmmeters that any change in battery voltage will cause a change in resistance reading, even if the meter is "zeroed" before each measurement. To minimize this effect, better electronic meters will include a voltage regulator to keep voltage constant. A regulator may be used with meters which operate on batteries or AC power or both. A typical regulator circuit is shown in Figure 69.

When power is applied to the circuit, E_{out} will stabilize at some value. Any change in load or supply voltage will be compensated for, to maintain that output.

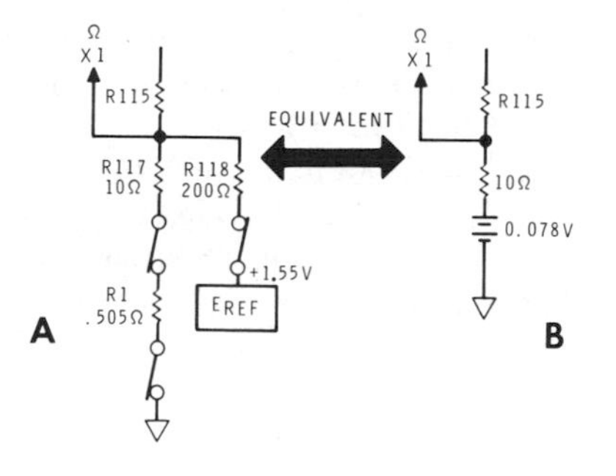

Figure 70

Assume that a load is placed at E_{out}. This would happen any time an external resistance is connected. More current will flow through R7 and E_{out} will tend to decrease. This decrease will be felt on the base of Q2, decreasing its conduction. This causes the emitter to become more negative. Since the emitters of Q1 and Q2 are connected, the emitter of Q1 will also be more negative, causing it to conduct harder. More current flows through R3 and R4, placing a more negative voltage on the base of Q3, causing it to conduct harder. Since the collector current of Q3 flows through the base-emitter junction of Q4, Q4 will conduct harder, drawing more current through R8 and bringing E_{out} back to normal. E_{out} is thus stabilized for any normal output conditions.

Low Voltage Supply In our typical meter, we have provided a low-voltage ohms range and a normal ohms range. The reference voltage E_{out} in our regulator circuit is used for both. Figure 70A shows the network used to reduce the output for low voltage. E_{ref}, R118, R117, and R1 make up a network which provides a low voltage of 0.078VDC with a source impedance of 10 ohms. Figure 70B shows an equivalent circuit of the network.

Ohms, Low-Voltage Figure 71 shows the circuit for the low-voltage ohms position. With the leads open, the full reference voltage is applied to pin 3 of the op-amp. The networks of R133-R135 and R145-R149 are adjusted for full-scale deflection. If an external resistance is connected to the leads, some current will flow through the circuit, dividing the reference voltage between the 10 Ω source resistance and the unknown resistance. The voltage drop across Rx is applied to pin 3 of the op-amp where it is converted to meter current which is calibrated in ohms. Switching to a higher resistance range increases the effective internal resistance of the reference source by a factor of 10. For example, when going from the Rx1 range to the Rx10 range, the resistance increases from 10 Ω to 100 Ω.

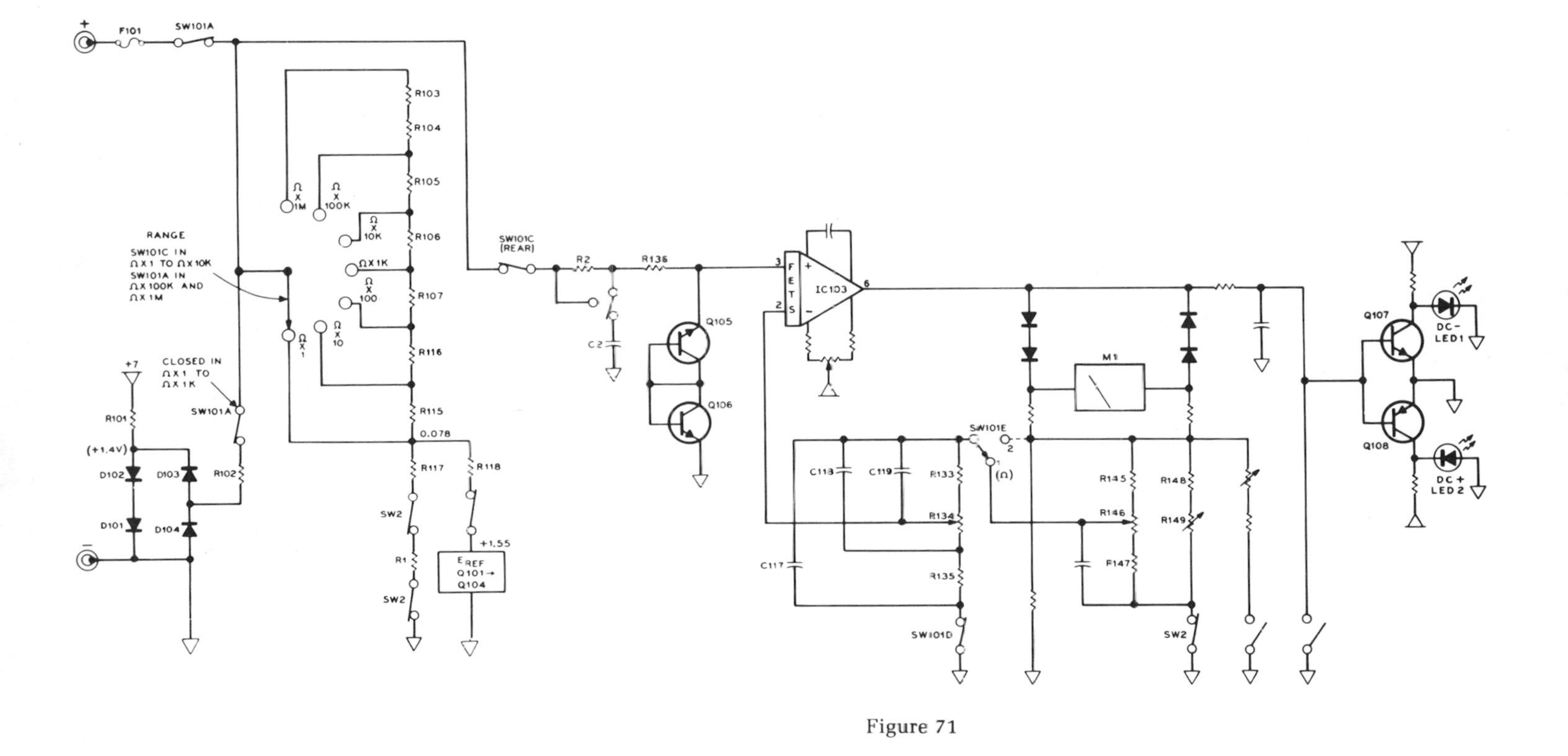
F101
SW101A
R103
R104
R105
R106
R107
R116
R115
0.078
R117
R118
SW2
R1
SW2
+1.55
E REF Q101→ Q104
RANGE
SW101C IN ΩX1 TO ΩX10K
SW101A IN ΩX100K AND ΩX1M
CLOSED IN ΩX1 TO ΩX1K
SW101A
R101
+7
(+1.4V)
D102
D103
R102
D101
D104
SW101C (REAR)
R2
R136
C2
Q105
Q106
IC103
FETS
M1
SW101E
(Ω)
C118
C119
R133
R134
R135
C117
SW101D
R145
R146
F147
R148
R149
SW2
Q107
Q108
DC- LED1
DC+ LED2

Figure 71

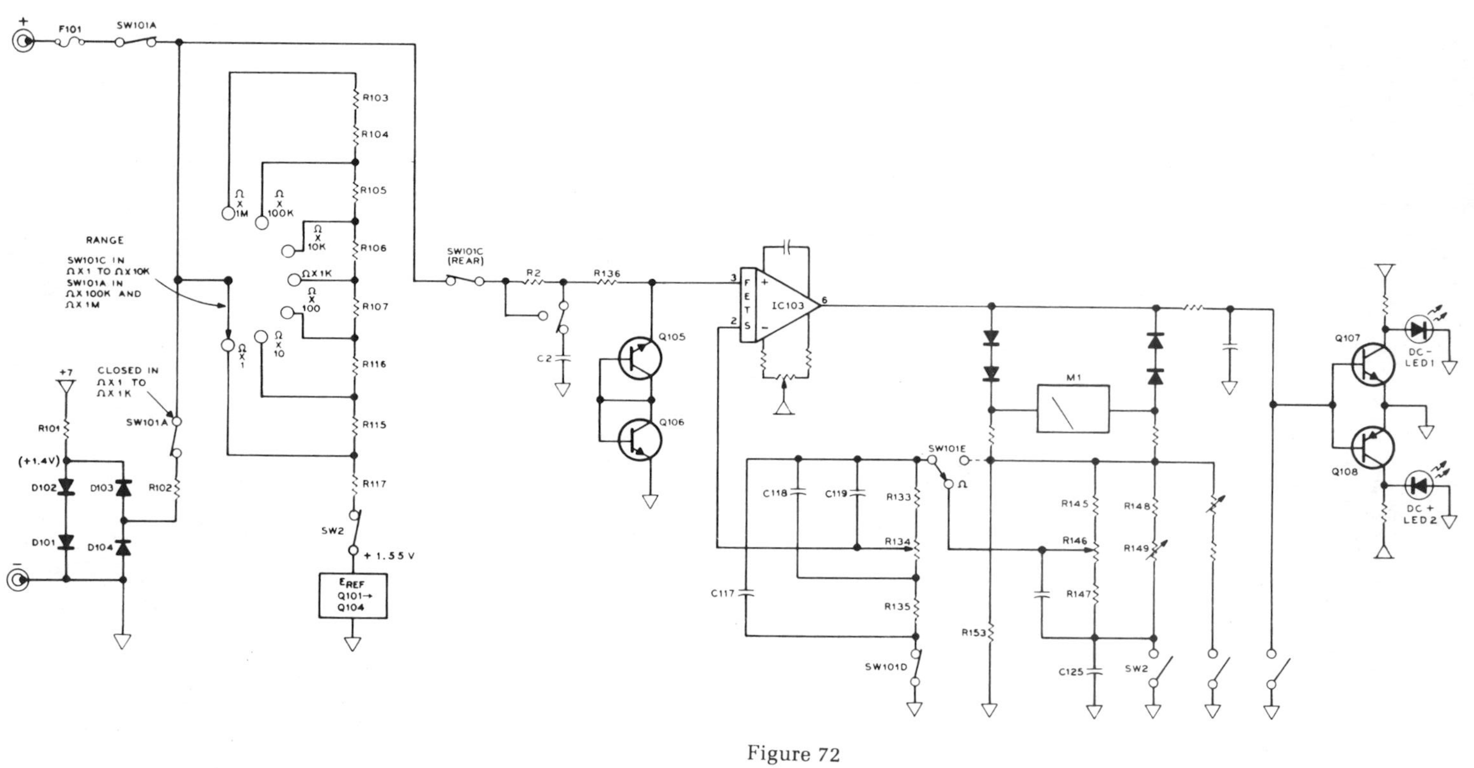

F101
SW101A
R103
R104
R105
R106
R107
R116
R115
R117
Ω X 1M
Ω X 100K
Ω X 10K
Ω X 1K
Ω X 100
Ω X 10
Ω X 1
RANGE
SW101C IN ΩX1 TO ΩX10K
SW101A IN ΩX100K AND ΩX1M
CLOSED IN ΩX1 TO ΩX1K
+7
R101
(+1.4V)
D102
D103
D101
D104
SW101A
R102
SW2
+1.55V
E REF Q101→ Q104
SW101C (REAR)
R2
R136
C2
Q105
Q106
F E T S
IC103
M1
SW101E
Ω
C118
C119
C117
R133
R134
R135
SW101D
R153
R145
R146
R147
C125
R148
R149
SW2
Q107
Q108
DC – LED1
DC + LED2

Figure 72

Ohms, High-Voltage The high-voltage ohms position works almost the same as the low-voltage ohms position, except that the full reference voltage of 1.55 V is now used and the scale factor of the meter is changed by switching R145-R149 out of the circuit, as shown in Figure 72.

Figure 73

Low Cost EVOM

Now, we'll discuss a simpler EVOM. Figure 73 is a photograph of the Heathkit IM-5284 low-cost, multifunction EVOM. The only significant difference in the controls of this EVOM and a passive VOM is the "zero adjust " which is used to electrically zero the meter on all ranges and functions. The "ohms adjust" is adjusted for "∞" with the meter leads not touching.

As with most electronic meters, the input resistance is high, 10 MΩ on DC and 1 MΩ on AC. Accuracy is ±3% of full scale for DC and ±5% of full scale for AC. Characteristics of this meter are similar to a passive VOM except for the high input resistance, which gives it an advantage when measuring low values of voltage in high resistance circuits.

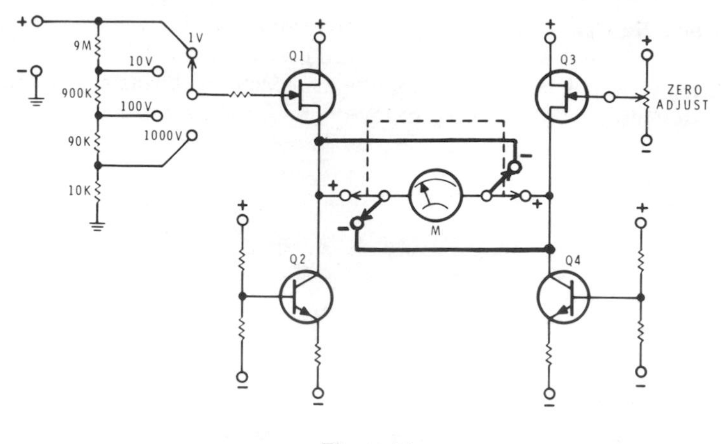

Figure 74

The meter is electrically simple, consisting of an input circuit and a meter bridge. The input circuit contains the function and range switches with multiplier and shunt resistors. The meter circuit is a balanced bridge which is unbalanced by the incoming signal. The bridge arrangement reduces DC drift which is normally found in single-ended amplifiers. Look at the simplified schematic of the +DC voltage function of this meter shown in Figure 74.

+DC CIRCUIT

With no signal applied to "+" input terminal, zero volts is felt on the gate of Q1. The gate of Q3 is connected to a variable resistor which can be adjusted from some positive voltage through zero to some negative voltage. The meter is connected between the sources of Q1 and Q3. Q2 and Q4 are current regulators which maintain the currents through Q1 and Q3 approximately equal.

When the "zero adjust" is set so that there is no difference of potential between the source of Q1 and the source of Q3, no current will flow through the meter. However, when a positive potential is applied to the gate of Q1, an imbalance will occur which will cause current flow through the meter from the source of Q3 to the source of Q1. In this case, the design is such that 1 volt at the gate of Q1 causes full-scale deflection. Input is through a voltage divider which has 1:1, 10:1, 100:1 and 1000:1 ratios available.

–DC CIRCUIT

The –DC circuit is identical to the +DC circuit with one exception. With a negative applied to the gate of Q1, the imbalance in the bridge is opposite to that discussed earlier and current will flow from the source of Q1 to the source of Q3. A switch is installed in the meter circuit to keep current through the meter in the same direction. Switch connections for –DC are shown by the heavy dark lines in Figure 74.

AC VOLTAGE

When measuring AC volts, the incoming signal is first rectified by the half-wave rectifier shown in Figure 75. The positive voltage thus developed is applied to the input of the voltage divider shown in Figure 74. The measurement is then made just as the positive DC measurement was made.

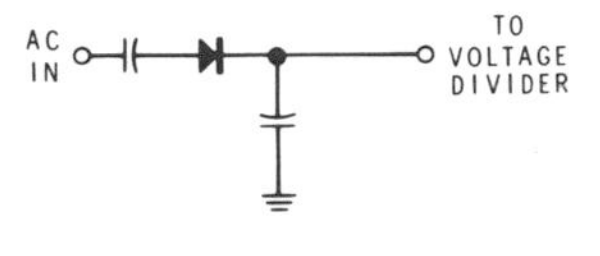

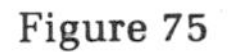
Figure 75

DC CURRENT

When measuring current, the meter reverts to a typical passive milliammeter. A simplified schematic of it is shown in Figure 76. On the 1 mA range, the shunt resistance is such that 100 μA flows through the meter circuit and the remainder through the shunt. Operation is similar to that of the Multimeter discussed earlier in that when switching to a higher range, a portion of the resistance is connected in series with the meter movement. For instance, in the 10 mA range, the meter resistance is increased by 125 Ω and the shunt resistance decreased by a like amount. Of course, the full-scale current of the meter doesn't change. It depends on the movement being used. The arrows show the current paths on the 100 mA range. This meter has no provision for measuring AC current, which is typical of low-cost meters.

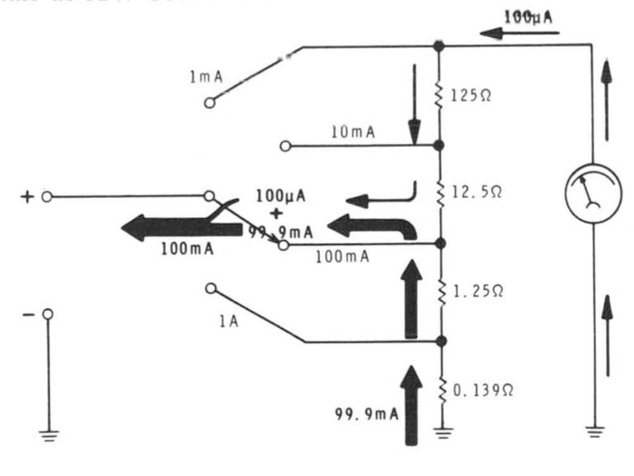

Figure 76

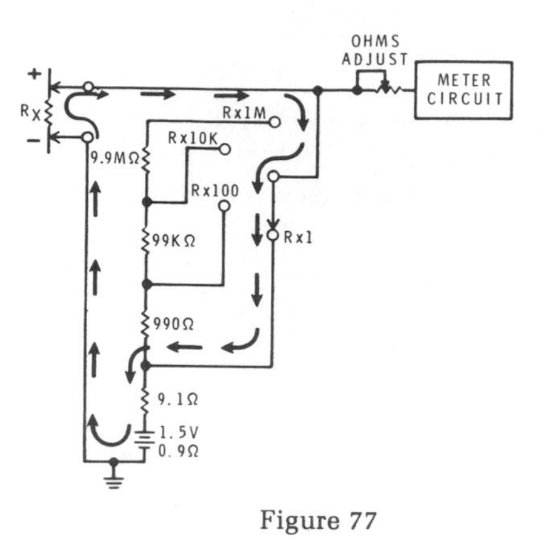

Figure 77

RESISTANCE

When measuring resistance, a battery is connected in series with the unknown resistance (R_x) and one or more of the multiplier resistors as shown in Figure 77. The voltage across R_x is measured by the meter circuit and converted to an ohms indication. The meter circuit is identical to that used for measuring voltage. The "ohms adjust" is set for full-scale deflection of the meter with an R_x of infinity (meter leads open). The current path for the R × 1 range is shown by the arrows in Figure 77. As you can see, the operation of the ohmmeter section of the meter is very similar to that of the basic ohmmeter discussed earlier.

COMPLETE METER

Figure 78 shows the complete schematic of the IM-5284 Multimeter. To help you visualize the circuit connections, the shaded line traces the signal path on the +DC volts function and the 0-to-10-volt range. Since this meter is simple, we will leave the tracing of other ranges to you.

EVOM Characteristics

POWER SUPPLIES

Power to drive the meter and associated circuits may be derived from a battery or an AC power source. It is very common today to find nickel-cadmium batteries for portable operation which may be recharged by plugging the meter into an AC source.

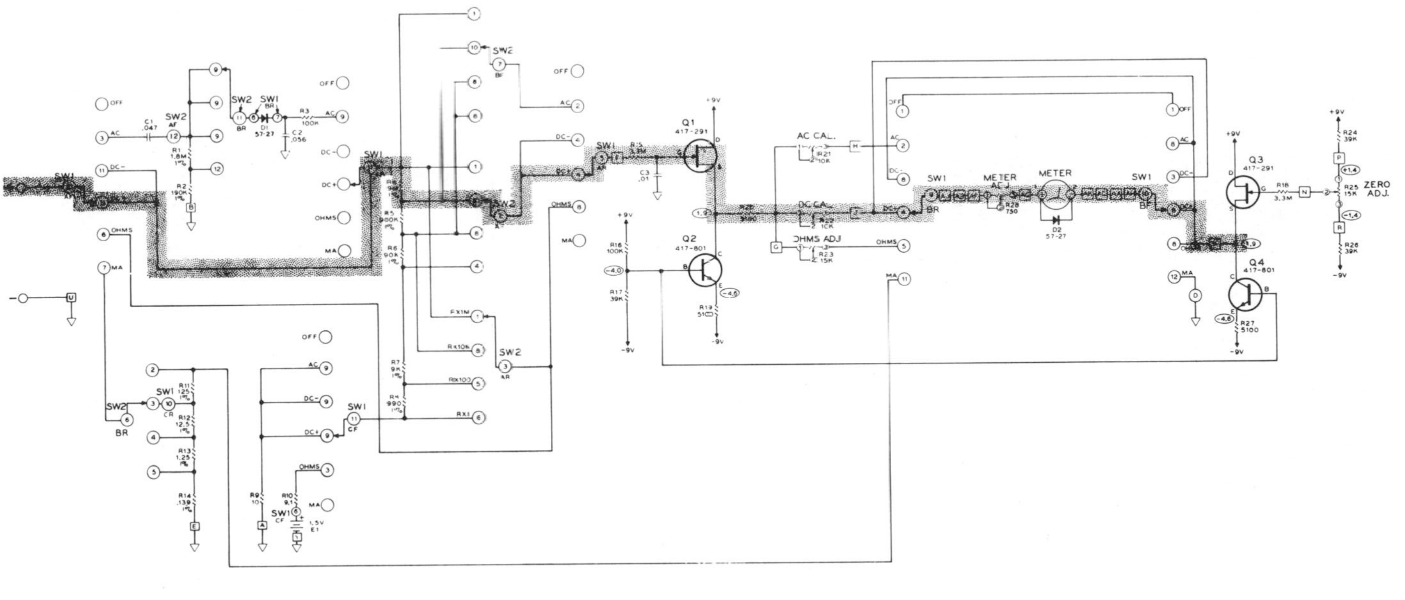

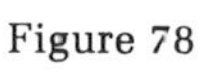

Figure 78

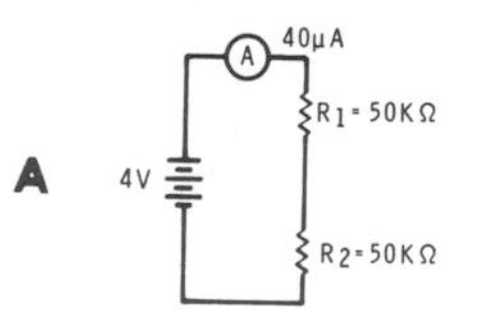

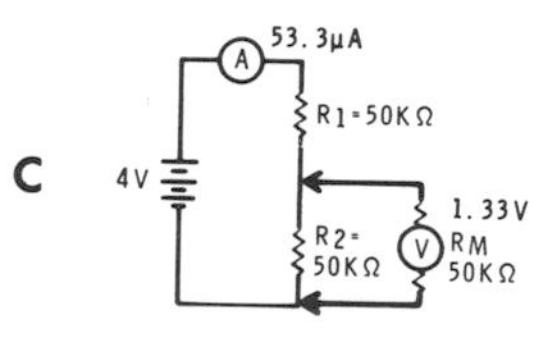

Figure 79

ACCURACY

The accuracy of an electronic meter is usually slightly better than that of a non-electronic meter. Accuracies near ±2% are common, while the typical VOM has an accuracy of 3 or 4%. Proper design will allow the electronic meter to maintain its accuracy over a wide range of frequencies and environmental conditions. However, accuracy is not usually the primary reason for choosing an electronic meter.

SENSITIVITY

Sensitivity is a very good reason for selecting an EVOM. The EVOM which you studied had a full-scale range of 0.1 volt, while the VOM had a minimum range of 0.25 volt. The EVOM is calibrated such that a resolution of 0.002 volts is possible, while the non-electronic VOM is capable of .005 volts resolution.

This does not mean that you can accurately measure a voltage of 2 millivolts with the EVOM. It means that you can detect a difference of as little as 2 millivolts, or detect the presence of 2 millivolts but not measure it accurately. Here, sensitivity and accuracy get tied together. Remember that accuracy is normally measured in percentage of full-scale deflection. Therefore, on the 100 millivolt range, a meter having an accuracy of 2% could be off up to two millivolts at any point on the scale. When attempting to measure a voltage as small as two millivolts, the error could be up to 100%. Therefore, if small voltages must be measured accurately, a very sensitive meter is required.

CIRCUIT LOADING

Reduced circuit loading is one of the major advantages of an electronic meter. You have already learned that placing an excessively high load on a circuit not only spoils the accuracy of the measurement but may cause the circuit to stop operating properly. Next, we will compare the results of measurements taken with a typical VOM and with a typical EVOM. The VOM has a resistance of 20 kΩ per volt and the EVOM has an input impedance of 10 MΩ. To demonstrate this effect, we will use the circuit shown in Figure 79.

Before a meter is connected to the circuit, the total resistance is 100 kΩ, 40 μA of current will flow, and the voltage across either R1 or R2 will be 2 volts.

Now assume the EVOM is placed with its 10 MΩ impedance across R2 as shown in Figure 79B. The resistance of R2 and Rm combined is 49.75 kΩ, the new current is 40.1 μA, and the voltage which will be measured by the meter is 1.99 volts. This reading is well within the normal tolerance of the circuit; therefore, the 10 MΩ meter has no significant effect on the circuit.

Next, assume the 20 k ohm-per-volt meter is connected in the circuit as shown in Figure 79C. Since we are measuring about 2 volts, select the 2.5 volt range of the meter. This will give a total meter resistance of 50 kΩ. The combined resistance of R1 and Rm will be 25 kΩ, circuit resistance will be 75 kΩ for a current of 53.3 μA. The voltage measured by the meter in this circuit is 1.33 volts, which is only 67% of the actual voltage. This is too much error for most applications. You can, therefore, see that for low voltage measurements in high resistance circuits, a high input impedance is necessary. The impedance of the EVOM is high enough to prevent loading all but the most sensitive circuits.

SELECTION

Buying the ideal meter can be rather difficult. Remember that a meter must do two equally important jobs. It must measure and communicate. It is just as important to be able to easily and properly interpret the meter scales as it is for the meter to be sensitive and accurate. Therefore, you should be sure that the scale is not confusing.

Accuracies vary from 0.5% of full scale to 5% or worse. An electronic meter with accuracies of 3% to 5% may cost less than forty dollars. Meters with accuracies of 1.5% are available for less than two hundred dollars. A word of caution here; an accurate meter must be calibrated regularly by a precision measurement equipment laboratory to maintain this accuracy.

The sensitivity of an electronic meter varies considerably. Low ranges of 50 mV full scale are easy to find. The smallest voltage you plan to measure should cause at least half-scale deflection.

Input impedance of a modern FET meter will be from 10 MΩ to 15 MΩ which is more than enough for most applications. To keep the loading effect of the meter to less than 2%, the input resistance of the voltmeter should be at least 25 times the resistance you are measuring across.

You cannot make a good meter selection until you have defined your actual application. Specification and selection will be discussed in the section on digital meters.

Unit 2

DIGITAL METERS

INTRODUCTION

Digital meters measure analog quantities and display them in digital form. This requires the analog input to be converted to a digital format. Therefore, an analog-to-digital (A/D) converter is required. There are several types, and we will discuss some of the more common ones. Also, several types of display are available, and we will discuss the more common of these.

When we have completed our study of digital meters, we will compare analog and digital meters. We will also cover some practical accessories and discuss factors to consider when selecting your own meter.

ANALOG-TO-DIGITAL CONVERSION

A digital meter must take an analog quantity and convert it to an equivalent digital readout. A practical meter will have the ability to measure a wide variety of values. It should be able to measure both AC and DC voltage from a few millivolts to several hundred volts. Often, the meter is needed to measure an unknown resistance or current. The basic functions required to accomplish this conversion are shown in Figure 1.

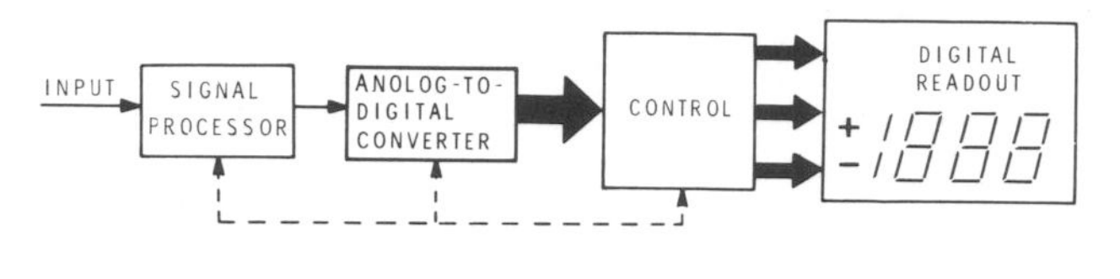

Figure 1

The unknown quantity to be measured is first applied to the signal processor. The signal processor must convert the incoming signal to a form usable by the analog-to-digital (A/D) converter. Since the A/D converter requires a DC voltage of limited range, the signal processor may be called on to amplify or attenuate the input, convert AC to DC and current to voltage. The signal processor must also provide a current source for measuring resistance.

The A/D converter does only one thing; takes the DC voltage from the processor and converts it to a digital number. A/D converters are single range devices. These ranges vary from device to device with some typical full-scale ranges being 200 mV, 1V, 2V and 10V. All inputs to the A/D converter must be a DC voltage within the proper range. Many of the meter characteristics are determined in these first two stages. The input impedance, sensitivity, number of ranges and number of digits are all determined here.

The control block manages the flow of information within the meter. In very simple meters, it may do little more than transfer the output of the A/D converter to the readout. In other meters, very complex timing functions are controlled here.

The digital readout or display is rather obvious. It provides a visible result of the meter's work.

We will take a more detailed look at some of these blocks starting with the analog-to-digital converter.

There are two basic techniques of analog-to-digital conversion, integrating and non-integrating. Each technique has certain advantages and disadvantages. The choice depends primarily upon intended application. In both techniques, there are several ways to accomplish the conversion. We will begin with a discussion of integrating techniques.

Integrating Techniques

Integrating techniques could also be called averaging techniques or charging techniques. Basically, integrating converters measure the time required to charge a capacitor to a given reference voltage. Thus, the charge on the capacitor will equal the average input voltage for that period of time.

A major advantage of the integrating technique of conversion is the reduction of the effects of noise. Because the noise is usually AC and is averaged in with the measured signal, it affects the measurement less. In fact, you will see that with some techniques, it is reduced to near zero.

There are several methods of performing the integrating techniques of A/D conversion. Some are very simple and some are more complex. We will start our discussion with one of the more simple methods.

SINGLE-SLOPE A/D CONVERSION

The A/D converter shown in Figure 2 is a single-slope type. It requires a crystal-controlled oscillator as a clock, a precise voltage source as V reference, and a precision value of C and of R_1. The accuracy of the meter depends on the stability of these components.

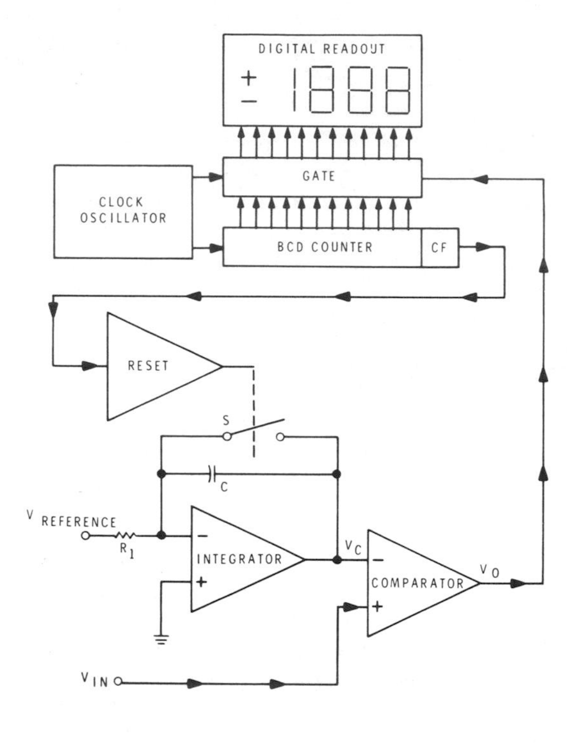

Figure 2

The BCD counter continuously counts the output of the clock oscillator. Each time the counter reaches a full count, CF (the counter-full flip-flop) will change states. This change is applied to the reset amplifier and changes the position of S. While shown here as a manual switch, S is some form of electronic switching device. When the output of CF is high, S is closed and capacitor C is discharged. Waveform R in Figure 3 shows the output from CF. The switch will be open during the time from T_1 to T_3, allowing C to charge.

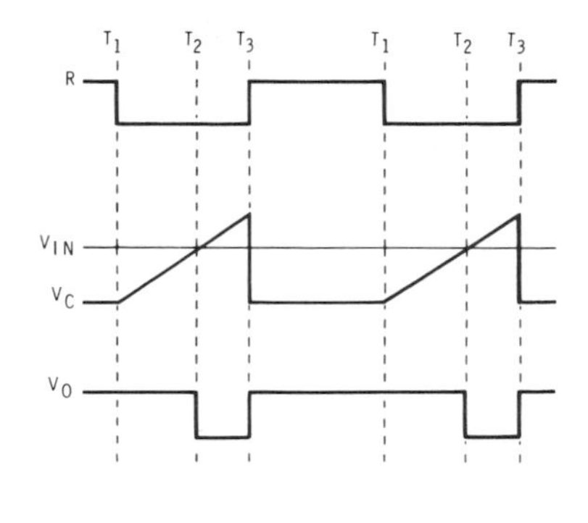

Figure 3

Anytime the switch is open, C is charged by the constant current source composed of V reference and R_1. This causes a linear increase in V_C as shown in Figure 3. As shown in Figure 2, V_C is applied to the inverting input of the comparator. The unknown voltage from the signal processor (V_{in}) is applied to the non-inverting input of the comparator. The output of the comparator (V_O) will be high until V_C exceeds V_{in}. As soon as V_C exceeds V_{in}, as shown at T_2 of Figure 3, V_O will go low. V_O will stay low as long as V_C exceeds V_{in}.

V_O is applied to the gate circuit. When V_O goes low, the gate will open and pass the number in the BCD counter to the digital readout. Since the BCD counter has been counting clock pulses at a constant rate and C has been charging at a constant rate, there is a direct relationship between V_C and the number in the counter.

Figure 3 shows that the counter starts counting at T_1, and V_C starts increasing. At T_2, when V_C equals V_{in}, the contents of the counter are displayed. Therefore, the number transferred to the readout is determined by the amplitude of V_{in}. With the proper calibration, the digital readout can indicate the value of V_{in}.

A 3-1/2-digit readout, as shown here, requires 2,000 clock pulses for each reading. A 3-1/2-digit readout can display up to 1,999. If the display incremented by one count for each pulse counted, 1,999 pulses are required for a full count. An additional pulse will then be required to reset the counter and prepare for the next conversion. Thus, a minimum of 2,000 pulses is required for each conversion. This is the **minimum** number of pulses required. The designer of a meter often adds refinements to the circuit which require additional pulses.

The time required for a conversion is a factor of clock frequency. For instance, 2,000 pulses from a 10 MHz clock require 200 microseconds, which would give 5,000 conversions per second. However, with a 1 MHz clock frequency, 2,000 pulse require 2 milliseconds and give 500 conversions per second.

There are several other factors which affect the number of readings per second that a meter can make. The RC time of the integrator is a big factor, and must be kept such that the ramp is linear. You will see in the section on dual-slope integration that certain integration periods have inherent advantages.

The frequency response of the signal processor is a limiting factor in speed of conversion. Until a change in amplitude can be processed and sent to the converter, it cannot be converted. Therefore, the time it takes to settle down at a new value must be added to the conversion time.

While it is theoretically possible for a single-slope converter to perform 5,000 conversions per seconds with a 10-MHz clock, it is extremely unlikely that anywhere near this rate would be used with a shop meter. In fact, most shop meters use a conversion rate of less than 10 per second.

The main advantage of the single-slope method of integration is simplicity. It is not the most accurate of conversion methods; however, accuracies of 1 percent are easily obtained, making this an attractive choice for economical meters.

The primary drawback of the single-slope method is that the stability of the clock oscillator, the reference voltage, R_1, and C must be very good. A variation in either of these directly affects accuracy.

DUAL-SLOPE INTEGRATION

The dual-slope integration method of A/D conversion is probably the most common in use today. With this method, the capacitor in the integrator is allowed to charge for a fixed period of time. The charge on the capacitor at the end of this time depends solely on the amplitude of the input. At the end of the fixed charge time, the capacitor is discharged by a constant current source. The time required for discharge is counted, and the count displayed as a voltage or current.

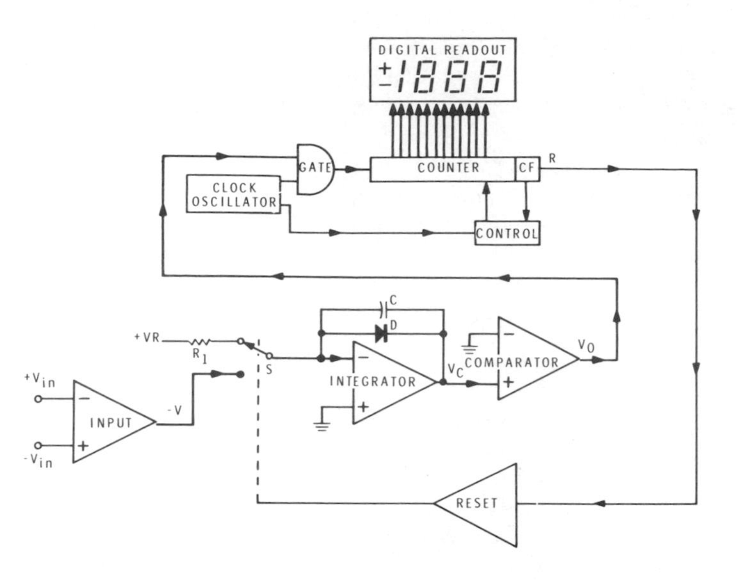

Figure 4

Figure 4 shows a simplified block diagram of a dual-slope integrator. We will start our explanation with no input. The reset signal shown as R in Figure 5 is at a high level. The switch S is in the V_R position. The positive voltage reference applied to the inverting input of the op-amp will charge C such that V_C will be a negative value. However, it is limited by diode D to approximately −0.5 volts. The output from the comparator will be low and the gate will inhibit the clock pulses to the counter.

At T_1 (Figure 5), the control circuit clears the counter and resets R (Figure 4) to a low state. When R is low, S will be in the −V position. Since −V is derived from V_{in}, C will charge at a rate proportional to the amplitude of the voltage being measured. The polarity of the charge on C is such that V_C will be positive. As V_C passes through zero volts at T_2 (Figure 5), V_O goes high and enables the gate. Clock pulses are now fed to the counter which begins a count. Keep in mind that the information in the counter is not fed to the digital readout until directed by the control. At this time, the readout is displaying the results of the previous conversion.

The count continues for a fixed period of time. This time is determined by the control circuits and is in no way affected by the input voltage. Since C is allowed to charge for a fixed period of time, the charge on C at the end of that time is directly proportional to the input voltage.

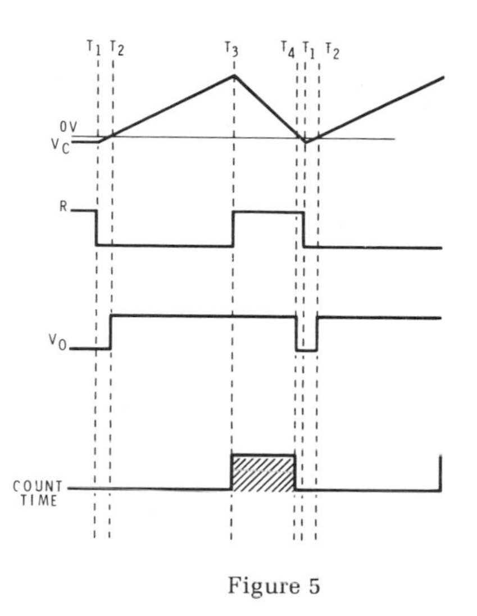

Figure 5

At the end of this period, which corresponds to T_3 of Figure 5, CF signals the control unit and places a high on reset line R. The control resets the counter to zero, and the reset voltage switches S to the $+V_R$ position. The counter once more starts a count from zero as C starts to discharge.

V_R is a precision reference voltage. Along with R_1 it provides a constant current source which discharges C at a constant rate. Thus, the time required for C to discharge is determined solely by its charge which was, in turn, determined by the voltage being measured. When V_C passes through zero volts, V_O will go low, inhibiting the gate and stopping the count.

The count time from T_3 to T_4 is the time required to discharge the capacitor. This time is proportional to the input voltage. Therefore, the count that is transferred to the digital readout is proportional to the input voltage. This display will be maintained until another cycle is complete and another count is made.

Since C is common to both charge and discharge, any long term changes in the value of C will have little affect on conversion accuracy. If, for instance, C increases in value, it will charge to a lower voltage during a given period of time. However, it will also take longer to discharge with a given current. Thus, the two changes tend to cancel each other and have little effect on accuracy.

The dual-slope method of conversion is also relatively immune to oscillator drift. Since the time which the capacitor is allowed to charge is determined by the number of clock pulses counted, and the number of pulses counted during capacitor discharge come from the same clock oscillator, the actual oscillator frequency has little or no effect on the measurement. Of course, if the oscillator were to change frequency while a conversion was in progress, the accuracy of that conversion would suffer. However, long term drift is well compensated for.

Dual-slope integration effectively eliminates noise that enters with the input signal, provided that the period of the noise is a multiple of gate length. One of the most common sources of noise is commercial power lines. This line-related noise is 60 Hz in the United States and 50 Hz in much of the rest of the world.

To see how line-related noise may be eliminated, look at Figure 6. Here we have chosen a gate length (capacitor charge time) of 1/10 second. During this time, exactly six cycles of 60-Hz noise will have occurred.

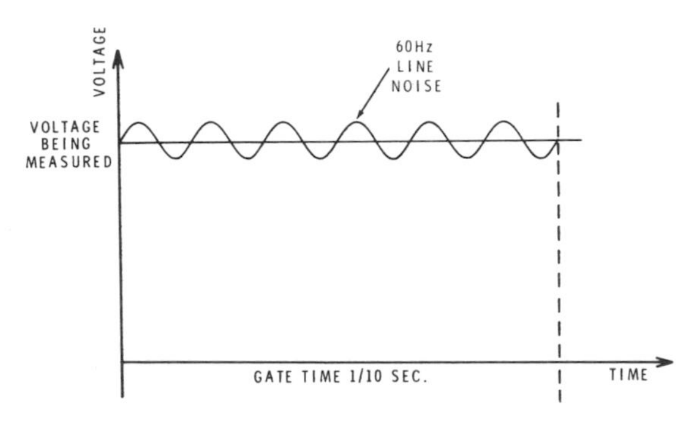

Figure 6

In basic AC, you learned that the average of a complete sine wave was zero. Therefore, the average of any number of complete sine waves will be zero. Thus, by selecting a gate length that is an exact multiple of line frequency, the effect of line related noise can be eliminated, as is any harmonic of that frequency.

Good accuracy and noise rejection can be obtained with conversion rates of ten samples per second or less, making the dual-slope one of the slowest methods of conversion.

VOLTAGE-TO-FREQUENCY

A simple voltage-to-frequency method of A/D conversion is shown in Figure 7. The voltage to be measured is converted to an equivalent frequency, which is counted and displayed. The conversion could be as simple as driving a voltage-controlled oscillator; however, this would have many limitations. A more useable method is shown in Figure 8.

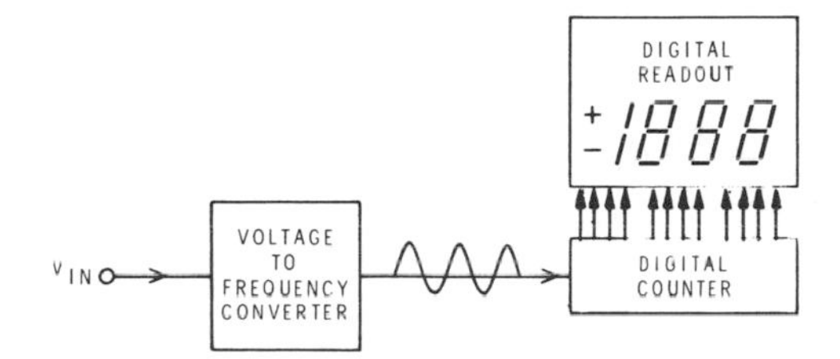

Figure 7

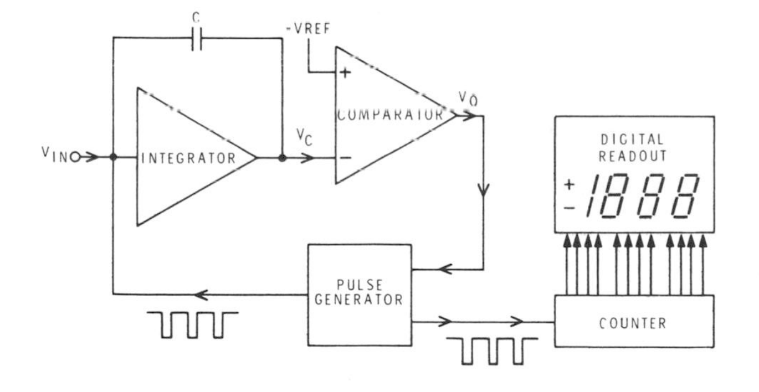

Figure 8

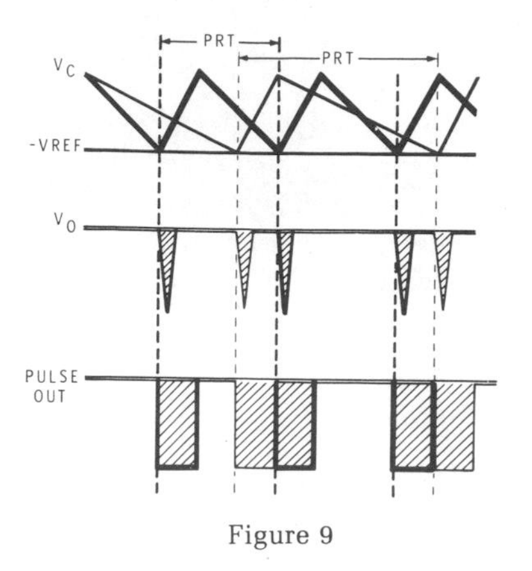

Figure 9

The input voltage causes the capacitor to charge such that V_C is negative as shown in Figure 9. When V_C reaches the potential of $-V_{ref}$, the comparator will trigger the pulse generator. When triggered, the pulse generator will produce a pulse which will return the charge on the capacitor to zero. At the end of the pulse, the capacitor will once more start charging and the cycle will repeat.

With a relatively low input voltage, V_C will take longer to charge to V_{ref} as shown by the black lines in Figure 9, resulting in a relatively low pulse-repetition frequency (PRF). If the voltage is increased as shown by the dark bold lines in Figure 9, V_C will reach $-V_{ref}$ in less time resulting in a higher PRF. Thus, the PRF is representative of the input voltage.

Each pulse generated is fed to an up-down counter. The result is then displayed on the digital readout.

An interesting feature of the voltage-to-frequency technique is the way resolution and reading rate are related. Assume that the maximum frequency available from the pulse generator is 100 kHz. Actually, frequencies of up to 300 kHz are possible. A 100-kHz pulse rate gives the possibility of 100,000 counts per second. Therefore, if we make the frequency counter gate one second in length — that is, make one measurement per second — we have the capability for 5-digit resolution. However, if we desire a faster reading rate — say 10 per second, the counter gate length will only be 0.1 seconds. In 0.1 seconds, the counter can accept 10,000 pulses for a 4-digit resolution. We can see that if we are willing to accept 3-digit resolution, a reading rate of 100 measurements per second is possible.

CHARGE BALANCE

Another method of analog-to-digital conversion that uses a form of voltage-to-frequency conversion is the charge-balance method. The method is so named because a charge placed on C (Figure 10) by V_{in} is, in turn, balanced by V_{ref}.

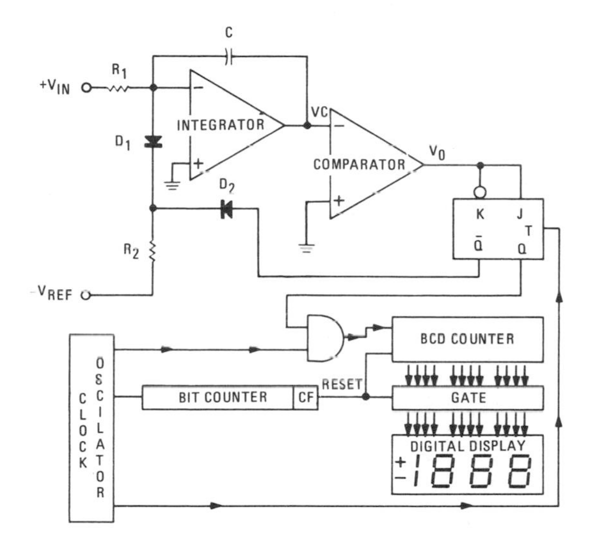

Figure 10

For explanation, we will use a 3-1/2-digit meter which, as you learned earlier, is capable of displaying up to 1999. Thus, in order to make full use of the display, a conversion cycle must be 2,000 pulses in length.

To keep the numbers simple and easy to handle, we will assume that the signal processor has converted the incoming voltage to a range of 0 V to +2 V and that R_1 has a value of 2 KΩ. V_{ref} is −2 V and R_2 is also 2 KΩ.

Assume that the J-K flip-flop is reset; that is, Q is low and $\overline{Q}$ is high. The high at $\overline{Q}$ is slightly less than 5 V. The voltage divider action between $-V_{ref}$ and $\overline{Q}$ will cause a potential of about +3.5 V to be felt at the junction of D_1, D_2, and R_2. This potential, which is more positive than V_{in} can ever be, will reverse bias D_1 and allow only V_{in} to feed the integrator.

At the same time this is happening, the low at the Q output of the J-K flip-flop will inhibit the AND circuit, preventing any pulses from reaching the BCD counter. Of course, the bit counter will continue to count until it reaches full count, which in this case, is 2,000. At that time, the contents of the BCD counter will be gated to the digital display and the BCD counter and the bit counter will be reset to zero to start another conversion cycle. This is a free-running system, so the bit counter will count to 2,000, reset, count, reset, etc., continuously as long as power is applied.

We will start the conversion cycle with some charge on C that causes V_C to be positive and V_O to be low. V_O would have to be low in order for the J-K flip-flop to be in the reset condition.

For the first measurement, assume that V_{in} is the smallest value which can be measured. In this case, 1 mV. With 1 mV at V_{in} and R_1 equal to 2 kΩ, the charge current applied to C will be 0.5 μA. The small amount of charge current will cause V_C to very slowly go in a negative direction (Figure 11).

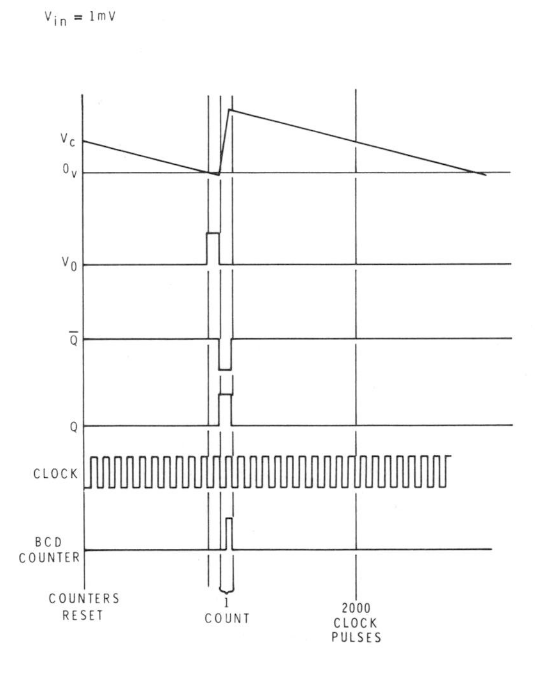

Figure 11

When V_c crosses the comparator threshold, V_O will go high. The comparator threshold is determined by the potential applied at the non-inverting input of the comparator. In this case, it is ground or 0 V. This places a high on J and a low on K of the J-K flip-flop. At the next clock pulse applied to T, the flip-flop will change states. Q will go high and $\overline{Q}$ will go low.

A high at Q will enable the AND gate and allow the BCD counter to start counting clock pulses. Simultaneously, the low at $\overline{Q}$ reverse biases D_2, which allows the $-V_{ref}$ to be felt through D_1 to the integrator.

Remember that $-V_{ref}$ is -2 V and R_2 is 2 kΩ; therefore, the current available to the integrator is 1 mA minus the current from V_{in} which was .5 μA. Thus, the current pushing V_C in a positive direction is 1 mA $-$.5μA or 0.9995 mA. With this current, V_C will charge to maximum positive very quickly.

As soon as V_C passes through the threshold, which happens almost immediately, V_O will go low. The J-K flip-flop will reset on the next clock pulse, inhibiting the AND gate and stopping the BCD counter.

Thus, the BCD counter is enabled for one count out of 2,000, which is correct for 1 mV out of 2 V.

When V_{in} is increased to 1 V, the current through R_1 increases to 0.5 mA. This current will cause C to charge much more rapidly as shown in Figure 12. The current available from $-V_{ref}$, which is $I_{ref} - I_{in}$, or 1 mA −0.5 mA, is the same as the input current. Thus, the positive-going slope of V_C is the same as the negative-going slope. Therefore, V_O will be high the same amount of time it is low, as will Q and $\overline{Q}$. This means that every other clock pulse will be gated to the BCD counter.

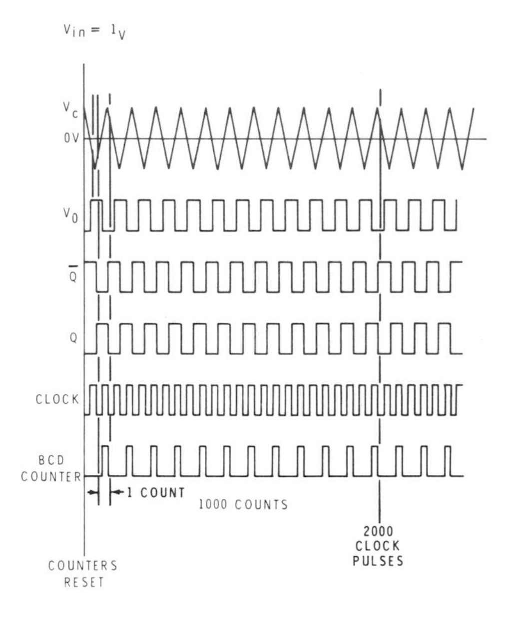

Figure 12

The BCD counter is, thus, enabled for 1,000 out of 2,000, which is correct for 1 V out of 2 V.

Looking now at Figure 13, V_{in} has been raised to 1.999 V, the maximum which can be converted by this circuit. Here is a situation that is opposite that in Figure 11. The charge current from V_{in} is 0.9995 mA, while current from V_{ref} can only be 1 mA − 0.9995 mA or 0.5 μA. This means that V_C will be negative for a much longer period of time than it will be positive. The ratio is the same as the current ratio. Therefore, the BCD counter will be gated on for 1999 counts out of 2000.

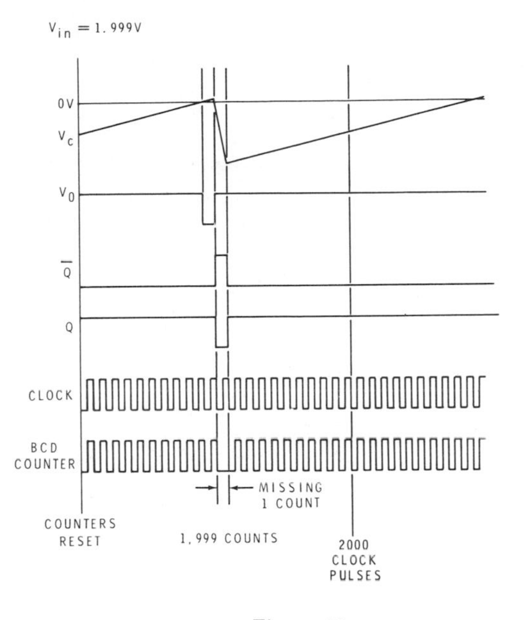

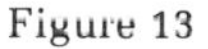
Figure 13

If V_{in} is raised to a higher value, V_C will never cross the 0 V threshold, and the BCD counter will continuously count. Once this happens, there is no way of knowing the value of V_{in}. Therefore, some form of over-range indication is provided to alert the operator.

Accuracy of the charge-balance technique depends primarily on the stability of V_{ref} and the accuracy of R_1 and R_2. With a 10 kHz clock, each reading requires approximately 0.2 seconds for a rate of 5 readings per second. Reading rate can be increased somewhat by using a higher frequency clock.

All of the A/D conversion techniques we have used so far have been integrating techniques which have the advantage of simplicity and reduction of the effects of line noise. The major disadvantage is in reading rate, which reaches a practical maximum of about 60 per second.

Non-Integrating Techniques

With integrating techniques, the unknown voltage was used to charge a capacitor. The conversion to digital was made by counting the number of pulses either to charge or discharge the capacitor. In the non-integrating technique, the capacitor is done away with and the conversion is done in a more direct method.

By eliminating the integration cycle, the conversion process can be speeded up; however, the inherent noise rejection is lost. Perhaps the simplest of the non-integrating techniques is the linear-ramp. We will use this method for an introduction to non-integrating techniques.

LINEAR-RAMP CONVERSION

A block diagram of the linear ramp method is shown in Figure 14. A ramp generator produces a linear swing from maximum positive to maximum negative as shown in Figure 15. The output from the ramp generator is fed to two comparators where it is compared with V_{in} and ground. As long as the ramp voltage is more positive than V_{in}, the output of comparator 1 will be low, and as long as the ramp is more positive than ground, the comparator 2 output will be negative. Therefore, at the start of the sweep, the outputs from both comparators are low, inhibiting both of the AND gates and keeping both inputs of the OR gate low.

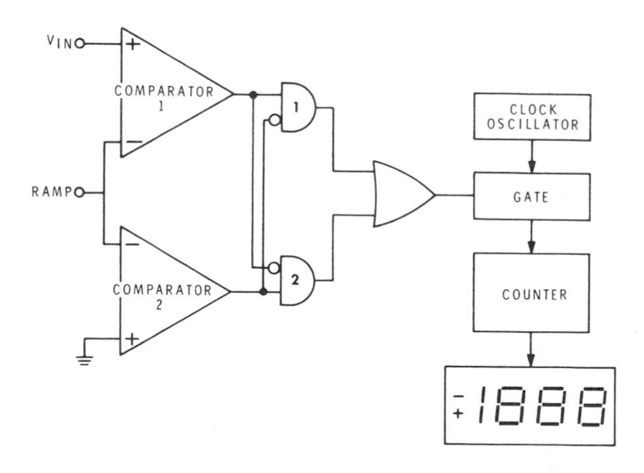

Figure 14

Assume that +5 volts is applied to V_{in}. The ramp starts its sweep from +15 volts in a negative direction. As the ramp voltage passes through +5 volts, the output of comparator 1 goes positive. The output of comparator 2 is still low; thus, the output of AND gate 1 goes high. This high is applied to the oscillator gate and the counter starts counting. As long as a high is applied to the oscillator gate, the count will continue.

As the ramp voltage passes through zero, the output of comparator 2 goes positive. This positive causes the output of AND gate 1 to go low. As long as the output of comparator 1 is positive, AND gate 2 will have a low output. The oscillator gate will thus close, stopping the count. Thus, we have counted a number of pulses proportional to the ramp time between +5 volts and ground. The time from T_1 and T_2 is the correct count time.

If the input voltage were negative, comparator 2 would go positive first, starting the count and comparator 1 would stop the count when the ramp voltage equals the negative input. Autopolarity is easily incorporated by sensing which comparator goes positive first.

A major disadvantage of the linear-ramp method of conversion is its susceptibility to noise. Noise on the input can cause a premature opening of the count gate as shown in figure 15. To achieve acceptable accuracy in this converter, a large amount of input filtering is required.

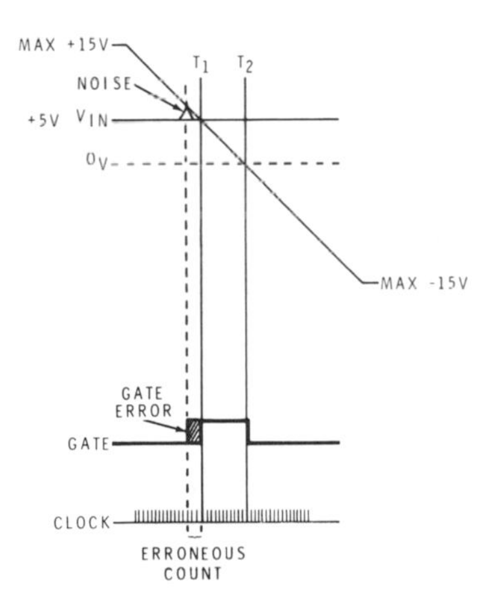

Figure 15

SUCCESSIVE APPROXIMATION

The successive approximation technique generates four reference voltages for each digit. The voltages are generated in a binary-coded-decimal sequence. How does this work in a 3-1/2-digit meter with a range of 10 volts? The reference voltages are sampled in an 8, 4, 2, 1 sequence for the most-significant digit, then decreased by a factor of 10 and sampled 0.8, 0.4, 0.2, 0.1, for the next digit, etc., until each digit has been determined. At start time, 8 volts will be sampled. If it is greater than the voltage being measured at E_{in}, it will be dropped. However, if it is less than E_{in} as shown in Figure 16, it will retain or add to the reference level. Next, 4 volts will be tried. If 4 volts causes the reference to exceed the input, the voltage will be dropped and the reference will remain at the previous level as shown in Figure 16. Two volts is also excessive, so that is dropped. However, one volt is retained, which raises the reference to 9 volts. Each segment will be sampled in turn and either added or dropped until the final reference voltage equals E_{in} as nearly as possible.

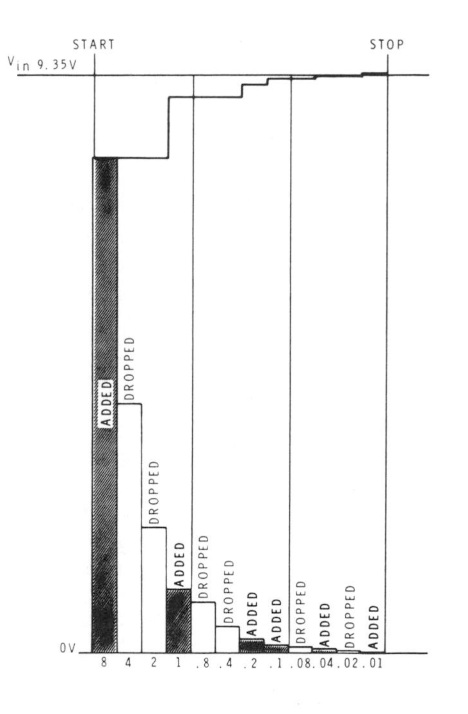

Figure 16

Figure 17 is a simplified block diagram of a successive approximation A/D converter. It allows the sampling sequence described in the preceding paragraph. The voltage being measured is applied to the inverting input of the comparator where it is compared with a feedback voltage (V_f) from the digital-to-analog (D/A) converter. As long as V_{in} exceeds V_f, the output from the comparator will remain high.

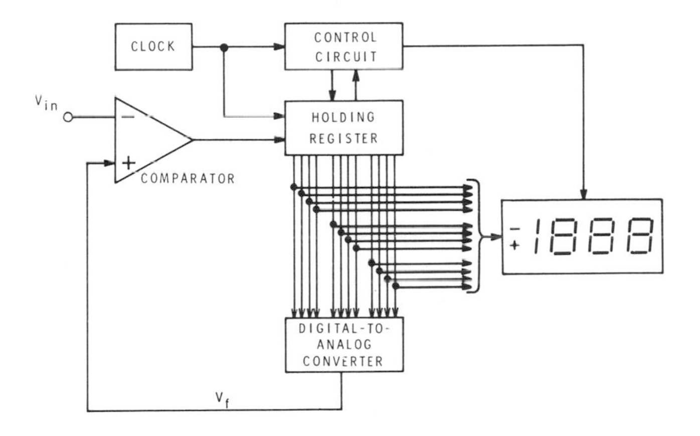

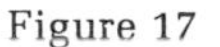
Figure 17

The control circuit starts the conversion by placing a 1 in the most-significant bit position of the holding register. All other bits will be 0. Thus, the BCD number sent to the D/A converter is 8.00. If V_{in} is 9.35 volts, as in the previous example, V_f will be less than V_{in}, the output from the compartor will be high, and the 1 will be held in the register. The register contents now read 1000 0000 0000.

The control circuit sequences through all possible bit combinations, holding or dropping each bit depending on the comparator output. When the conversion sequence is complete, the register will contain 1001 0011 0101 for the input of 9.35 V. This number is gated to the digital readout; the register is reset and another conversion begins.

Instead of the 2000 or more clock pulses required for a conversion by the integrating converters, the successive approximation technique requires 12 clock pulses. Actually, a few more pulses may be required to reset the register and gate the readout to the display. Even so, the successive approximation A/D converter is one of the fastest available.

Reading rate is fixed and depends on the switching rate of the A/D converter. A 3-digit meter is capable of making over 1,200 readings per second. An accuracy of ±0.02% is practical. However, this method is also sensitive to noise, and considerable filtering is required.

Summary

We have discussed some of the more common methods of A/D conversion. There are many more that we have not mentioned. There are as many modifications to the techniques as there are meters, but they all use either an integrating or non-integrating technique.

The integrating techniques have the inherent advantage of noise rejection, but are rather slow. On the other hand, non-integrating techniques are fast but may require a lot of input filtering, which reduces speed. There is no clear advantage for either category in accuracy or resolution. This depends more on the individual meter design.

SIGNAL PROCESSOR

The analog-to-digital converter is capable of handling only a limited range of input voltage. Typically, the voltage must be DC and the range may be from 0 volts to 1.0 volt or less, full scale. The integrator is often an op-amp with a low impedance.

It is the signal processor (Figure 18) that must provide the voltage division necessary for a number of useful ranges. The processor must also change AC to DC and current to voltage as well as providing a source of current for resistance measurement. The input impedance must also be provided here.

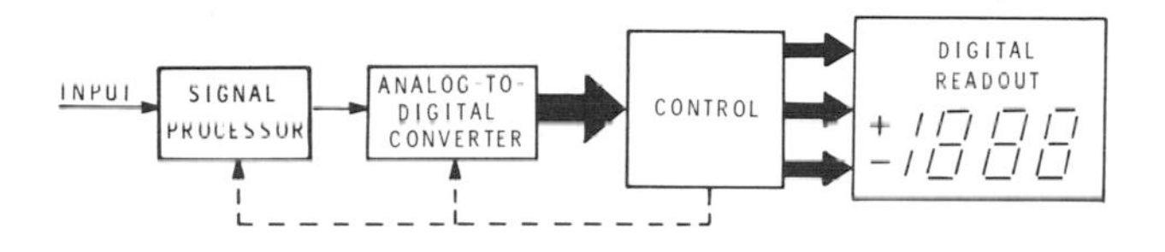

Figure 18

DC Voltage

A simplified illustration of the DC voltage input is shown in Figure 19. Assume that the maximum voltage that can be passed to the integrator is 0.1 volt. This input is taken from an FET connected in a source-follower configuration which provides impedance matching to the low impedance input of the integrator. Input to the FET is from a voltage divider that has a total resistance of 10 M ohms. The input resistance of the FET is in the neighborhood of 100,000 M ohms. Therefore, the position of the switch will have no noticeable effect on the input resistance, and the input resistance on all ranges will be 10 M ohms.

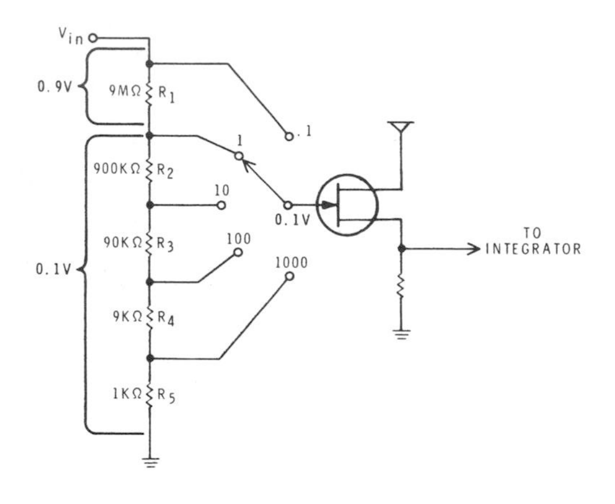

Figure 19

The numbers on the range switch indicate the full-scale voltage which can be measured. At the 0.1 volt position, the full voltage at V_{in} is applied to the gate of the FET. However, when the switch is moved to the 1-volt position as shown in Figure 13, 0.9 volt will be dropped by R_1 and 0.1 volt will be developed across the rest of the network. Thus, 0.1 volt is applied to the gate of the FET with 1 volt at V_{in}. A similar voltage division occurs on all other ranges. Thus, the input to the FET with full-scale voltage at V_{in} on any range will never exceed 0.1 volt. In practice, provision is usually made for "overranging." What this means is that instead of measuring 1.000 volts on the 1 volt range, we could measure 1.999 volts. We will discuss "overranging" in more detail later in the lesson.

Another way to change the range of a Digital Voltmeter (DVM) is by varying the gain of an op-amp as shown in Figure 20. If the integrator can accept an input from 0 volts to 1 volt, then the output of the op-amp must be within this value for all ranges. We will look at three separate ranges to see how this is accomplished. They are 0.01 volt full scale, 1 volt full scale, and 100 volts full scale.

On the 0.01 volt range, the op-amp must increase the input voltage at full scale from 0.01 volt to 1 volt which requires a gain of 100. Remember that the gain of an op-amp is $A = \frac{R_f}{R_{in}}$, where A is the gain of the amplifier, R_f is the feedback resistance and R_{in} is the input resistance, as shown in Figure 20. The "−" sign merely indicates a phase reversal. Therefore, to obtain a gain of 100, the ratio of R_f:R_{in} must be 100:1. If we assign a value of 100 k ohms to R_{in}, R_f must equal 10 M ohms to give us 1 volt out for 0.01 volt in.

For the 1 volt range, a 1:1 ratio is required; therefore, R_f must equal R_{in}. A switching arrangement similar to that shown in Figure 21 could be used. In this figure, the 10 M ohm resistor is always in the circuit. When switching to the 1 volt range, S1 is closed, placing a 101 k ohm resistor in parallel with the 10 M ohm resistor for a total resistance of 100 k ohms. Thus, a gain of 1 is achieved.

When changing to the 100-volt range, S2 is closed with S1 remaining closed. The three resistors combine for a total resistance of 1 k ohm. The ratio of R_f:R_{in} now becomes 1:100 for a gain of 0.01, reducing the 100 volt input to 1 volt at the output.

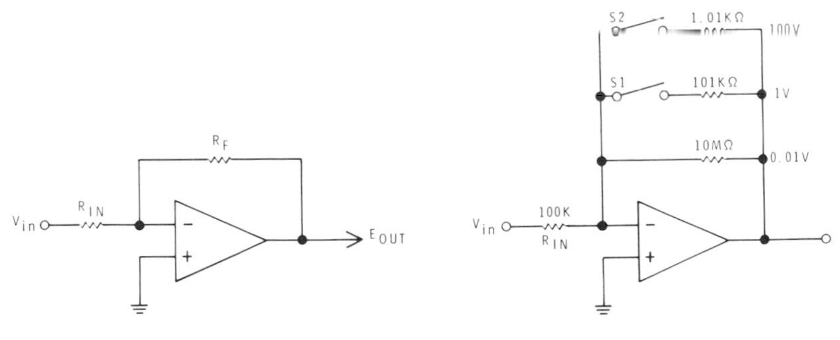

Figure 20

Figure 21

By proper selection of feedback resistances, it is possible to design a meter with any combination of ranges. Also, by designing the digital counting unit such that an output is available any time the input goes "overrange," we could have automatic switching of the resistors, and, thus, achieve "auto-ranging" where the meter selects the proper range. We would no longer have to worry about selecting the proper range as long as we did not exceed the maximum range of the meter. Figure 22 shows how this is done in a typical auto-ranging DVM.

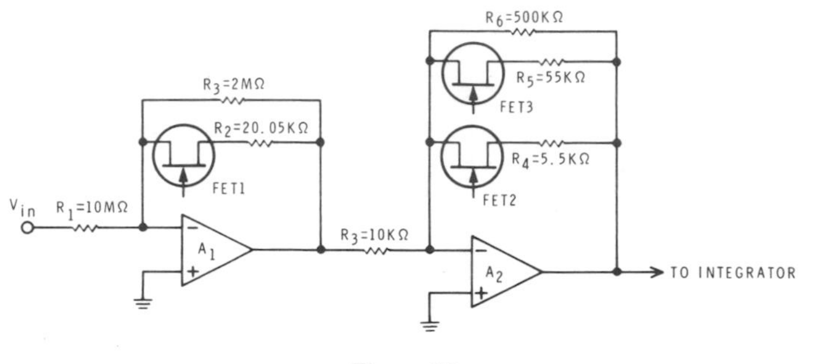

Figure 22

AUTO-RANGING

The switches are FET's and are controlled by the A/D converter. The converter turns on certain switches for a given range overflow. On the lowest range (0.2 V), none of the FET's are turned on. (See Figure 23.) With all switches off, A_1 has an R_f of 2 M ohms for a gain of 0.2 and A2 has an R_f of 500 k ohms for a gain of 50. Thus, the total gain is 10 and the output to the integrator is 2 volts for 0.2 volts in. The A/D converter used in this meter can accept an input of ±2 volts.

RANGE	A_1 GAIN	A_2 GAIN	TOTAL GAIN	FET_1	FET_2	FET_3
0.2V	0.2	50	10	OFF	OFF	OFF
2V	0.2	5	1	OFF	OFF	ON
20V	0.2	0.5	0.1	OFF	ON	ON
200V	0.002	5	0.01	ON	OFF	ON
2000V	0.002	0.5	0.001	ON	ON	ON

Figure 23

When switching to a higher range, one or more of the FET's will turn on and reduce the total gain. On the 2-volt range, only FET_3 will turn on, leaving FET_1 and FET_2 off. Thus, the gain of A_1 will be unchanged while the gain of A_2 will be decreased to 5 by reducing R_f from 500 k ohms to 50 k ohms. By comparing Figures 22 and 23, you can determine the gain of A_1 and A_2, the total gain, which FET's are on and which are off.

You can also calculate the total R_f for each op-amp.

In this application, the input voltage is fed to the input of the op-amp through a resistor network which sets the input resistance. The op-amp has an FET input much like that discussed in analog meters.

AC Converter

While the A/D converter will function properly only with a DC input, a digital meter is often required to measure both AC and DC. Therefore, to make the meter truly versatile, some method of AC to DC conversion must be incorporated.

The main purpose of any AC converter is to change the incoming AC to DC, which can be processed by the A/D converter. Basically, there are two types of conversion: average responding and true RMS conversion.

AVERAGE RESPONDING CONVERTER

The average responding converters are more common. As you learned in the section on analog meters, an average responding meter is so called because the output is proportional to the average of the incoming waveform. By using the proper scale factor, the readout can be calibrated to read RMS, peak, or any other value.

A simplified drawing of an average responding converter is shown in Figure 24. This particular converter uses a half-wave rectifier. A full-wave rectifier could be used, but since it is more complicated and offers no real advantage, the half-wave is more common.

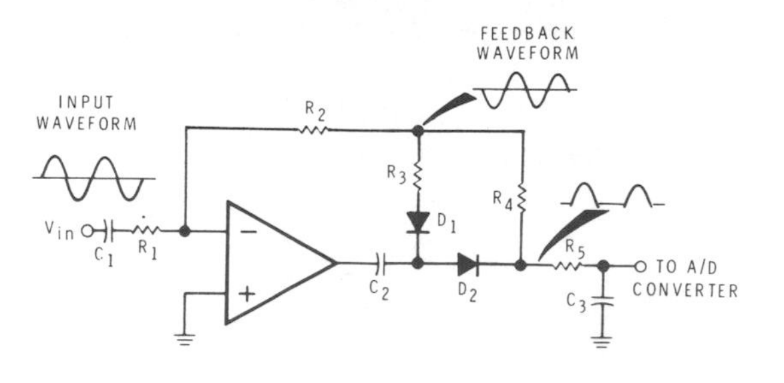

Figure 24

The AC signal to be measured comes from the input circuits through C_1 and R_1 to the inverting input of the op-amp. During the positive portion of the input, the output of the op-amp goes negative. This portion of the signal is coupled through D_1 and R_3 to the junction of R_3 and R_4 as a negative alternation of voltage. At the same time, the negative alternation is blocked by D_2 and prevented from reaching the output circuits.

As the input voltage goes negative, the output of the op-amp goes positive. This positive is coupled through D_2 and R_4 to the junction of R_3 and R_4 as a positive alternation of voltage. Thus, a feedback waveform is developed at the junction of R_3 and R_4 which is out of phase with the input waveform. This AC voltage is coupled through R_2, the feedback resistor to control the AC gain of the amplifier.

At the cathode of D_2, a half-wave rectified signal is present as shown in Figure 24. This pulsating DC is then filtered by R_5 and C_3 and a DC voltage is fed to the A/D converter.

D_2 is the rectifier which provides us with the output voltage. It also provides a feedback voltage. D_1 is a steering diode that provides a negative alternation of feedback that matches the positive alternation from D_2, giving a balanced feedback.

Since it is usually more desirable to know the RMS value of an AC waveform than it is to know the average value, the gain of the amplifier is usually adjusted so that the charge on C_3 represents the RMS value of the input. Since we know that the RMS value is 1.11 times the average value, the gain of the amplifier is adjusted so that the output is 1.11 times the input.

The major disadvantage to this type of circuit is that it is only accurate with a pure sine wave input. As discussed in Unit 1, the scale factor is different for each wave shape, making it impossible to calibrate the meter for all applications.

TRUE RMS CONVERTER

The true RMS converter gives an accurate RMS indication of the amplitude of any wave shape. The most common type of true RMS converter uses a thermocouple input. The AC input is fed through the heater element as shown in Figure 25. The DC output is fed to an op-amp, then to the A/D converter.

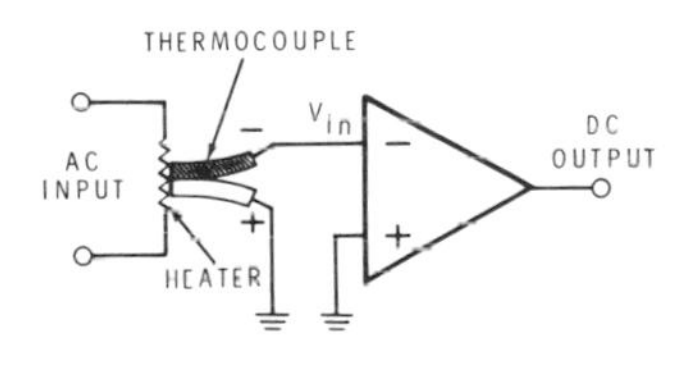

Figure 25

As mentioned in Unit 1, the thermocouple has a square law output. Also, changes in the surrounding temperature affect the output of the thermocouple and appear as errors in the reading. Since it is not desirable to have changes in ambient temperature affect the measurement, we must, in some way, compensate for these changes. Also, most A/D converters require a linear input. The square law output from the thermocouple does not meet this requirement. This, too, must be compensated for.

One way of overcoming these disadvantages is by using a DC feedback to drive a second thermocouple which is connected in opposition to the first, as shown in Figure 26. First, we will discuss temperature compensation. If the ambient temperature increases, the output from both thermocouples will increase by a like amount. Since the thermocouples are connected in series and opposing each other, the net change in output voltage will be zero. However, if the AC input were to increase, only thermocouple 1 would be directly affected. Therefore, the output of thermocouple 1 will increase, increasing the output of the amplifier.

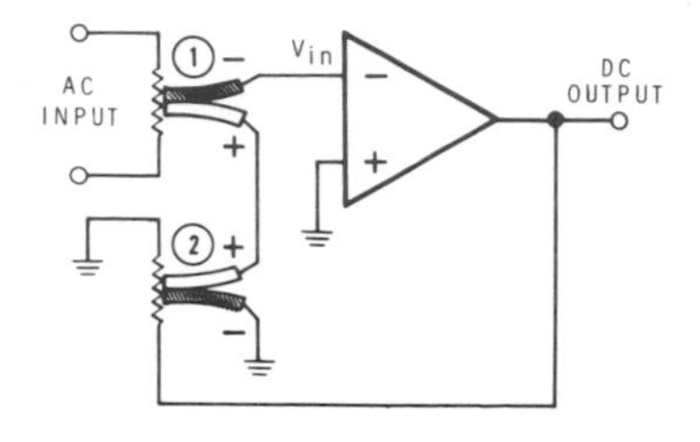

Figure 26

Now, how does this feedback arrangement compensate for the non-linear output of the thermocouple? Assume that 2 V RMS is connected to the AC input. This will produce a given DC potential from thermocouple 1. This potential is amplified by the op-amp. The amplified voltage causes a feedback current through the heater of thermocouple 2. The output of thermocouple 2 is a DC that opposes the output from thermocouple 1. From your study of op-amps, you know that the feedback voltage will be equal and opposite to the input voltage. Actually it will not be quite equal, but the extremely high gain of the op-amp will reduce V_{in} to a few microvolts. Thus, if the AC input is 2 V RMS, the DC output must be approximately 2 V DC.

Suppose the AC input is cut in half to 1 V RMS. Since the thermocouple is a square law device, the output from thermocouple 1 will be reduced to one-fourth of its original value. Therefore, the DC output will decrease until the output from thermocouple 2 is approximately equal to the output from thermocouple 1. For this to occur, the DC output must once more equal the AC input. Therefore, the DC output will decrease to 1V. Thus, the output of the op-amp will be linear and the effects of the temperature change will be minimum.

The true RMS Converter has superior performance to the average responding converter in all areas. It can measure any wave shape with better accuracy. The major disadvantage is cost, which is significantly higher than the average responding converter.

Ohms Converter

If a digital meter is to be a true multimeter, it must be capable of measuring resistance and current as well as voltage. The simplest and most common method of measuring resistance is to provide a constant current through the unknown resistance and measure the voltage developed. A simplified circuit is shown in Figure 27.

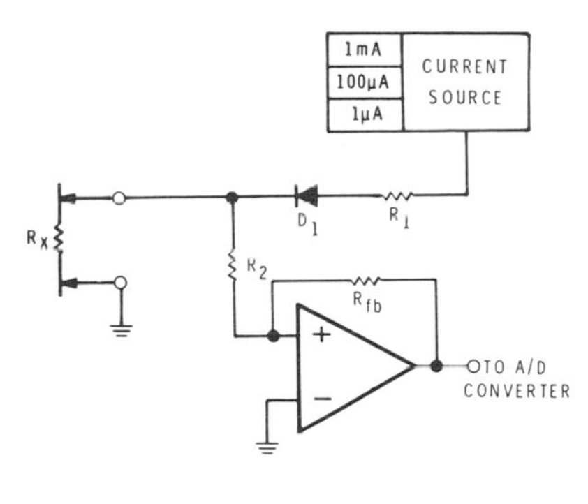

Figure 27

Assume that the maximum voltage that can be measured by the meter is 1 volt. A current source must be selected that will develop between 0 V and 1 V across R_X. If R_X is zero ohms, the voltage at the top of R_2 will be 0 V regardless of the current source selected. With the leads open, $R_X = \infty$, the voltage will exceed 1 V, and the meter will indicate an overrange condition. With the 1 mA current source selected, 1 V will be obtained when $R_X = 1{,}000\,\Omega$. To measure a higher resistance, a lower value of current must be selected. For instance, 10 k ohms can be measured with the 100 μA source, and 1 MΩ is the maximum for the 1 μA source. Additional ranges can be obtained by providing more current sources.

Current

The most common method of measuring current is to convert the current to a voltage drop across a known resistor and measure the voltage. Figure 28 is a simplified circuit showing how this is accomplished. As shown, the switch is in the lowest current range. Current flows through R_2, R_3, and R_4, developing maximum voltage. To measure higher values of current, the switch is moved so that there is less resistance for the current to flow through. The voltage thus developed is fed through R_1 to the integrator amplifier where it is handled as any other voltage measurement. With this arrangement, the value of R_{in} changes with each switch position. However, so does V_{in}, so with the proper selection of resistor values, a correct conversion will be obtained.

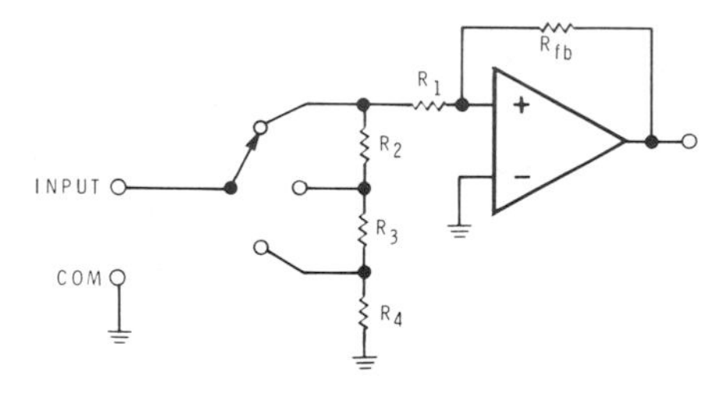

Figure 28

DISPLAYS

Once the input voltage or current has been converted into a series of digital pulses and the pulses counted, the results must be displayed. There are a number of different display types available, each having certain advantages and disadvantages.

Format

Each type of display is arranged into one of two basic formats, the segmented display and the dot matrix. The segmented displays are usually either 7 or 14 segments. The dot matrix can have any format, but for meter use, the 5 × 7 matrix is by far the most common.

In the **seven-segment format**, each digit is composed of from two to seven segments. Any digit can be formed by lighting the proper segments. To form a "1," segments b and c are lighted; for a "3" (as shown in Figure 29), segments a, b, c, d, and g are lighted; and an "8" is formed by lighting all segments.

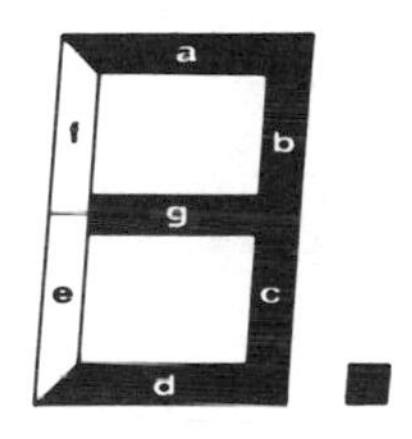

Figure 29

A 3-1/2-digit meter has a total of 23 segments. Depending upon the type of display used, each segment can require up to 50 milliamperes of current. If all the segments of the 3-1/2-digit number were illuminated simultaneously, a total current of 1.15 amperes would be required. This is a rather high amount of current and is far too much for a battery-operated meter.

To reduce current and, therefore, battery drain, the segments are **multiplexed**. First, the multiplexer will scan the "a" segments, turning on each required segment in turn. Next, the "b" segments are scanned, then the "c," etc. Each segment that is a required portion of the displayed number will be illuminated for a very brief period of time. The sequence is repeated many times per second so that they display appears to be on continuously. This technique significantly reduces the total current required and makes the battery operation practical. Multiplexing also reduces driving circuitry.

By far, the most common of the **dot matrix formats** is the 5 × 7 format shown in Figure 30. Each dot can be addressed individually, allowing any number of characters to be displayed. In Figure 30, a "3" is being displayed. This format is particularly useful when letters or symbols other than numbers must be displayed. However, for general meter use, the 7-segment display is used more frequently. Multiplexing is even more important in the dot matrix due to the large number of dots. A 5 × 7 matrix has 35 individual dots which require excessive circuitry unless multiplexing is used.

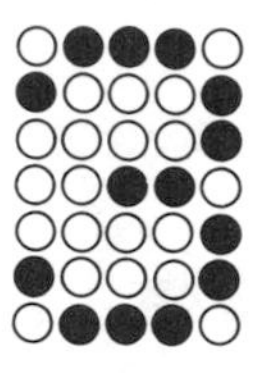

Figure 30

Types of Display

Several types of display are available to the designer of a meter. In addition, each type is available in different sizes. Brightness varies within types and from one type to another. A variety of colors is available in some types and severely restricted in others. Some of the more common types of displays are discussed here.

LIGHT EMITTING DIODE

By far, the most popular type of display is the light emitting diode (LED). The LED is a PN junction constructed from a material that emits light when electrons combine at the junction as shown in Figure 31. The LED has the advantage of relatively low cost combined with high reliability and long life. LED's are low voltage devices, but they require an average current of about 20 milliamperes per segment. Comparatively speaking, this is a moderate amount of current; however, some of the smaller units may need as little as 3 mA.

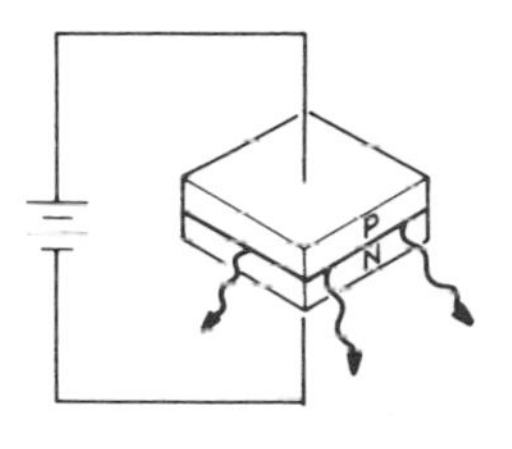

Figure 31

Almost any format of presentation is available with LED's. Some of the more common are the 7-segment numeric, the 16-segment alphanumeric, and the 5 × 7 dot matrix. All of these are available as single characters or in groups of up to 12 digits contained in one package. In the smaller 7- and 16-segment displays, each segment is formed by a row of LED's usually encapsulated in plastic. The larger displays use plastic light pipes, fiber optics, and plastic reflectors to enhance brightness and to give each segment a more uniform appearance. Of course, in the dot-matrix display, each dot is formed by an individual LED.

Many sizes of characters are available, starting with 0.135 inch, which usually has a built-in magnifier, and going up to about 0.6 inch. For characters over about one-half inch, another type of display is usually more practical. Four colors of LED are currently available: red, yellow, orange, and green. Red is the most economical to construct, and is thus the most common. However, most people find red hard to look at for an extended period. Orange is aesthetically the most pleasing, while the human eye is more sensitive to green.

In general, the brightness of LED's is very good. Most are readable in normal room light, and some red and yellow units are readable in bright sunlight. Filters are used to increase contrast, but they can't change the color.

LED's are temperature sensitive. However, most will operate over a fairly wide range of environmental conditions. Temperature ranges of −20°C to +70°C are common; however, operating above 85°C may cause premature failure.

LIQUID CRYSTAL DISPLAY

The liquid crystal display (LCD) uses less power than any other type of display. The older type of LCD is the dynamic scatter type. This type depends upon a small amount of current through the crystal to align the molecules. Only a few microwatts of power are required.

The dynamic scatter LCD presented a frosted image with either a clear or a mirror background. Contrast was a problem, as was viewing angle. The display could not be read from wide viewing angles or in very dim light conditions.

A more recent development with liquid crystal displays is the field-effect LCD. The field-effect LCD is definitely a power miser, using only nanowatts of power. This low power is possible because the LCD does not emit light. It transmits light or reflects light.

The reflective LCD also requires a fairly high level of ambient light. If there is sufficient light to read the printed page, there is enough light to read the reflective LCD. The reflective LCD is constructed of two thin glass plates, each having a very thin coating of electrode material where the segments of the 7- or 14-segment display are located. The electrode material is so thin, it appears transparent. Between the plates is the liquid crystal material. Under static conditions, the liquid crystal is clear. (Figure 32A) However, when a potential is applied to the electrodes, the material becomes opaque (Figure 32B). Since opacity is localized to the vicinity of the applied voltage, the segments are individually addressed. The rear plate is mirrored and reflects light from any part of the display that is not energized.

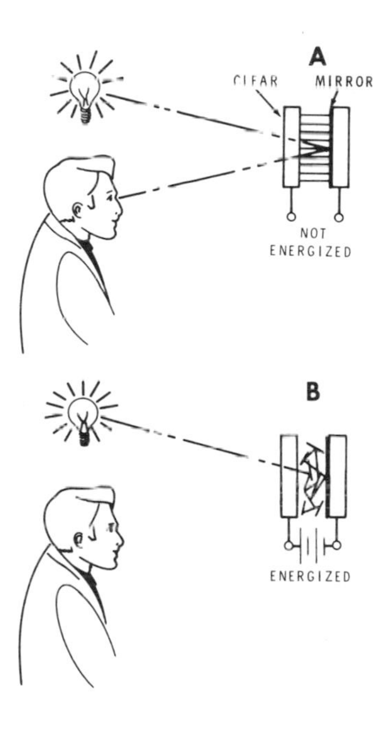

Figure 32

The transmissive LDC is constructed in the same manner as the reflective LDC, except that the rear plate is not mirrored. Thus, light from behind the display can pass through any portion that is not energized (Figure 33A). Light is prevented from passing through an energized portion (Figure 33B). Backlighted units can be made almost any color by filters.

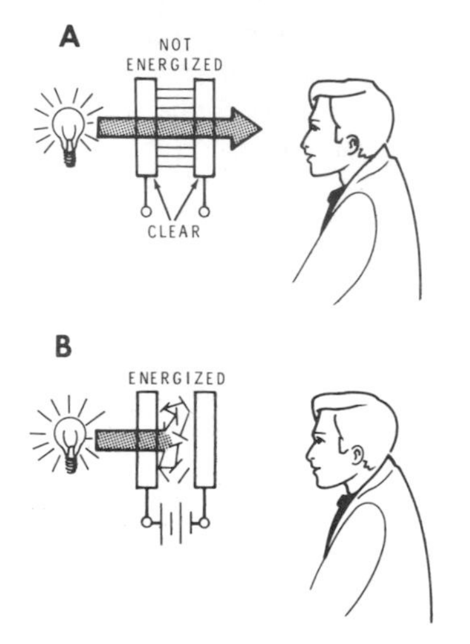

Figure 33

Either type of LCD can be made with light letters on a dark background or with dark letters on a light background. Since the LCD does not emit light, there is no rating for brightness. Instead, a contrast ratio is given, which indicates how well the display stands out against the background. A contrast ratio of 15 or 20 to 1 is good.

LCD's are available in a wide variety of sizes from the very small ones used in watches to the characters several inches in height. Power consumption does not increase significantly with size, especially with the field-effect LCD. LCD's operate over a moderate range of temperatures from +5° to 80°C. At temperatures lower than those, a heater is required for reliable operation. At higher temperatures, failure may occur. Also turn-on and turn-off times vary considerably with temperatures.

Because of its low power drain and low cost which is competitive with the LED, the LCD is becoming fairly common in battery-operated meters.

VACUUM-FLUORESCENT

Another display that is becoming more popular is the vacuum-fluorescent. This display is available in segmented format and as a dot matrix. Both are made in several sizes as single or multiple character units.

The vacuum-fluorescent display is constructed similar to a triode vacuum tube (Figure 34.) Each segment or dot has a cathode, a control grid which is used to turn the segment on and off, and a phosphor-coated anode which is the visible portion of the segment or dot. The cathode is an especially selected material that emits large quantities of electrons when heated. A cloud of electrons will thus form around the cathode.

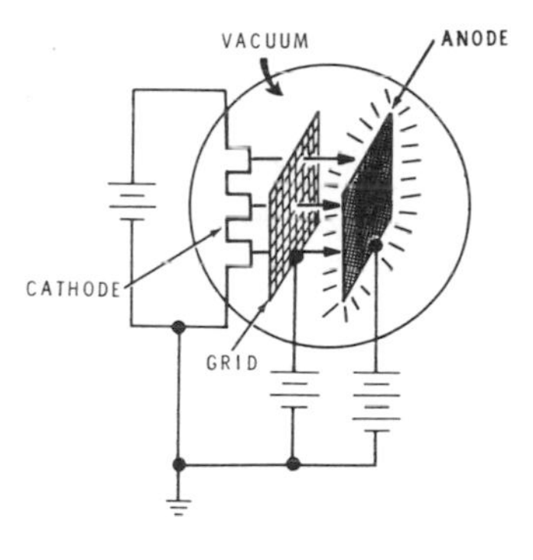

Figure 34

The control grid is a wire mesh mounted between the cathode and anode. If the potential on the grid is negative, the electrons surrounding the cathode will be repelled by the grid and forced to remain in the vicinity of the cathode. However, if the grid is made positive, it will attract the electrons from the cathode.

Most of the electrons will pass through the mesh of the grid and be attracted to the much higher potential of the anode. The anode is coated with a phosphor material that emits visible light when bombarded by electrons. Because of the construction, the voltage required is rather high, being in the neighborhood of 30 to 40 volts. However, power consumption is low, about one-half that of the LED.

The vacuum-fluorescent display has a high brightness and a pleasing blue-green color. This, combined with its relatively low cost, makes it an attractive selection for a designer.

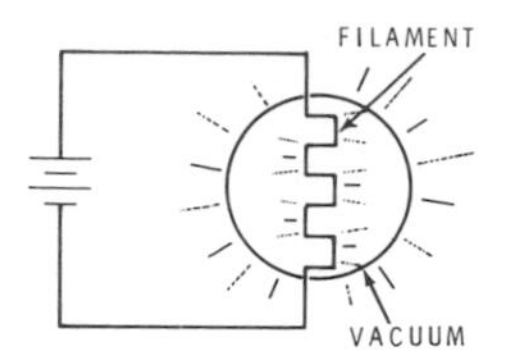

Figure 35

INCANDESCENT

The brightest of all displays is the incandescent. It consists of a filament located in a vacuum (Figure 35). The filament is heated by a current until it glows. The higher the current, the brighter the glow. This display requires the most power of all, with 20 to 30 mA per segment being common.

Both segmented and dot matrix displays are available. Each segment will be composed of an individual filament. With the matrix display, each dot is usually a separate bulb. Since the basic filament emits a white light, any color of display is available by the use of filters. There is a wide variation in sizes available, from 0.34 inch to 3.375 inches.

Several meters are available with incandescent displays. These are especially useful in conditions of high ambient light where portable operation is not required.

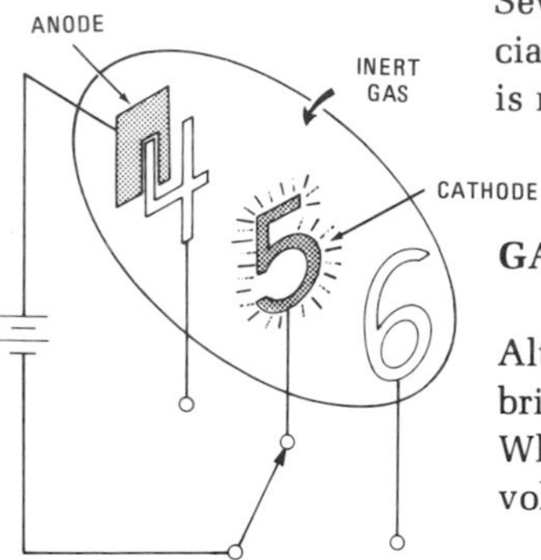

Figure 36

GAS DISCHARGE

Although not as bright as the incandescent, the gas discharge display is brighter than the LED, making it the second brightest display available. While the gas discharge does not require a large amount of power, a high voltage of approximately 200 V is needed.

The gas discharge display consists of a cathode and an anode inside a glass container which is filled with an inert gas. A potential difference is applied between the anode and the character to be displayed. (Figure 36) A high enough potential will cause current to flow, ionizing the gas near the negative element. The character which we wish to display is electronically selected. When formed characters are used as shown in Figure 36, they are stacked one behind the other. Thus, the outline of other characters may be seen even when they are not illuminated.

Formed characters, segmented and dot matrix formats, are available in sizes from 0.25 inch to more than 1 inch in height. The predominant color is orange; however, green is also available.

Although fairly high in cost, the gas discharge display is found in several meters.

SPECIFICATIONS

Many of the specifications given for digital meters mean the same as those specifications for analog meters. However, there are some unique terms which will be explained in this section.

Range

The term range on a digital meter is interpreted somewhat differently than it is on an analog meter. The "full-scale" range of a digital meter is the maximum value that can be measured without going "overrange." This value is directly related to the number of "full" digits. Full digits are those which are capable of registering from 0 to 9. A 3-digit meter will have three digits capable of registering from 0 through 9. Thus, the maximum range of a 3-digit voltmeter would be 999 V.

Almost all meters today have "overranging." With "overranging," the A/D converter and counter circuits are capable of processing a value larger than the "full-scale" reading. A typical meter will have an overrange capability of 100%, meaning that a 3-digit meter can measure from 0 V to 1999 V. This is not a rule, and overranging varies from 20% to 200%, depending on meter design.

Overranging is indicated by the addition of a digit in the most-significant position. The additional digit is called a half digit. Therefore, a 3-digit meter with 100% overranging is called a 3-1/2-digit meter.

Here is an example. A 3-digit meter is indicating 99.9 V. If the voltage is increased to 100 V, and there is no half digit, we will have to turn to a higher range for the measurement. The meter will then indicate 100 V, but we will have lost the tenths indication. By going to the higher range, we have reduced the accuracy and sensitivity of the meter by a factor of 10.

By including overranging and adding the half digit, the indication will be 100.0 V, retaining the sensitivity and accuracy of the lower range. Thus, a 3-1/2-digit meter on the 100 V range is capable of measuring up to 199.9 V.

Resolution

There is often confusion between the terms **resolution**, **accuracy**, and **sensitivity**. They are related but do not mean the same thing. Resolution is the ability of a meter to display the difference between values.

For instance, the Heathkit IM-2202 digital multimeter is a 3-1/2-digit meter with several ranges. The lowest range is 100 mV and the highest is 1000 V. On the 100 mV range, the most-significant digit indicates 100 mV. The next most-significant digit is 10 mV, then 1 mV, and the least-significant digit represents 0.1 mV. Thus, the smallest voltage that can be displayed on the 100 mV range is 0.1 mV. The highest is 100 mV. It is often said that the resolution is 0.1 mV. Actually, this is not entirely correct.

Resolution is not a value of current or voltage. It is a ratio of the minimum value that can be displayed to the maximum value that can be displayed on a given range. Overranging is disregarded when specifying resolution, so on the 100 mV range, resolution is 0.1 to 100 or .1%. On the 1000 V range, the resolution will be the same. The most-significant digit is 1000 V and the least-siginificant digit is 1 V, which is a ratio of 1 to 1000 or .1%. This does not necessarily mean that the meter has an accuracy of .1%. The accuracy cannot be better than the resolution; however, it can be much worse.

Sensitivity

We said earlier that on the 100 mV range, the meter could measure or sense a difference as small as 0.1 mV. We, therefore, say that the meter has a sensitivity of 0.1 mV. The sensitivity of the meter is the smallest change in voltage that the meter can respond to. It can be found by multiplying the lowest range times the resolution. Thus, the sensitivity of a 3-digit meter with a 100 mV range is $.001 \times 100$ mV = 0.1 mV. Of course, this only applies if the measuring and counting circuits are capable of the indicated sensitivity accuracy.

Accuracy

Accuracy is an indication of the maximum error that can be expected between the actual voltage being measured and that indicated by the meter. In digital meters, it is usually given as a percentage of reading plus a percent of full scale. The percent of full scale is often expressed as plus or minus one digit. For instance, the accuracy of the Heathkit IM-2202 is given as ±0.2%, ±1 digit. Remember that when measuring a voltage, a gate is opened for a period of time proportional to the absolute value of the input voltage. While the gate is open, pulses are being counted. The nature of the measurement is such that the gate can close between pulses or during a pulse. It is, therefore, possible for the least-significant digit to change back and forth between two adjacent digits while the meter is measuring a constant input.

In order for a manufacturer to claim a given accuracy for his meter, the accuracy must be traceable to a legal source. In the United States, the National Bureau of Standards (NBS) keeps the official standard for all measurements. Therefore, measurements are compared to the "legal volt" maintained by NBS. When you purchase a new meter, it should meet the accuracy specification. However, from that point on, accuracy will start to decrease. It is the user's responsibility to maintain the standard.

The individual user will seldom have the equipment and knowledge necessary to calibrate test equipment. Since calibration is required at intervals of 30 days to one year, some provision must be made. Usually the meter is sent to a Precision Measuring Equipment Laboratory (PMEL) or returned to the manufacturer at regular intervals. The equipment required for PMEL is expensive and skilled technical help is required. Therefore, the cost of calibration can be significant and should be considered in your purchase decision.

Some companies such as Heath have provided built-in calibration references, allowing you to do your own calibration. Accuracy is not as good as with laboratory standards. However, 0.5% or better can be obtained. With laboratory standards, some meters are capable of 0.01% or better accuracy.

Error Factors

There are several factors which contribute to the accuracy of a meter. In this section, we will discuss some of the factors which can cause errors.

QUANTIZING ERROR

Quantizing error is an error that occurs because a digital meter can only measure steps of voltage or current. The steps can be very small, but they are discrete steps. As an example, assume that a 3-1/2-digit meter is measuring 65.3 volts on the 100 V range. There is an ambiguity in the least-significant digit in that we have no way of knowing if the actual voltage is 65.30 or 65.39 since the reading will not change until the input reaches 65.40. Thus, there is a possible error of up to 0.09 volts on the 100 V range.

A proportional amount of error occurs on any other range. The error is up to nine tenths of the least-significant digit. In better meters, the circuit is designed such that the changeover occurs halfway between digits. With this method, the error is no more than one half of the least-significant digit instead of up to a full digit.

NORMAL-MODE ERROR

Normal-mode error is caused by noise and is frequently referred to as normal-mode noise. Normal-mode noise is always present. It is picked up from power lines and electromagnetic fields and is generated in the meter itself. In many cases, the noise has a higher amplitude than the signal being measured. The effects of this noise must be reduced.

Different methods must be used for different types of noise. Using an integrating type of A/D converter can reduce line noise by a significant amount. The reduction due to integration will be at multiples of the reading rate. For instance, the Heath IM-2202 meter makes five readings per second. Thus, higher rejection will occur at 5 Hz, 10 Hz, 15 Hz, 20 Hz, and all multiples of 5 Hz. This reading rate reduces the effects of line noise at either 50 Hz or 60 Hz.

Frequencies other than those which are multiples of the reading rate must be reduced by filtering. In meters which do not use an integrating converter, filtering must be used for all noise reduction.

Normal-mode rejection of 30 dB or more is adequate for most applications; however, in very noisy environments, much more attenuation may be required.

COMMON-MODE ERROR

Common-mode noise is noise that is present at both input terminals of the meter. Remember, noise is any unwanted signal. It may be a DC potential, an RF field or line related.

Figure 37 shows a situation where common-mode noise should not be present. If the negative terminal of the meter and circuit ground are at the same potential, all of the current is caused by the voltage being measured and current will follow the path indicated by the arrows.

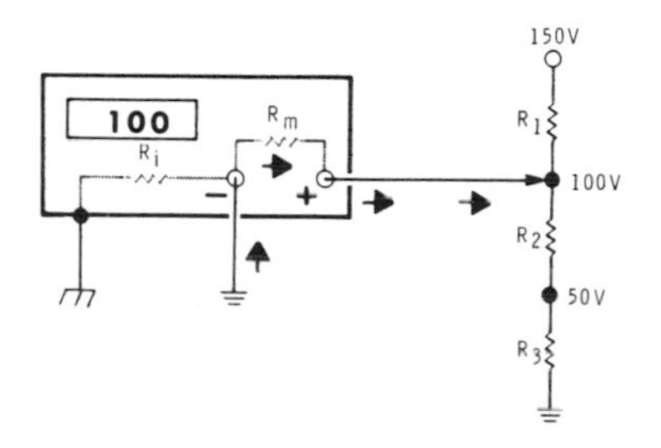

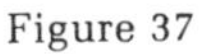
Figure 37

However, when a floating measurement is made as in Figure 38, a potential difference is established across R_i. This places the common terminal of the meter above ground potential.

In this example (Figure 38), the common terminal is at 50 V and the positive terminal is at 100 V. We would like for the meter to measure the 50 V drop across R_2. We can see then that the 50 V across R_3 is common to both leads of the meter.

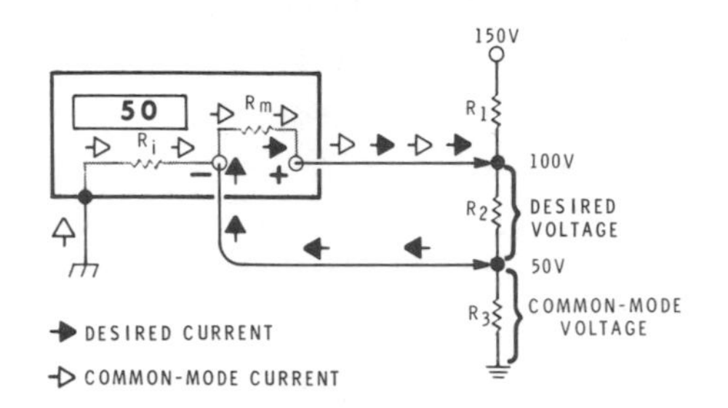

Figure 38

A meter with a differential input will indicate the proper voltage due to its common-mode rejection capability. Remember that the formula for finding the voltage out of an op-amp (V_{out}) is:

$$V_{out} = A\ (V_1 - V_2)$$

Where A is the amplifier gain and V_1 and V_2 are the inverting and non-inverting inputs, respectively, as shown in Figure 39.

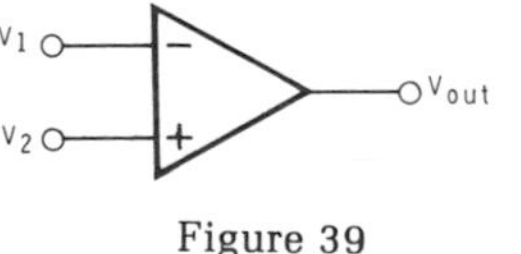

Figure 39

Assume that the gain of the amplifier in the meter in Figure 38 is 0.1. Thus,

$$V_{out} = 0.1\ (100 - 50)$$
$$= 5\ V$$

The output is proportional to the voltage we wish to measure.

Of course, this only works if both halves of the operational amplifier are perfectly balanced. Since this is impossible in actual practice, any common-mode signal will produce some output which is called common-mode error.

How well a meter eliminates common-mode error is the common-mode rejection (CMR) of the meter. CMR may be found by the formula:

$$\text{CMR} = 20 \log \frac{V_{out}}{\frac{V_{in}}{A}}$$

and is measured in dB.

Thus, when 1 V is applied to both terminals simultaneously, the gain of the amplifier is 100, and the meter indicates 1 mV, the CMR is:

$$\text{CMR} = 20 \log \frac{.001}{\frac{1}{100}}$$

$$= 100$$

A good meter should have a CMR of **at least** 60 dB. More is better.

In laboratory meters which must measure very small values in the presence of high common-mode noise, special techniques are used to reduce the noise reaching the terminals. In the average shop meter, these techniques are usually not needed and, therefore, not incorporated.

There are some steps which you can take to reduce common-mode error in your measurements. Using shorter meter leads can reduce stray signal pickup. Operating a portable meter on batteries should also extend the CMR somewhat.

STABILITY

Another factor which affects the overall accuracy of the meter is stability. There are two types of stability which may be specified; long-term and short-term.

Long Term Over an extended period of time, the components in the meter age and change value. These changes cause a corresponding change in reading for a given input. The period of time referred to usually the calibration interval, and the error is stated as ± a percent of reading or of full scale for that time period.

Short Term Short-term stability is essentially repeatability. Good short-term stability does not infer good accuracy. However, in order to have good accuracy, the meter must have good short-term stability. Short-term stability is a measure of how well a meter responds to successive measurements of the same voltage. The time interval is usually 24 hours; however, there is no standard. Once more, the error is expressed as a ± percentage.

TEMPERATURE

Each meter is designed to maintain its basic accuracy within certain temperature limits. Many of the laboratory meters are specified at 25°C ±5°. Others give a range of temperature such as 10°C to 40°C. Often, the manufacturer's specifications don't state the temperature range.

The effects of temperature variation are in addition to the error specified for normal accuracy. They are given in different ways, such as a percentage of overall value (±0.01% reading and ±0.01% full scale/C°) or in parts per million per C°. For shop use, the change may be insignificant, but for field use in extreme weather conditions, it can be a significant factor.

SPEED

There are two things to consider when discussing the speed of a meter. They are conversion time and reading rate. Conversion time is the time required for the A/D converter to change the input voltage to a digital readout. Basically, this can be broken down into two segments; settling time and digitizing time.

Whenever the input voltage to the meter changes, it takes the amplifier converter a certain period of time to sense or respond to the complete change. This is the settling time. It is longer if an AC to DC conversion is required or if heavy filtering is used. If the meter has auto-ranging capability, the settling time may be much longer.

Next, the A/D converter must digitize the measurement. Digitization time may be fixed or variable, depending upon the type of converter used. The time from the start of a measurement to the digitization of the signal is the conversion time. If the meter is to be used as part of a "system," this time is very important. However, for normal bench use, conversion time by itself doesn't tell much. For bench use, reading rate is more important.

Reading rate is the number of readings per second, including trigger and display time, which conversion time excludes. Specifications for reading rate should be carefully considered. Meters using the successive approximation technique of A/D conversion may be capable of several hundred readings per second under ideal conditions. However, when the necessary filtering is added for practical applications, it may be slower than a good integrating meter. Reading rates of from 5 to 10 per second are reasonable.

ANALOG VS DIGITAL

In Unit 1, the characteristics of several types of analog meters were discussed. Thus far in this Unit, the characteristics of digital meters have been covered. Each type of meter has certain advantages and disadvantages. A comparison of analog meters with digital meters will aid in determining where the advantages lie.

The digital meter is superior to the analog meter in almost all areas. What are some of these advantages? One major advantage is the display. The number is in easily-read digits, complete with decimal point and polarity sign. You don't have to worry about scale factor, parallax, interpolation, or which way the leads are connected.

Another advantage is resolution. About the best you can read with an analog meter is one part in 100. That is 1%, so even if your analog meter has 0.5% accuracy, you can't read it. A 3-digit digital meter has a resolution of one part in 1,000, and a 4-digit meter has one part in 10,000. This is a resolution of 0.1% and and 0.01%, respectively.

A digital meter is usually many times as accurate as an analog meter. If you look hard and pay a lot, you can find an analog meter with ±0.5% accuracy. A 3-digit digital meter which costs about the same will have an accuracy of ±.1%. The best digitals have accuracies up to ±0.005% and the resolution to match.

Reading time for an analog meter is usually a second or more. A slow digital makes five readings per second. In some applications, the reading rate may be several hundred per second. Thus, the digital is much faster.

The analog meter is relatively immune to circuit noise; however, it may be more susceptible to radio magnetic interference. With filtering, the digital can do as well or better here too.

At one time, the analog had the advantage of portability. Today's digitals are very portable and go where the job is. They can be built very rugged, as the absence of moving parts makes them hard to break.

The only advantage left to the analog is cost. If you require a low-cost meter with an accuracy of 3% or 4%, you can find it in an analog; however, for a few dollars more, you can have a digital with an accuracy of ±1%. When the final decision is made, a lot will depend on individual preference.

PROBES

All meters, both analog and digital, have limitations on the measurements which can be made. There is a maximum voltage, a maximum frequency, or some other limit beyond which we may occasionally want to go.

In this section, we will discuss how probes can be used to extend the usefulness of a meter. Three of the most useful probes are; the RF probe, the AC current probe, and the high-voltage probe.

RF Probe

Any meter, digital or analog, has a limited frequency response. In a general-purpose meter, this range will usually have a low of about 50 Hz and a high of anywhere from 2 kHz to 10 kHz. To measure higher frequency voltages, a radio frequency probe is required. A typical RF probe is shown in Figure 40. The incoming RF is coupled through C1, which blocks any DC from entering the probe, to D1. D1 rectifies the RF signal and produces a DC voltage across R1 which is equal to the peak RF input. This DC is then applied to the meter and measured just as any other DC voltage would be. If an RMS indication is desired, a dropping resistor can be inserted in series with the input to scale that input to 0.707 of the peak.

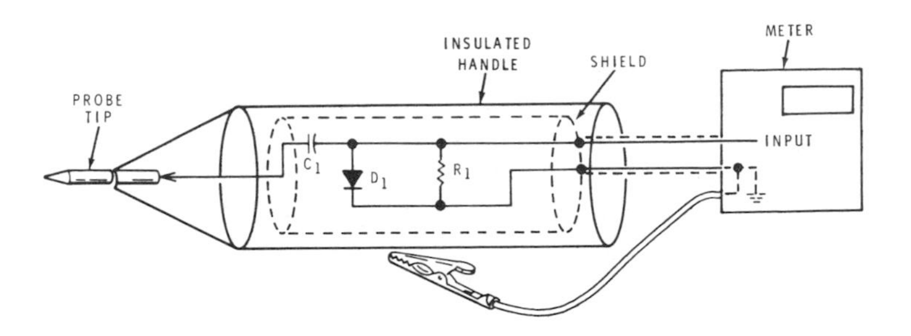

Figure 40

Another method of measuring the RMS value of RF is with a thermocouple. The thermocouple is especially good if the waveform of the RF voltage is something other than a sine wave. Special thermocouple probes are available that enable a meter to measure RF in the VHF and UHF range.

Current Probe

You have learned that when measuring current, the circuit must be broken and the ammeter inserted in series with the line. You have also learned that many lower-cost meters, especially electronic meters, have no provision for measuring AC. If the meter does have AC capability, the additional resistance can interfere with proper circuit operation.

One of the ways to overcome these problems is with a current probe. The probe operates in the same manner as the clamp-on meter discussed in an earlier lesson. The probe consists of a ferrite core which can be opened and placed around the current-carrying wire. A number of turns of wire are wound on the core as shown in Figure 41. The ends of this wire are connected to the input of an AC voltmeter.

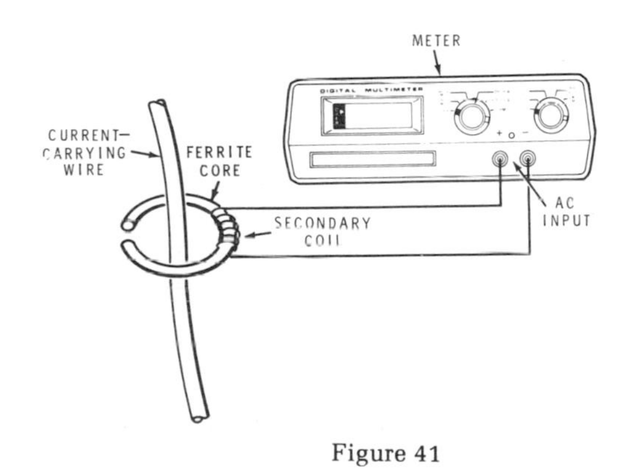

Figure 41

Thus, we have a transformer. The primary is the current-carrying wire and the secondary is the coil that is connected to the input of the voltmeter. The turns ratio is usually such that 1 mA of current produces 1 mV of output. Therefore, the meter can be read directly. Of course, this type of probe can be used on AC circuits only.

High-Voltage Probe

The typical digital or electronic voltmeter has a maximum voltage range of about 1000 V. To measure voltages higher than this, a high-voltage probe is required.

Electrically, the high-voltage probe is nothing more than an additional multiplier resistor which extends the ranges of the meter. Figure 42 shows how the probe works. Usually, the probe is designed so that 1/100 of the voltage at the probe tip is applied to the meter input. Thus, for 20 kV at the probe tip, the meter will indicate 200 V.

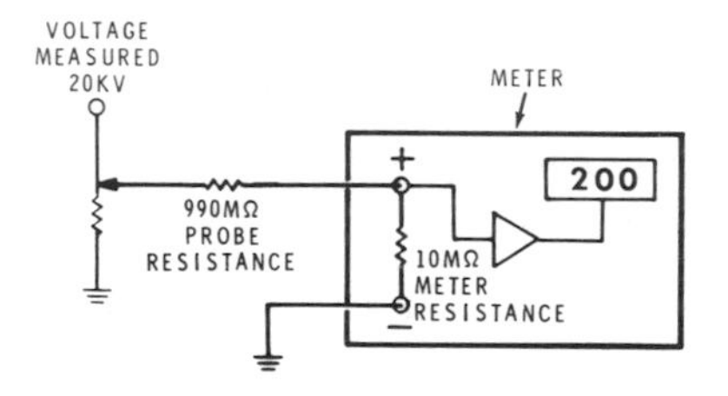

Figure 42

The resistance of the probe is dependent upon the input resistance of the meter being used. A ratio of 99 to 1 will provide the proper voltage division. Therefore, a meter with a 10 M ohm input resistance requires a probe resistance of 990 M ohms.

While it is possible to connect the proper resistance in series with your normal meter lead, it is extremely unwise. Potentials of several thousand volts can give you a shock that your survivors will remember for a long time. A high-voltage probe should be designed to minimize shock hazard and the possibility of arcing. Figure 43 shows how such a probe is usually constructed.

The handle and shield are made of a sturdy insulating material. The hand shield keeps the hand from slipping down the barrel of the probe. The barrel is long to minimize the possibility of arcing. Also, to prevent arcing around the resistors, the resistance is physically a very long resistor or a string of resistors in series.

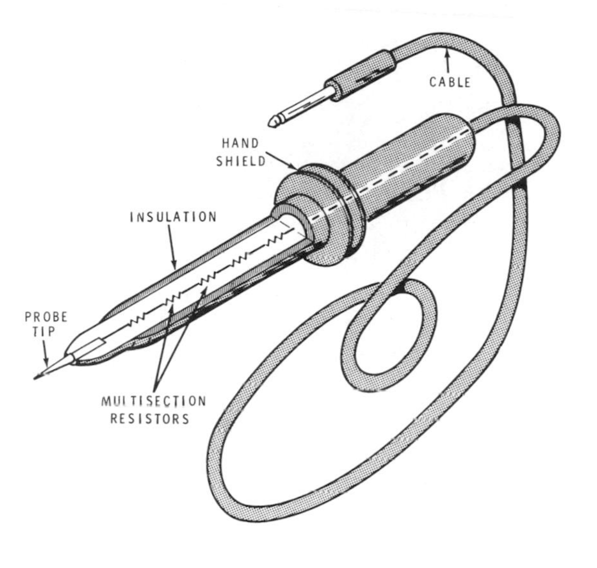

Figure 43

High-voltage probes are usually designed to operate with a particular meter. In most cases, the probe can be used with any meter having the same input resistance, provided the connectors fit. If the probe is used with a meter that has an input resistance other than what the probe is designed to work with, measurements will be inaccurate.

For those who have a meter for which no probe is available, there are several high-voltage probe meters available. These are constructed similar to a high-voltage probe with an indicator built in. The indicator may be either a digital readout or an analog movement as shown in Figure 44. Thus, the versatility of almost any meter can be improved by the use of appropriate accessories.

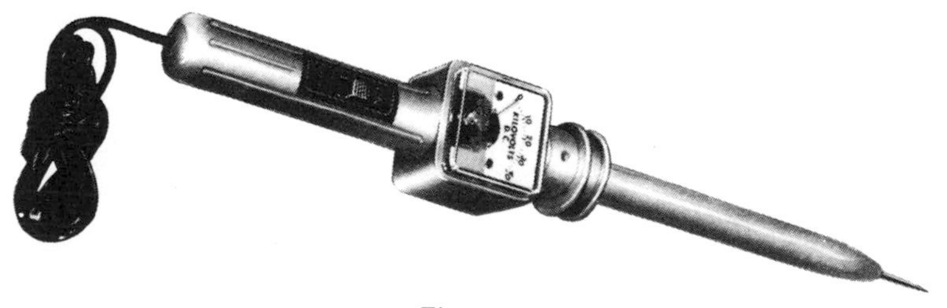

Figure 44

Unit 3

OSCILLOSCOPES

INTRODUCTION

Oscilloscopes are very useful test instruments for testing and designing in electronics. There are an almost infinite number of wave shapes that can exist. There are sine waves, square waves, triangular waves and combinations of these and other waves. In a defective circuit, the wave shapes may contain undesirable voltages. These voltages can be detected by a meter. If the shape of the voltage waveform is known, the average, peak, rms, or peak-to-peak value of voltage can be determined. However, there is no way to determine with a meter whether the voltage measured is caused by the desired waveform or by noise.

Also, in many instances, it is not the amplitude but the shape and timing which are most important. Only with the oscilloscope can we look at the shape and spacing of electrical signals. Thus, with an oscilloscope, we can see what is happening inside our television or computer.

In this section, we will discuss the components and circuits which make up the oscilloscope. We will also discuss the function and operation of the various controls and how to interpret various scope presentations.

CATHODE RAY TUBE

The heart of the oscilloscope is the cathode ray tube (CRT) just as the picture tube is the heart of your television set. The CRT is similar in many respects to the picture tube in a black and white television and could, in fact, be used to display a TV picture. There are three basic parts to the CRT: the phosphor screen, the deflection plates, and the electron gun as shown in Figure 1. The phosphor screen can be thought of as a piece of electronic graph paper upon which such details as amplitude, phase, frequency, distortion, pulse duration, and shape may be plotted. The screen is composed of a phosphor coating on the inside of the face of the CRT.

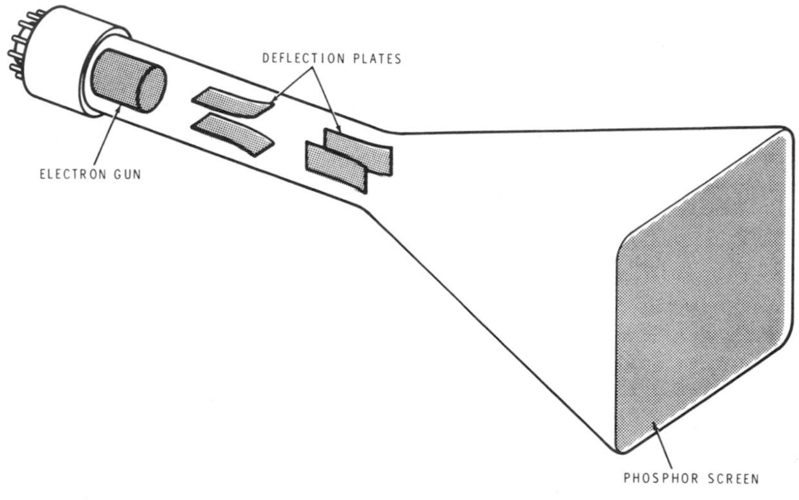

Figure 1

Phosphors are materials which give off light when struck by electrons. There is light emission during electron bombardment, which is called **fluorescence**. There is also a continued emission after bombardment stops, which is known as **phosphorescence**. The length of time that a phosphor continues to glow after bombardment stops is the **persistence** of the phosphor.

There are approximately forty different phosphors used for coatings on CRT's. Only a few of these are suitable for use with oscilloscopes. Others may be used for radar, television, computers, and various other displays.

Phosphors are available in white, yellow, orange, red, green, blue, and combinations of these. Persistence varies from less than a microsecond to more than a minute.

In selecting a phosphor, the designer must consider the characteristics of the signal being displayed and the response of the human eye. The signal that is displayed by an oscilloscope may be one that is changing very slowly which suggests a long persistence, or it may be changing rapidly and need a shorter persistence. In practice, a compromise is made, with persistence times being between fifty microseconds and fifty milliseconds. Of course, designers may deviate from these limits depending on the application.

The human eye responds best to some form of green light. Phosphors in this color group that fulfill the persistence requirement are green and yellowish green. These are the colors most often seen in oscilloscopes.

The Electron Gun

The electron gun provides a very fine beam of high velocity electrons that bombard the screen. The electron gun is shown in Figure 2. The electron gun is composed of five sections. The cathode (k) and the four grids are shown as they are mounted in the tube. In this particular CRT, the elements are held in place by insulated supporting rods. Although they are not considered a part of the electron gun, the vertical and horizontal deflection plates are shown here to illustrate their physical placement in the tube.

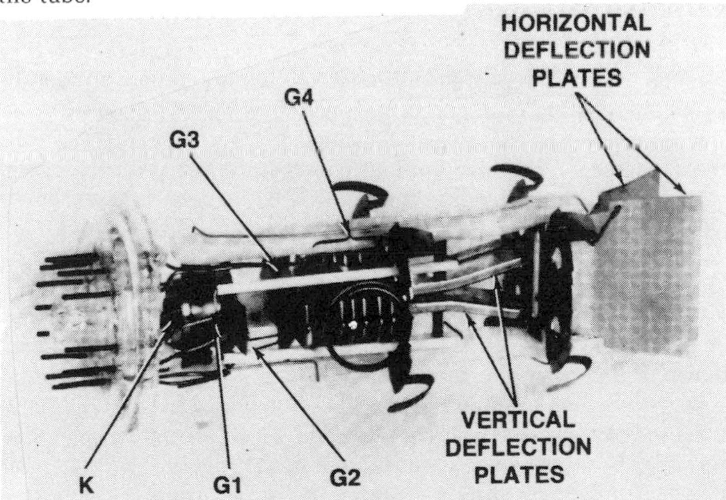

Figure 2

An exploded view of the electron gun is shown in Figure 3. The cathode is a cylindrical metal disc coated with barium sulphate and heated by the heater. When heated inside a vacuum, barium sulphate releases some of its electrons into the surrounding space. These electrons form a **cloud** around the cathode. This cloud of electrons is known as a **space charge** and becomes the source of electrons for the CRT. The control grid (G_1) often called a "Wehnelt cylinder" partially encloses the cathode. Electrons are allowed to escape through a small hole opposite the cathode.

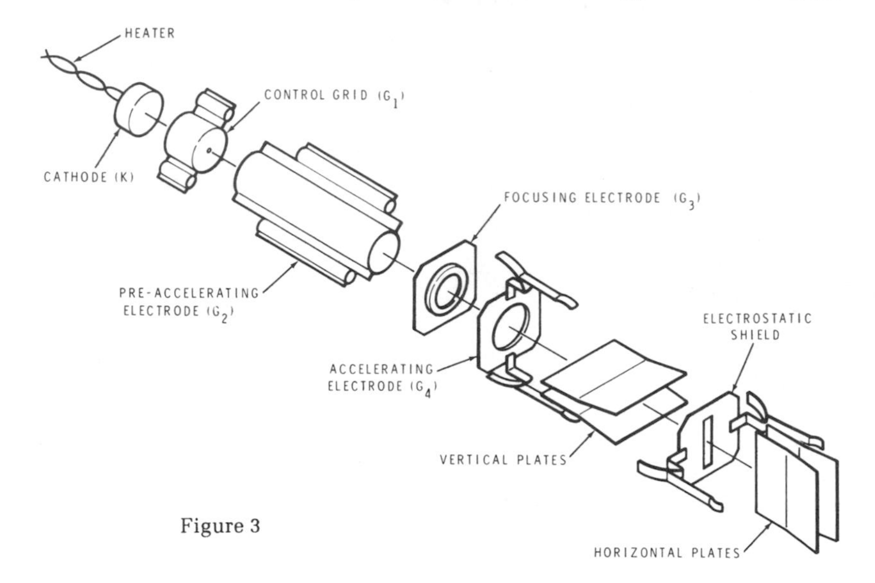

Figure 3

Next, is the pre-accelerating electrode (G_2) which is another cylinder. It has a small hole in one end and is fully open at the other. The end with the small hole is facing the control grid. The focusing electrode (G_3) is a metal ring with a negative potential applied which forms the electrons into a very thin beam. The accelerating electrode (G_4) then imparts a final burst of speed to the electrons before they pass through the deflection plates.

The inside surface of the CRT has an aquadag coating (this is a conducting material, usually graphite) and is connected to a high positive potential. When the electrons strike the phosphor screen, they not only cause the screen to glow, they also cause secondary emission. Secondary emission occurs when an electron beam strikes a material and causes other electrons to be knocked off from that material. These secondary electrons are picked up by the aquadag coating and thus returned to the cathode via the power supply, providing a complete path for current through the CRT.

The effect of the various grids on the electron beam can be seen in Figure 4. The heated cathode emits a cloud of electrons. The negative voltage on G_1 forms the electrons into a slightly divergent beam. These electrons are attracted by G_2, which is positive. Electrons which pass through the aperture of G_2 are accelerated toward the screen in a divergent beam. The negative voltage on G_3 causes the electrons to converge such that they arrive at the screen as a very fine point. G_4 aids in the convergence of the beam and gives a final acceleration to the electrons.

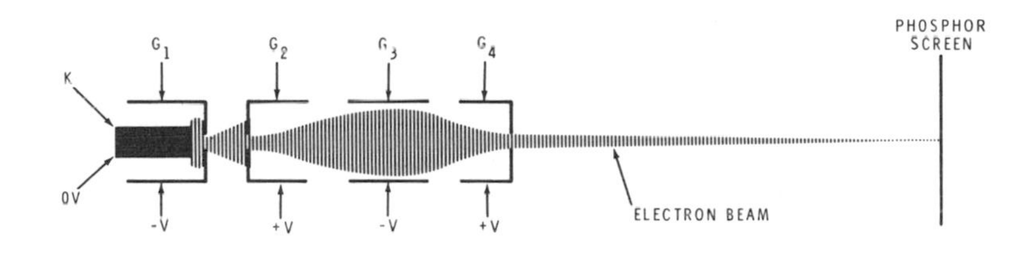

Figure 4

The acceleration provided by the electron gun is adequate for a low frequency scope with a maximum bandwidth of about 5 MHz. At higher frequencies, the beam moves across the phosphor screen at such a speed that the trace brightness decreases. Therefore, the energy of the electrons striking the screen must be increased.

More energy could be given to the electrons by increasing the potentials on the electron gun. However, this will require increased voltages or increased area for the deflection plates. Either of these can cause a decrease in frequency response of the scope. It is better to accelerate the electrons after they pass the deflection plates. This is called "post deflection acceleration."

The easiest method of providing post deflection acceleration is by a resistive helix wound inside the tube as shown in Figure 5. With a high positive voltage applied to the screen end of the helix, electrons will be progressively accelerated until they strike the screen. The main drawback to this method of acceleration is increased CRT length.

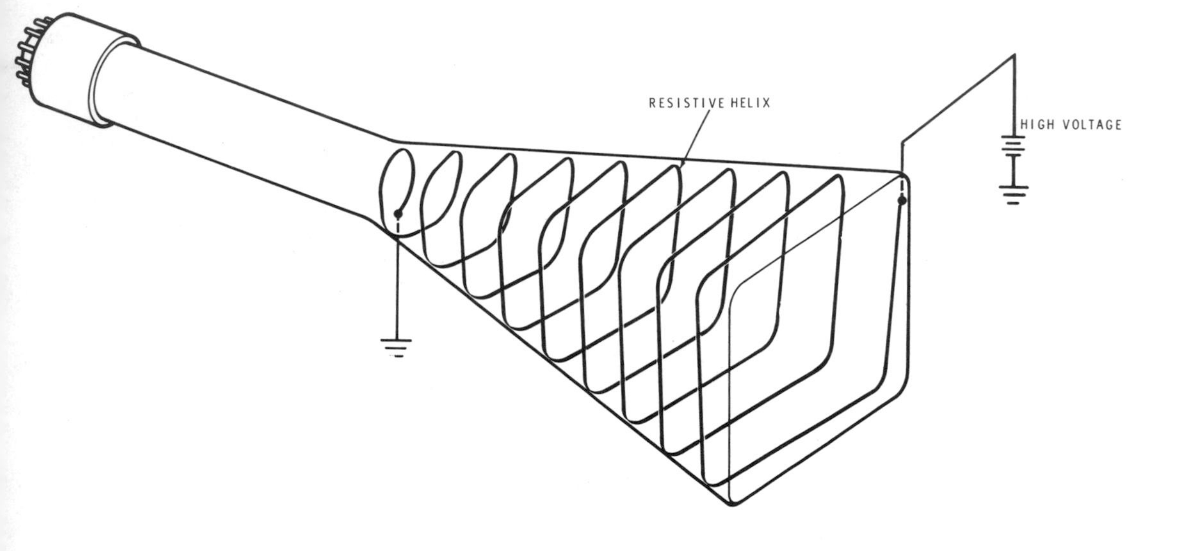

Figure 5

A method of providing considerable electron acceleration in a tube which is much shorter than one using the helix is the high-contoured expansion mesh designed by Hewlett-Packard. The mesh is mounted on the screen side of the deflection plates and connected to electron gun potential as shown in Figure 6. The inside surface of the CRT is coated with a conductive material.

When a high positive potential is applied to the conductive material, a strong electrostatic field is established between the mesh and the wall of the CRT. This field is depicted in Figure 6. Notice that the field is spherical in shape so that, while the electrons are accelerated, their direction is not changed. This type of post deflection acceleration requires a tube less than one half as long as is required for the same size display using a helix.

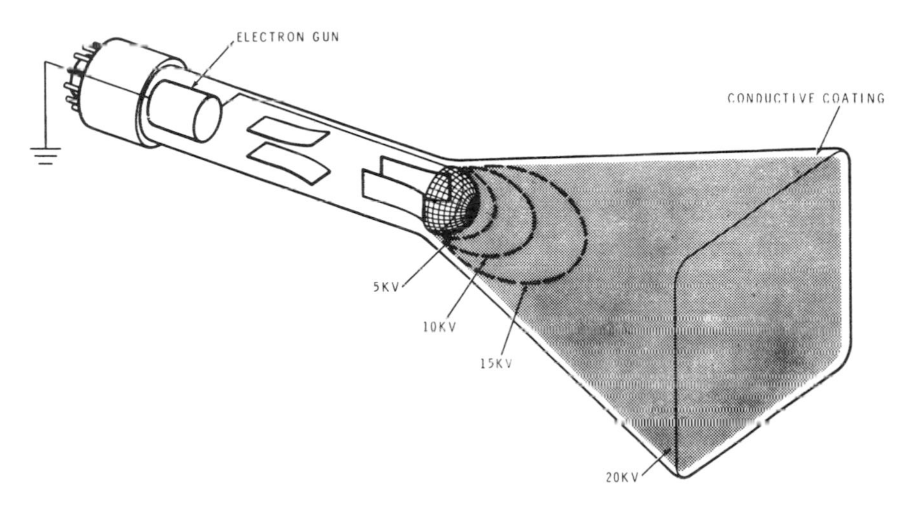

Figure 6

Brightness can also be improved by insuring that all of the light emitted by the phosphor is seen by the viewer. This may be accomplished by **aluminizing** the screen.

Figure 7A shows what happens with a screen that is not aluminized. The electron beam strikes the phosphor, causing it to emit light. Here, the light is emitted in all directions with only a small portion going toward the viewer. However, when the tube has been aluminized, by vapor depositing a very thin layer of aluminum over the rear of the screen, most of the light is reflected toward the viewer as shown in Figure 7B. This gives a significant increase in brightness.

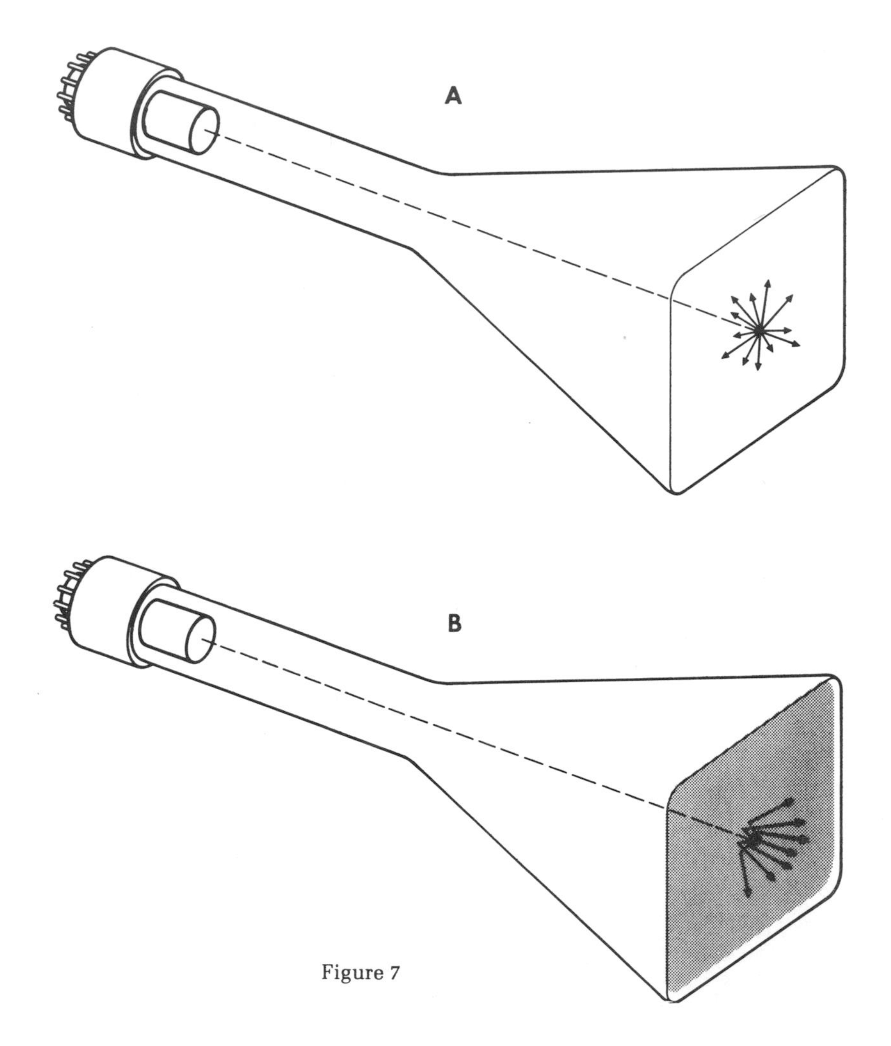

Figure 7

Beam Deflection

The beam from the electron gun travels in a straight line and forms a small dot on the screen. With no external influence, the beam should strike the center of the screen. For the scope to be useful, the beam must be moved or deflected in both horizontal and vertical directions.

There are two methods of beam deflection in a CRT: electromagnetic and electrostatic.

ELECTROMAGNETIC DEFLECTION

Electromagnetic deflection is the type used with TV picture tubes. The beam is deflected by magnetic fields produced by coils arranged around the outside of the tube. The size of the deflection coils increases as frequency increases, and becomes prohibitive at higher frequencies. While magnetic deflection gives better resolution, it is seldom used in oscilloscopes due to an upper frequency limit of approximately 5 MHz.

ELECTROSTATIC DEFLECTION

Electrostatic deflection, which is used in most oscilloscopes, works on the principle that unlike charges attract and like charges repel. To accomplish this, a set of deflection plates is installed inside the tube just after the electron gun. The vertical plates usually come first, followed by the horizontal plates. The relationship of the vertical and horizontal deflection plates is shown in Figure 8. With no difference in potential across the plates, the beam will strike the center of the screen as shown.

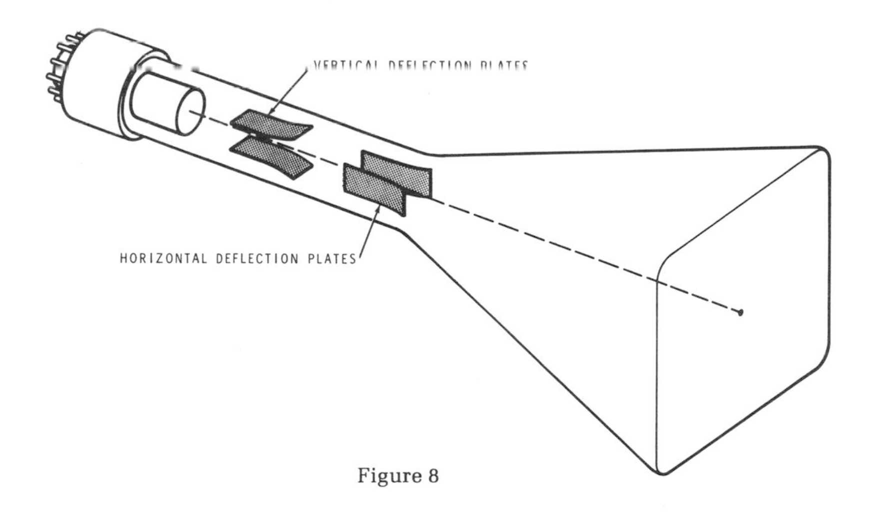

Figure 8

When a potential is connected across the vertical plates, as shown in Figure 9, the electron beam will be deflected away from the negative plate toward the positive plate. The amount of deflection is proportional to the voltage applied to the plates.

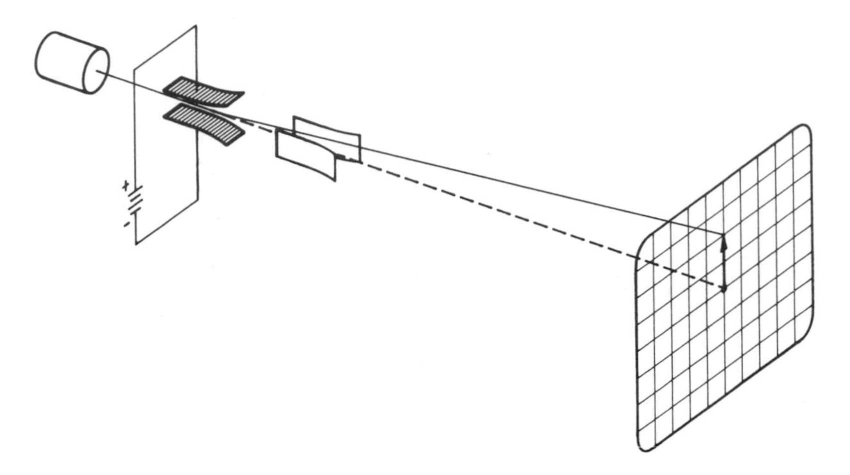

Figure 9

Notice that one plate is positive while the other is negative, giving an overall potential of 0 V for the plates. This is necessary to avoid changing the acceleration of the electrons with the deflection voltage; thus changing the brightness of the trace.

If the voltage is applied across the horizontal deflection plates as shown in Figure 10, the beam will be deflected in a horizontal direction.

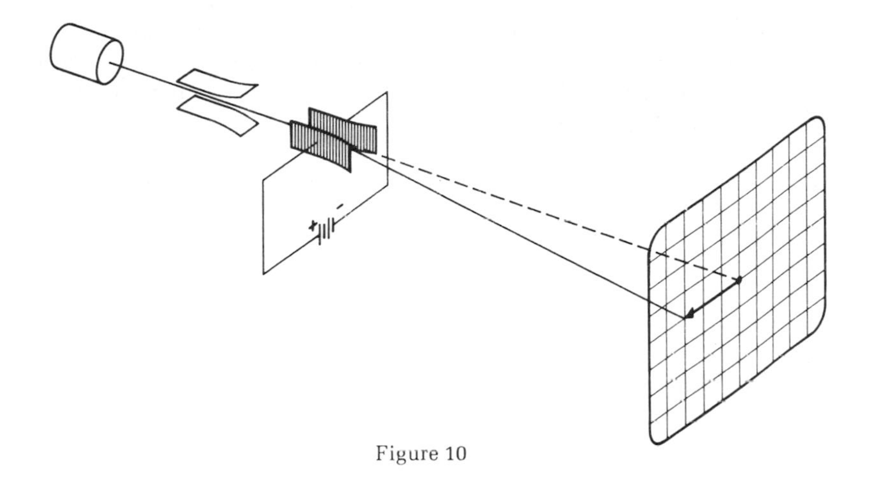

Figure 10

DEFLECTION SENSITIVITY

If we look at the placement of the deflection plates within the CRT, we find that the vertical deflection plates are nearer the electron gun and farther from the screen than the horizontal plates. Thus, a smaller angular deflection is required by the vertical plates for a given distance of spot travel on the screen than is required by the horizontal plates.

A "typical" CRT requires from eight to twelve volts of potential difference between the vertical deflection plates for a deflection of 1 cm on the screen. The voltage required for a 1 cm deflection is known as the deflection sensitivity; therefore, we could say that the vertical sensitivity of a typical CRT is 10 V/cm. This does not mean that the vertical sensitivity of the oscilloscope is 10 V/cm. Amplifiers are used to increase the sensitivity far beyond that of the CRT.

We have already seen that the horizontal sensitivity is less than the vertical sensitivity. Horizontal deflection usually requires between 15 and 20 V/cm.

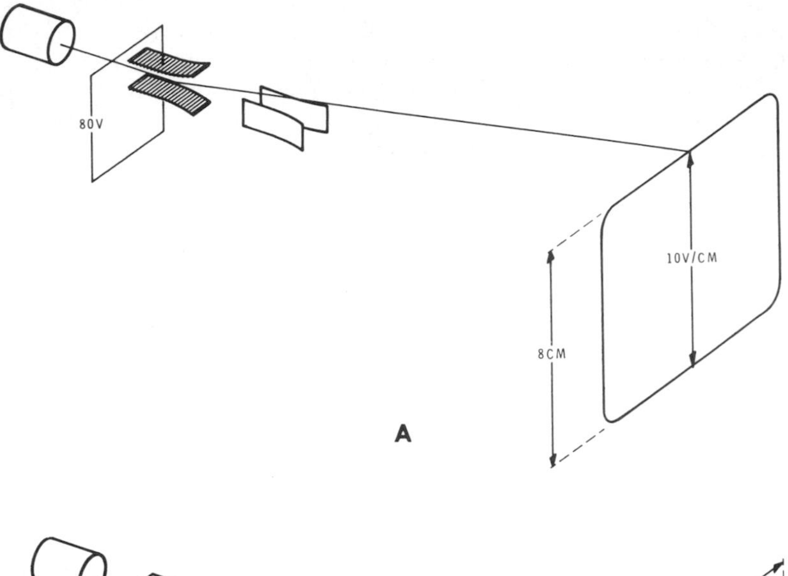

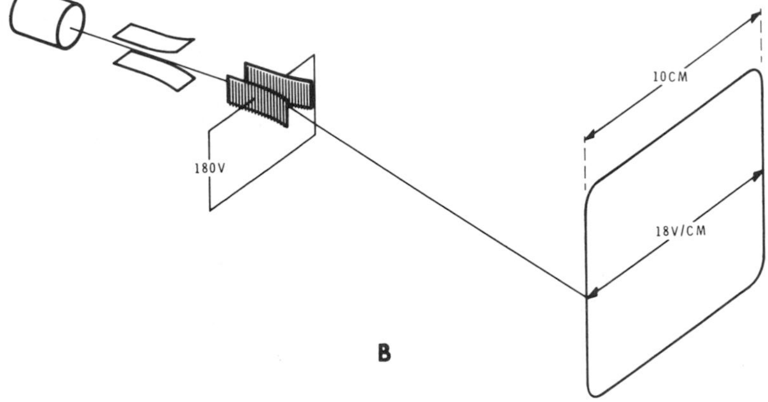

Figure 11

A common display size is 8 × 10 cm. Thus, if the vertical deflection sensitivity is 10 V/cm, 80 V must be applied to the vertical deflection plates for full deflection as illustrated in Figure 11A. As shown here, the top plate will be positive 40 V and the bottom plate negative 40 V, for a difference of 80 V. It is common for both plates to be at some positive potential. The actual potential is unimportant as long as the difference is 80 V.

Using the same size display with a horizontal deflection sensitivity of 18 V/cm, a potential of 180 volts must be applied to the horizontal plates as shown in Figure 11B. Once more, the actual potential on the plates is irrelevant as far as deflection is concerned. It is the potential difference between the plates that counts.

The actual potential on the plates does have some effect on focus and brightness. Therefore, when the deflection potential goes positive on one plate, it must go negative by a like amount on the opposite plate so that the overall potential remains constant.

We have discussed the construction and use of the CRT, and used illustrations which show the physical and electrical relationship of the tube components. As we develop the circuits which work with the CRT to make an oscilloscope, we will be primarily concerned with the electrical characteristics only. Schematically, the CRT is represented as shown in Figure 12. This representation will be used as we continue.

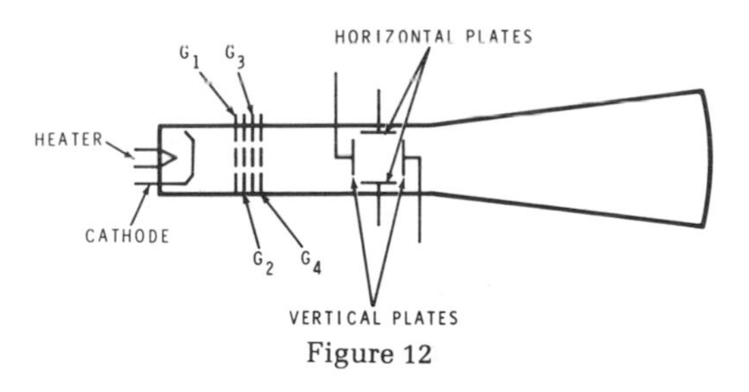

Figure 12

DEFLECTION CIRCUITS

Before we start our discussion of deflection circuits, we will review some necessary basics.

If a capacitor is connected across a power source as shown in Figure 13A, the capacitor will charge to the value of E. The time required for the capacitor to charge to a given value is a function of the value of R and C. This time may be calculated by the formula:

$$T = RC$$

Where T = time in seconds

R = resistance in Ohms

C = capacitance in farads.

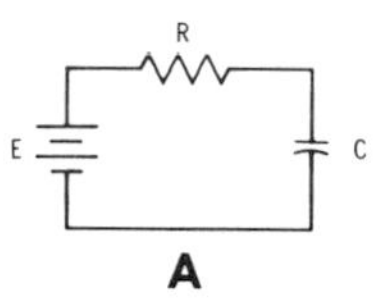

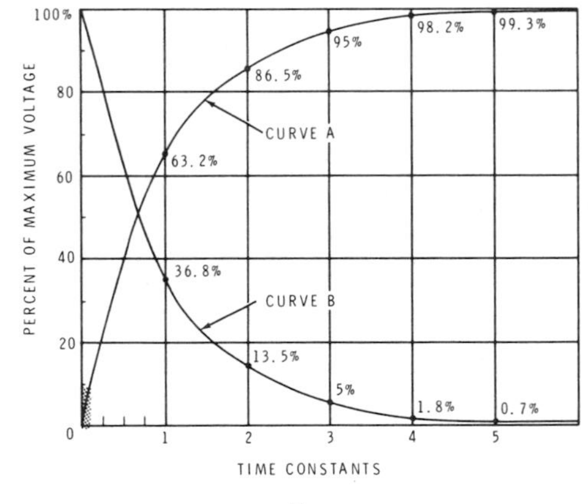

Figure 13

In one time constant (T), the capacitor will charge to a 63.2% of E as shown in Figure 13B curve A. Thus, for a capacitor to charge quickly, the value of R and C must be made as low as possible. Another way to reduce the time required to reach a given voltage is to make the supply voltage much greater than the desired voltage on C. If E is 10 times the value that C is required to charge to, then C will reach the required voltage in only 0.1 time constants. Also, the change will be in the linear portion of the charge curve as shown in Figure 13B.

If we relate the circuit shown in Figure 13A to an oscilloscope, C becomes the capacitance of the plates and R becomes the effective resistance of the vertical or horizontal amplifier. In a general-purpose oscilloscope, the deflection plates have a capacitance of about 5 pF. As you will see in the section on rise time, in a 15 MHz scope, the vertical amplifier must be capable of charging that capacitance to the desired voltage within about 25 nS.

The requirements of the horizontal amplifier are similar to those of the vertical amplifier except that a **higher** voltage is required on the horizontal plates.

It is also possible to view the deflection circuits in the frequency domain instead of the time domain. The amplifier can be treated as an AC generator with the output applied to an RC circuit as shown in Figure 14A.

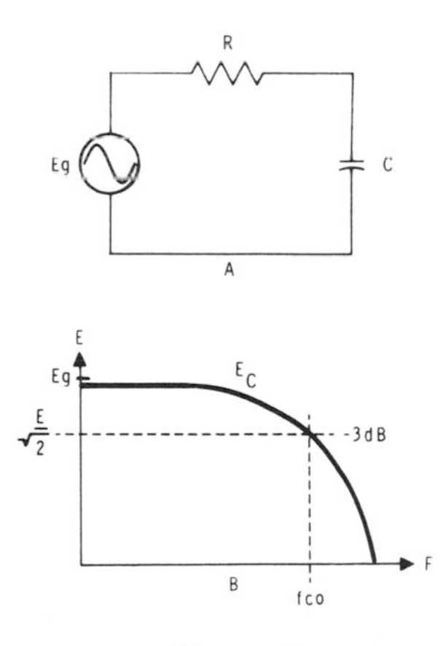

Figure 14

At the lower frequencies, almost the entire voltage is developed across C. However, as frequency increases, X_c decreases, causing the resistance to drop a higher portion of the voltage. When the voltage across the capacitor decreases to the half power point (−3 dB), the cut-off frequency of the scope (f_{co}) has been reached. This is the upper frequency limit of the scope. The voltage applied to the deflection plates at this frequency is .707 times the original voltage as shown in Figure 14B.

Thus, we can see that the design of the vertical and horizontal amplifiers affect both the bandwidth of the scope and the fidelity with which it displays certain waveforms.

Horizontal Sweep

The most basic display which can be obtained on an oscilloscope is a straight horizontal line across the screen. This is called a sweep or a trace. To see how this sweep is developed, look at Figure 15.

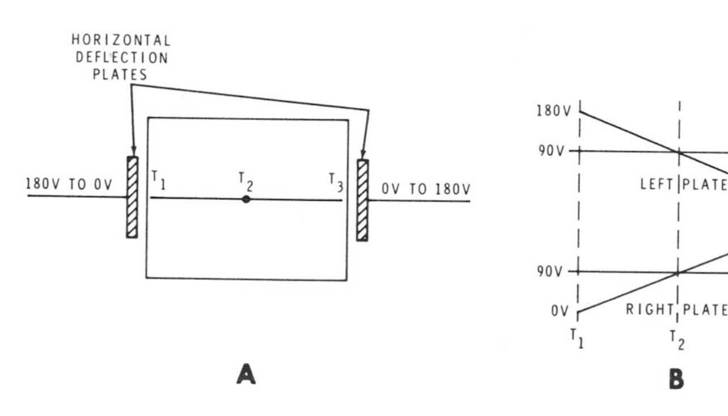

Figure 15

Each of the horizontal deflection plates has a voltage applied which can vary from 0 V to 180 V. The action begins with the left plate at 180 V and the right plate at 0 V. The electron beam will be deflected to the more positive potential. Since deflection potential is maximum, the deflection will be maximum, placing the beam at the extreme left of the screen as shown at T_1 in Figure 15.

As time passes, the sweep circuits cause the voltage on the left plate to decrease and the voltage on the right plate to increase. This causes the beam to be deflected to the right. At time T_2, both plates are at 90 V and there is no difference of potential between them; therefore, there is no deflection voltage and the beam is at the center of the screen as shown at T_2.

As time continues, the voltage on the left plate continues to decrease while the voltage on the right plate increases. At T_3, the left plate will be at 0 V and the right plate will be at 180 V, placing the beam at the right edge of the screen.

If, at the end of the sweep, we cause the deflection voltage to snap back to its original potential, we can start the sweep over again. In practice, the sweep is repeated continuously, causing the sweep to appear as a constant line across the CRT.

Remember that the beam is always producing a small dot on the screen. However, if the beam is swept across the screen at a high enough speed, the persistence of the phosphor coating and the characteristics of the human eye cause it to appear as a solid line or trace on the scope. As sweep speed becomes slower and slower, the trace will first begin to flicker. Then at even slower speeds, it becomes possible to discern the spot of light moving across the screen.

Do not stop the beam in its movement across the scope. This will concentrate the energy in one spot and possibly burn the phosphor. If the phosphor is burned, a dark spot will appear on the screen and remain there forever.

Vertical Deflection

The vertical deflection plates work very much like the horizontal deflection plates. Figure 16 illustrates how a vertical trace is developed.

As the action starts at T_0, both the upper deflection plate (plate A) and the lower deflection plate (plate B) have a potential of 50 V. There is no potential difference and the beam is centered on the screen. At this time, plate A starts going more positive while plate B becomes less positive. A potential difference is thus developed which causes the beam to deflect toward plate A. At time T_1, plate A will be at a maximum positive potential and plate B will be at 0 V. The beam will be deflected in the maximum upward direction and a line will be drawn from Y_0 to Y_1.

As time continues, the voltage on plate A starts in a negative direction and the voltage on plate B in a positive direction. The beam starts to move from Y_1 back toward Y_0. At T_2, both plates have returned to 50 V and the beam is back in the center of the screen. The plate potential continues to change, with plate B going positive and plate A becoming negative. This deflects the beam toward Y_2 and, at T_3, a line will have been drawn to the bottom of the screen. At T_4, the beam is back at the center.

As long as this signal is applied to the vertical deflection plates, the beam will continue to draw a vertical line on the screen.

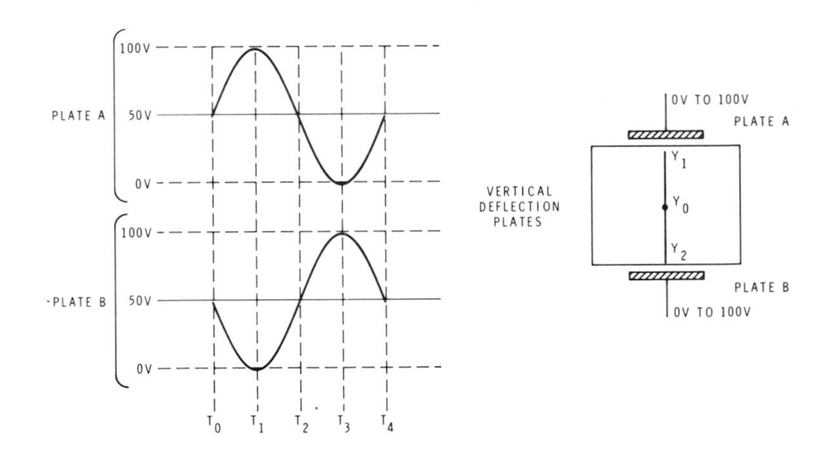

Figure 16

Developing a Presentation

We have seen how a horizontal sweep or a vertical trace can be developed on an oscilloscope. However, neither by itself does any good. For the scope to be useful, it must present some information about the incoming signal.

One of the desirable characteristics of an oscilloscope is its ability to display the frequency or period of a signal. To do this, there must be some method of comparing the incoming signal to time. Figure 17 shows how this is done.

At T_0, the horizontal deflection voltage is at its maximum negative value, which places the electron beam at the extreme left of the screen. At the same time, the sine wave which is applied to the vertical deflection plates is at 0 V.

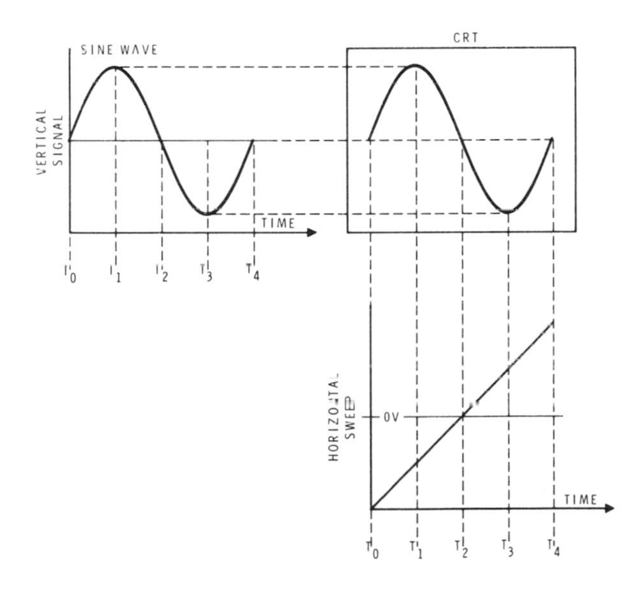

Figure 17

At T_1, the horizontal deflection voltage is at a more positive potential, which moves the beam from left to right. However, the vertical deflection voltage is now positive, which moves the beam toward the top of the screen. At T_2, the horizontal trace is passing through the center of the screen and the sine wave is at the 180° point. As the horizontal voltage continues to sweep in a positive direction moving the trace from left to right, the sine wave continues its excursion. At T_4, the horizontal voltage has reached maximum and the sine wave has completed one cycle. Thus, we can see that if the time of the horizontal sweep is exactly equal to the period of one sine wave, then we can display one sine wave on the screen.

If we know how long it takes the horizontal plate to deflect the beam from left to right, we can calculate the frequency of the waveform on the vertical plates by the formula

$$f = \frac{1}{T}$$

By controlling the time required for horizontal deflection, we can display any number of cycles of a given frequency applied to the vertical plates. In Figure 18, the sweep time has not changed, but the frequency of the vertical signal has doubled. Therefore, we will display two cycles on the CRT. If we wish to display only one cycle of the new frequency, we will have to cut the horizontal deflection time in half.

By controlling the horizontal sweep time, any number of cycles can be displayed.

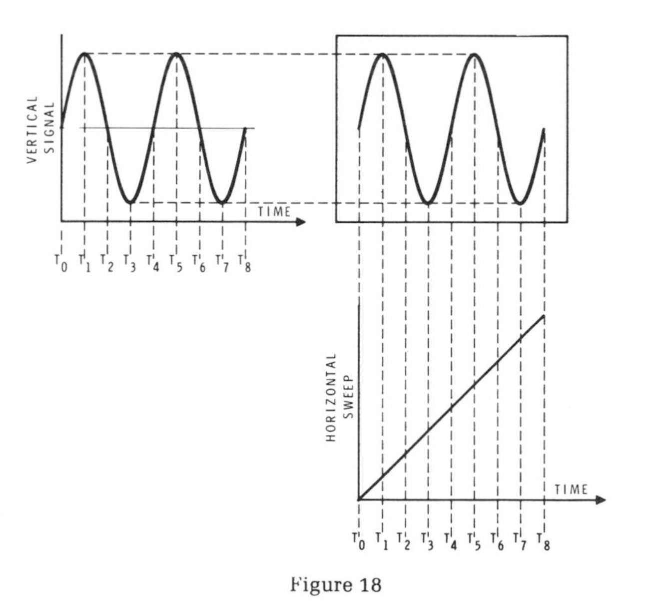

Figure 18

OSCILLOSCOPE CIRCUITS

Figure 19 is a block diagram of a basic oscilloscope. It contains the cathode ray tube with its vertical and horizontal deflection plates. These plates are driven by amplifiers which increase the input from as low as 1 mV to a value that will cause deflection.

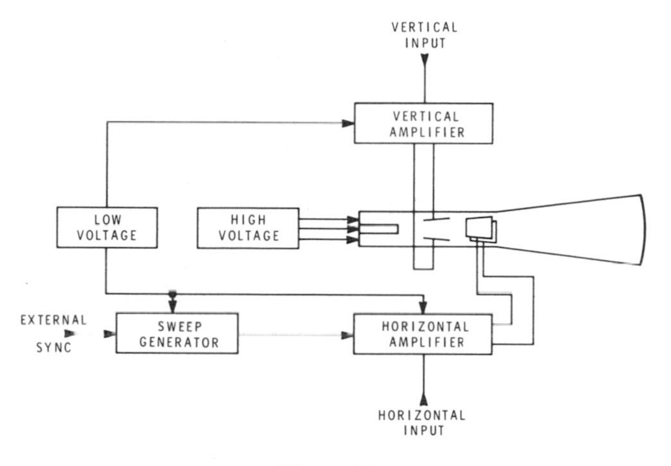

Figure 19

The horizontal amplifier has two inputs. One, from the sweep generator, is used to produce a horizontal trace across the screen. The other allows an external input to be displayed on the horizontal axis.

The vertical amplifier has only one input. The signal to be displayed is applied to this input. Thus, a quantity can be compared to an internally generated sweep or to another external input on the horizontal plates.

Sweep Generator

We have seen that a sweep is a horizontal trace on the face of the CRT that occurs in a selected period of time. The sweep generator generates a waveform that increases at a linear rate with respect to time until it reaches a predetermined value, at which time it suddenly drops to zero. It is this waveform that produces the horizontal trace or sweep on the CRT.

A basic sweep generator is shown in Figure 20. We will start our study with the switch closed. There will be no charge on the capacitor. As soon as the switch is opened, the capacitor will start to charge toward battery voltage. At first, C will charge at a linear rate. (See Figure 13B.) Therefore, if the switch is closed after a very short time, the capacitor will never reach the non-linear portion of its charge curve. Opening and closing the switch at the proper time will produce a continuous sawtooth of voltage.

Of course, it is impossible to operate a manual switch at the speed required by an electronic circuit. So, there is more to a practical sweep generator than is shown here.

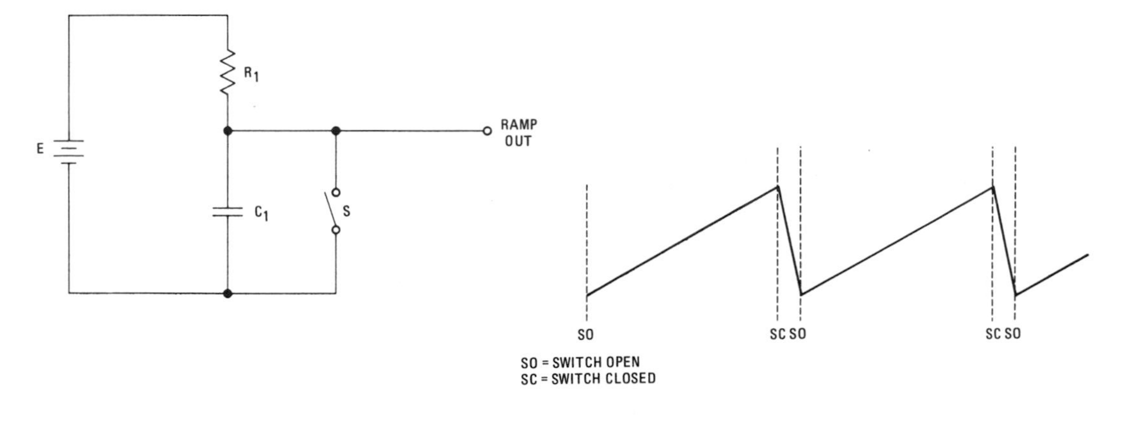

Figure 20

Figure 21 is similar to the ramp generator portion of the sweep generator circuits in a current Heathkit oscilloscope. When power is applied to the circuit, C begins to charge through Q1. Q1 is a constant current source that insures that C charges at a linear rate. During this time, Q2 is cut off, allowing the full charge on C to be felt at the ramp out. The ramp voltage is also applied to IC1 where it is compared to a pre-determined reference voltage. When the two are equal, a positive voltage is applied to the base of Q2, causing it to conduct and discharge C. Q2 corresponds to the switch in Figure 20. Thus, C discharges through Q2, which is almost a short circuit, giving a very fast discharge rate. Therefore, we have produced a ramp with a controlled rise time and a very short fall time. When the charge on C falls below a given level, forward bias is removed from Q2, starting the cycle all over again.

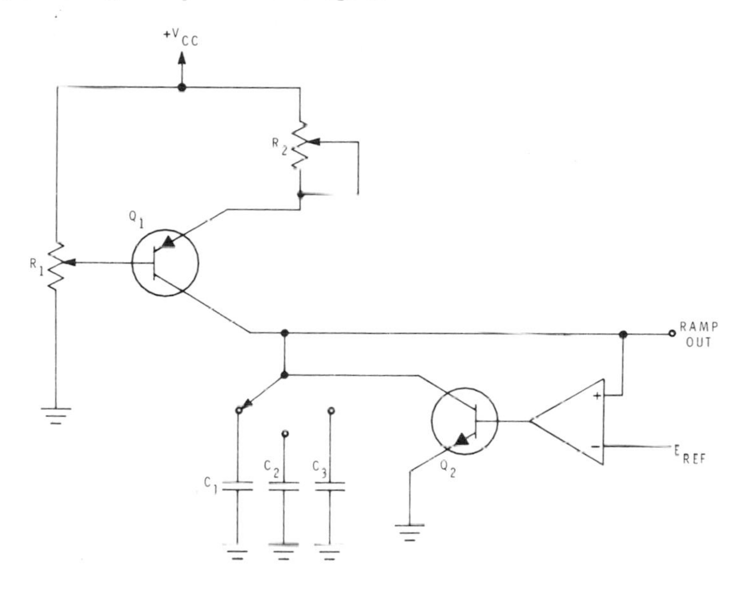

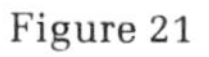
Figure 21

An operational scope places a number of demands on the sweep generator. The sweep must be linear if the display is to be a true picture of the incoming signal. The sweep must have a constant amplitude with no varying change in sweep time. The operator must also be able to control the sweep time when measuring frequency.

The precise control of the time it takes the sweep to move across the screen establishes a time base. It is this time base which allows the measurement of pulse width, pulse recurrence time, period of a waveform, etc. The time base will be calibrated in seconds per division. In almost all modern scopes, a division is equal to one centimeter. Thus, the calibration will read TIME/CM or SEC/DIV.

If the scope is to allow measurements over a wide range of frequencies, there must be some means of varying the time base. A common method is by switching to a different value of capacitance, as shown in Figure 21. In fact, anything that changes the charging current will change sweep time. Thus, we could also change sweep time by changing the resistance of R2 in Figure 21.

If an arrangement similar to that in Figure 21 is used, two types of control are available. There is a calibrated step control by switching capacitors and an uncalibrated vernier control by varying R2. Later, in the section on oscilloscope operation, you will learn how these controls are used.

Blanking and Unblanking

In looking at the sweep waveform (Figure 20), we see that the sweep goes slowly positive then rapidly negative. During the positive-going portion of the waveform, a horizontal trace is drawn on the screen from left to right. This portion of the signal has the proper timing as selected by the TIME/CM control.

At the end of the positive-going portion of the sweep, the voltage returns rapidly to its original level. In so doing, the electron beam will move from the right side of the screen to the left side. Unless some action is taken to prevent it, this **retrace** will be visible on the screen. It is undesirable to have a visible retrace; therefore, some means must be provided to **blank** the scope during retrace time.

In the section on CRT, you learned that the intensity of the electron beam can be controlled by controlling the voltage between the cathode and the control grid (G1) of the CRT. As the control gird is made more negative with respect to the cathode, the trace will become dimmer. When a sufficiently negative voltage is applied to the grid, the beam will be completely cut off, and the screen will be blank.

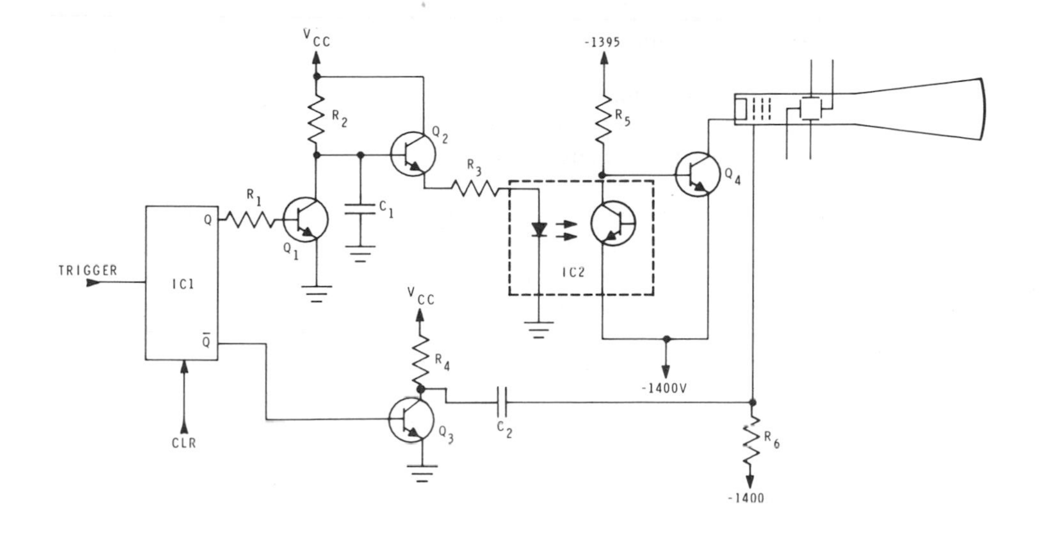

Figure 22

There are two ways to blank and unblank the CRT. The potential on the grid can be controlled or the potential on the cathode can be controlled. One Heathkit scope controls both. A simplified schematic of the blanking circuitry of this scope is shown in Figure 22.

IC1 is a J-K flip-flop which is set by the sweep trigger. We will discuss triggering later. When the flip-flop is set, the Q output is high and the $\overline{Q}$ output is low. With a high at Q, a positive voltage is applied to the base of Q1, causing it to conduct. This places the base of Q2 at essentially ground potential, holding it cut off. With Q2 cut off, no current flows through the diode portion of IC2.

IC2 is an optical isolator which consists of a light-emitting diode and a photosensitive transistor. As we know, a light-emitting diode emits light when it conducts, and a photosensitive transistor conducts when exposed to light. The arrangement is such that there is no electrical connection between the two. In this case, the optical isolator is used to isolate the high negative voltage of the cathode from the low voltage of the timing circuits.

If the diode portion of IC2 is drawing no current and emitting no light, Q4 will have −1395 V on the base and −1400 V on the emitter. Q4 is thus forward biased, which will allow cathode current to flow from −1400 V through Q4 to the cathode, through the CRT to the aquadag coating, back to the power supply. Thus, the cathode is **unblanked** during sweep time.

The $\overline{Q}$ output of the J-K flip-flop, which is low, is applied to the base of Q3, cutting off the transistor and causing the collector to go high. This high is coupled through C2 to the grid of the CRT, unblanking the CRT. Obviously, since this voltage is coupled through C2, it will start to decrease immediately. Therefore, unblanking the grid will only be effective for a short period of time. However, it is not necessary for the grid to be high in order for the tube to conduct. It is only necessary that it not be significantly more negative than the cathode.

We have seen how the CRT is unblanked during sweep time. Now, let's see how the blanking is accomplished. When the sweep voltage reaches a given value, a pulse is applied to the clear input of the J-K flip-flop. This resets the flip-flop, causing $\overline{Q}$ to go **high** and Q to go **low**.

When $\overline{Q}$ goes high, it causes the collector of Q3 to go low. This low is coupled through C2, driving the grid negative and blanking the CRT. As we saw previously, grid blanking as used here will only last for a short period of time. However, at the same time, $\overline{Q}$ went **high,** Q went **low**.

The low on Q drives Q1 to cutoff. The collector of Q1 will attempt to go high but will be delayed by a predetermined time by the RC time of R2 and C1. When C1 has charged to the necessary positive potential, Q2 will conduct, causing current to flow through the diode section of the optical isolator, IC2. This will cause light to strike the photosensitive transistor, allowing it to conduct. When the photosensitive transistor conducts, it places −1400 V on the base of Q4. With the same potential on both base and emitter, Q4 will cut off. Thus, no current can flow and the CRT is blanked.

Of the two blanking systems just discussed, the cathode circuit takes care of low frequencies of less than 60 Hz. The values of R2 and C1 are selected so that Q4 will never cut off at frequencies above 60 Hz. At these higher frequencies, all blanking is accomplished in the grid circuit.

We have shown you how blanking is accomplished in one Heath oscilloscope. There are many techniques used to blank a CRT. The important point is that the electron beam must be cut off during retrace time and turned on during sweep time.

Triggering

We mentioned earlier in this lesson that the sweep was started by a trigger. This is usually the case in modern scopes, but it is not universal. There are basically two types of sweep: the triggered sweep and the recurring or free-running sweep. Let's first look at the recurring sweep.

In the recurring mode, the sweep waveform will be as shown in Figure 23. The voltage will rise to maximum, drop back to the starting point, rise to maximum, etc. as long as power is applied to the circuit. The electron beam will sweep from left to right, retrace, sweep, retrace, etc.

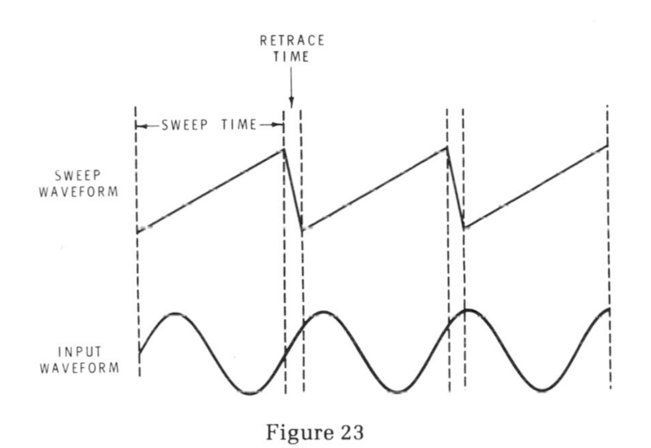

Figure 23

Suppose we wish to view a waveform, the period of which is exactly equal to sweep length. This waveform, a sine wave, is drawn to scale in Figure 23. Remember that we will display the portion of the sine wave that occurs during sweep time and blank out the portion that occurs during retrace time. Thus, during each sweep time, the portion of the input sine wave that occurs during that time will be displayed. Figure 24 shows how the display will appear with this input. The waveform shown here contains only three repetitions. Imagine what the display would look like with a large number of sweep cycles.

From this, we can deduce that for a display to be meaningful, the input waveform for each sweep must be identical to the previous one. In other words, each waveform must be superimposed on the one before. With the recurring sweep, this will occur only if the period of the incoming signal is an exact multiple of the sweep time plus the retrace time. Thus, we can see that it is impossible to view **one** complete cycle of the incoming

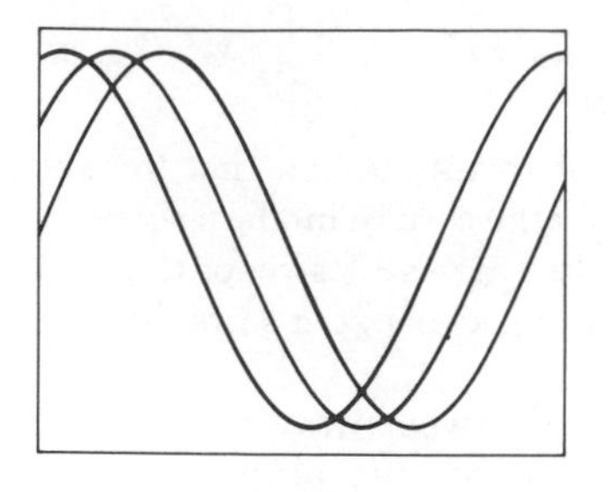

Figure 24

signal. It is also very difficult to synchronize or stabilize the display. If there is only a small difference between the incoming frequency and the sweep frequency, the display will appear to drift across the screen.

All of these problems can be cured by using a triggered sweep. A triggered sweep does not start until triggered. The most common method is to use the incoming signal to generate a trigger which starts the sweep.

Figure 25 shows how this works. When the incoming signal goes more positive than the trigger level, a trigger pulse is generated. The trigger pulse in turn starts the sweep generator. The sweep time can be whatever time the operator selects. In this case, it is equal to the period of one cycle of the input.

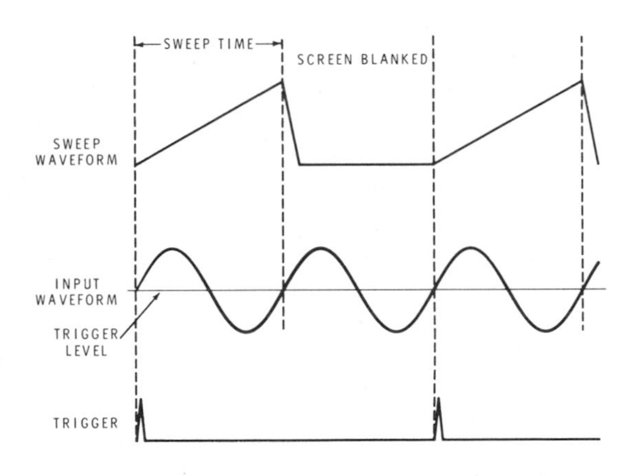

Figure 25

At the end of the sweep time, the screen is blanked and retrace occurs. The screen will remain blanked until another trigger is generated. With the triggered sweep, each sweep will start at the same relative point on the incoming signal. Therefore, each successive presentation will be superimposed on the one before. The display will appear as a distinct waveform as shown in Figure 26.

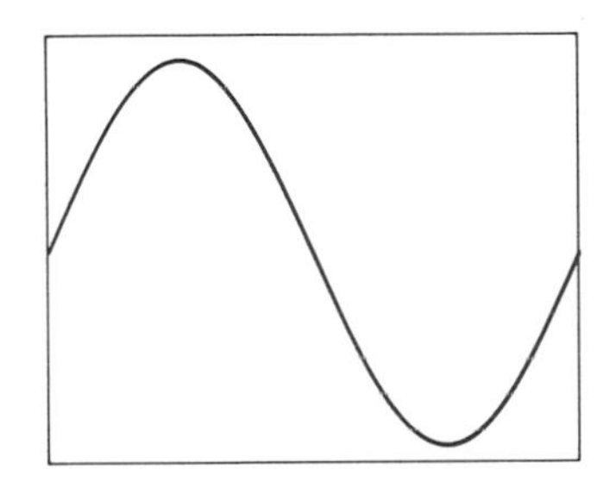

Figure 26

In this case, one complete cycle of the input waveform is shown. By varying the sweep time, it is possible to view a part of a cycle or any number of cycles.

The example in Figure 25 shows the trigger being generated when the input goes more positive than the trigger level. This is known as "triggering on the positive slope." Most scopes also have provisions for triggering on the negative slope, which means generating a trigger when the input goes more negative than the trigger level.

Figure 27A shows a slope detector circuit. The slope switch is in the positive position. This places a "low" on NAND gate A and a "high" on NAND gate B. The other input to the NAND gates comes from the comparator. This comparator has two outputs, one inverting and the other non-inverting. The non-inverting output is the one with the same sign as the input. For example, if the input signal is applied to the (−) input, the (−) output will not be inverted and the (+) output will be inverted.

Since the positive slope is selected, we will start our explanation by assuming that the input signal is more negative than the selected level. With a "low" into the comparator, the non-inverting output (−) will be "low" and the inverting output (+) will be "high." During this negative portion of the input signal, two "lows" are applied to NAND gate A; the output of A will be "high." At the same time, two "highs" are applied to the input of NAND gate B; the output of B will be "low." This places a "high" and a "low" on the input of NAND gate C; the output of C will be "high."

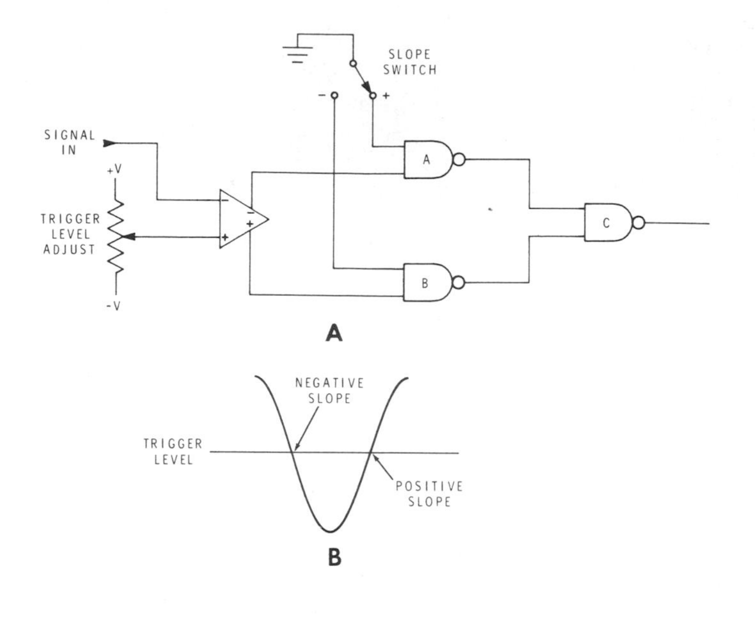

Figure 27

When the "signal in" passes through the trigger level in a positive direction, the outputs from the comparator will change. The non-inverting input will go "high" and the inverting input will go "low." This places a "high" and a "low" on A; the output of A remains "high." It also places a "high" and a "low" on B, causing the output of B to go "high." There are now two "highs" at the input of C; therefore, the output of C will go "low." When the output of C goes "low" a trigger will be generated which starts the sweep.

If the slope switch is placed in the (−) position, the same sort of action takes place. The difference is that the output of NAND gate C will be "low" when the **signal in** crosses the trigger level in a negative direction.

The major advantage of (+) and (−) triggering is that it allows the viewing of either the positive or negative portion of the input signal. A portion of the waveform can thus be expanded to better evaluate small portions.

Amplifiers

At least two very important oscilloscope specifications – bandwidth and sensitivity – are determined primarily by the amplifiers in the scope. It was mentioned earlier that the deflection sensitivity of a typical CRT was about 10 V/cm in a vertical direction and about 20 V/cm in the horizontal direction. Since the quantity to be measured or the output of the sweep generator is sure to be less than the amount required for sufficient deflection, amplification is required. Since amplitude is usually measured in the vertical channel, the vertical amplifier will be covered first.

SENSITIVITY

Vertical sensitivity is a measure of the amount of input voltage required to deflect the beam a given amount in a vertical direction. The amount of deflection is usually one centimeter. Thus, if 20 mV is required for 1 CM of deflection, the sensitivity would be stated as 20 mV/CM, or perhaps just 20 mV. Obviously, a scope which requires 5 mV to deflect the beam 1 division is more sensitive than one that requires 20 mV.

Sensitivity is very important when working with modern communications equipment. Signal voltages of 10 mV P-P or less are common. To display a 10 mV P-P waveform at a size suitable for proper interpretation requires a deflection of at least 2 divisions. Thus, a 5 mV/DIV sensitivity is required.

Some of the signals require a low-capacity probe to avoid loading the circuit or distorting the waveform. The low-capacity, or passive, probe is normally a ×10 probe. This probe attenuates the signal by a factor of 10. Thus, if a 10 mV P-P signal is to give two divisions of deflection, a 0.5 mV/DIV sensitivity is required.

Some of the more expensive scopes have a vertical expander, or magnifier, either ×5, or ×10, or both. The expander increases the effective sensitivity by a factor of 5, or 10, or whatever it is. Therefore, a scope with a sensitivity of 10 mV/CM and a ×5 magnifier has an effective sensitivity of 2 mV/CM.

On many scopes, when the expander is used, the bandwidth is decreased. Expansion by a factor of 5 reduces the bandwidth by a factor of 2. A ×10 expansion reduces bandwidth by a factor of 4. Thus, a scope might have a bandwidth of 20 MHz without expansion and 5 MHz with a ×10 expansion. Other scopes may not have an expander labeled as such but may reduce the bandwidth at the higher sensitivity.

Sensitivity means the ability to measure low voltages.

Measuring high voltages can be just as important as measuring low voltages. All of the in-between voltages are important too; so all scopes have some form of input attenuator. Usually, this will be a step attenuator with the steps calibrated to indicate the vertical sensitivity. The attenuation is usually stepped in a 1-2-5 sequence. Thus, the **VOLTS/DIV** switch on a good oscilloscope might read, 1 mV, 2 mV, 5 mV, 10 mV, 20 mV, etc. up to a maximum of 10 V, 20 V, or 50 V. If the maximum **VOLTS/DIV** setting is 20 V, full deflection of 8 divisions is utilized and a ×10 probe is used, a maximum voltage of $20 \times 8 \times 10 = 1{,}600$ volts can be measured. This is usually more than enough.

Sometimes it is desirable to set the vertical deflection to a fixed amount. Usually this is done when a comparison between values is more important than the actual value. Most scopes include a **VARIABLE** control for just this purpose. This control will have a calibrated or **CAL** position, which is normally at the fully clockwise position. In the **CAL** position, the setting of the **VOLTS/DIVISION** control are usually accurate within ±3 to 5 percent. With the **VARIABLE** control out of the **CAL** position, the height of the waveform is continuously variable, usually between two adjacent attenuator values. Normally, the signal size will be reduced as the **VARIABLE** is rotated counterclockwise.

BANDWIDTH

A vertical amplifier to meet the preceding requirements would be relatively simple if it were required to operate at only one frequency or even a narrow band of frequenices. However, this is not the case. An oscilloscope is often required to measure everything from a DC voltage to the RF output of a transmitter. Some scopes are capable of displaying a wider range of frequencies than others, depending on the vertical bandwidth of the oscilloscope. Bandwidth is a measure of the frequency range over which the oscilloscope faithfully displays the input waveform. Ideally, the vertical amplifier should amplify each frequency by the same amount. It is nearly impossible to obtain a perfectly flat amplification curve. Some frequencies are going to be amplified more than others. Normally, the manufacturer will keep the level within ±3 dB throughout the rated bandwidth. That means that the amplifier can have a voltage gain that varies from 0.7 to 1.4 times the nominal gain.

When looking at oscilloscope specifications, don't be misled by statements such as "usable to 30 MHz" or "down 3 dB at 10 MHz." The statements are no doubt true, but they leave some unanswered questions. What does usable mean and what happens between 0 Hz and 10 MHz? The type of specification you should be looking for reads "within 3 dB from DC to 10 MHz, DC to 5 MHz ±3 dB, or DC to 15 MHz (±3 dB)."

Beyond the rated bandwidth, there should be a 6 dB per octave rolloff rate. In other words, the amplitude should be cut in half everytime the frequency is doubled. Thus, sensitivity decreases and vertical calibration becomes less and less accurate as frequency increases.

It is possible to display a signal with a frequency above the bandwidth of the scope. Of course, you will be unable to measure the amplitude and the amplitude will be smaller than normal. If the waveform is other than a pure sine wave, it will have a particular harmonic content. Since the harmonics are at a higher frequency than the fundamental, they will be attenuated more. Therefore, the wave shape may not be displayed accurately.

In the section on CRT's, we mentioned that a given amount of deflection voltage is required at the deflection plates of the CRT. As frequency increased, more voltage was required. It is the responsibility of the vertical amplifiers to provide this voltage. At higher frequencies, the amplifiers will be unable to supply the necessary drive, and the display will be reduced in amplitude. Changing the sensitivity setting will have no effect, as the problem is lack of drive potential, not lack of input signal.

A typical oscilloscope will have a 3-position input switch; AC, DC, and GROUND. In the AC position, the DC component of the input signal will be blocked. However, the reactance of the capacitor will also attenuate the low frequency component of the signal. The −3 dB point for the low frequency end of the bandwidth usually occurs at between 2 and 10 Hz.

In the DC position, the signal is coupled directly from the input to the vertical amplifier, and there is no low frequency limit.

The GROUND position places the input at chassis ground and is normally used to establish a 0 V (zero-volt) reference for measuring DC voltages.

RISE TIME

Rise time (Tr), or how fast the vertical channel can react, is very closely related to bandwidth. Rise time is defined as the time required for a waveform to change from 10% of its maximum value to 90% of its maximum value as shown in Figure 28.

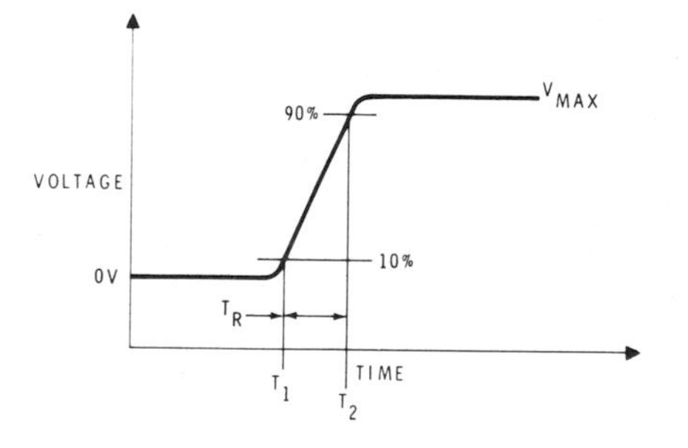

Figure 28

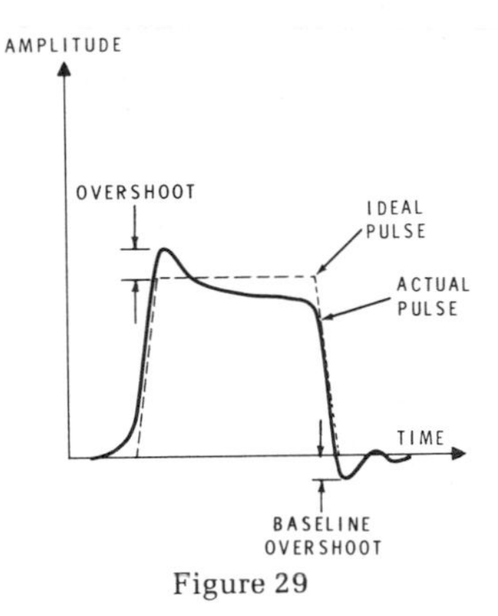

Figure 29

One would expect that a faster rise time would be better. However, this is not always the case. Unless the vertical amplifier has sufficient bandwidth, too fast a rise time may cause distortion in the form of overshoot as shown in Figure 29. The ideal pulse shape is shown by broken lines and the actual pulse shape is shown by a solid line.

Some amount of overshoot is inevitable. Therefore, a compromise must be made. The best trade off occurs when rise time multiplied by bandwidth equals approximately 0.35. Thus, given the required rise time, the necessary bandwidth can be calculated by the formula

$$BW = \frac{0.35}{T_r}$$

If we view the vertical amplifier and the deflection plates as a simple RC circuit as shown in Figure 30A, the half-power (−3 dB) point can be found by the formula $F_{co} = \frac{1}{2\pi RC}$

Since the lower frequency limit of an oscilloscope is usually extended to DC or 0 Hz, the upper limit (F_{co}) will determine the bandwidth (Figure 30B). Therefore, we can substitute BW for F_{co} and arrive at

$$BW = \frac{1}{2\pi RC}$$

with R being the effective resistance of the amplifier and C the capacitance of the deflection plates.

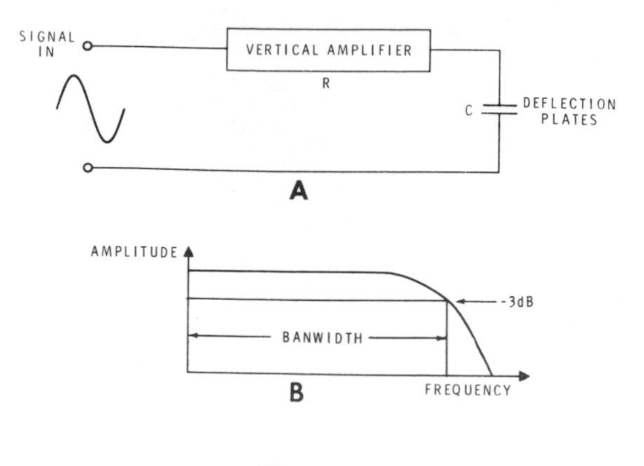

Figure 30

Still using the RC circuit analogy, but this time considering the time domain instead of the frequency domain, we see that with a square wave applied, a certain period of time is required for the plates to charge (Figure 31A). We have already defined rise time as being the time required for the plate voltage to change from 10% of maximum to 90% of maximum. Therefore, we can see that rise time (T_r) can be found by $T_r = t_2 - t_1$ (Figure 31B).

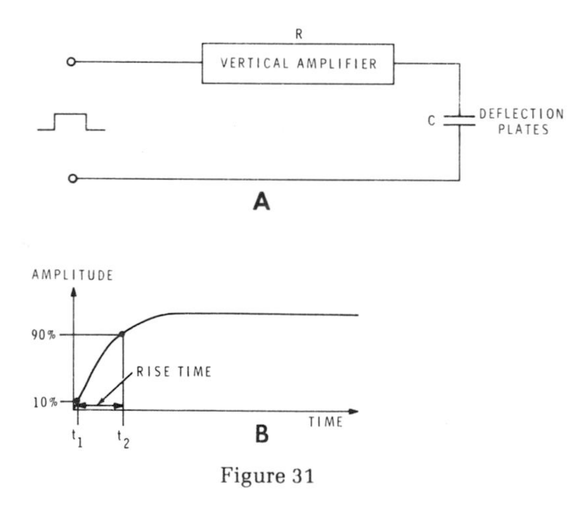

Figure 31

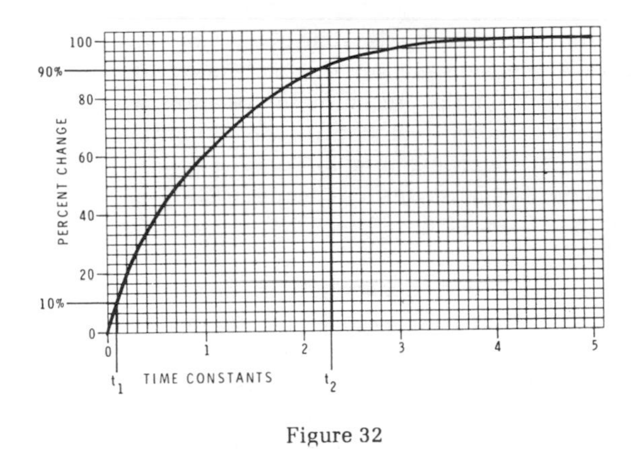

Figure 32

Looking at the universal time constant chart (Figure 32) we see that $t_2 - t_1$ is equal to 2.2 time constants. We know from previous lessons that one time constant is equal to the product of R times C. Thus, $T_r = 2.2\ RC$.

If the formula $BW = \dfrac{1}{2\pi RC}$ is solved for RC, we have

$$RC = \frac{1}{2\pi BW}$$

In the formula $T_r = 2.2\ RC$, substitute $\dfrac{1}{2\pi BW}$ for RC and we have

$$T_r = \frac{2.2}{2\pi BW} \quad \text{or} \quad T_r = \frac{0.35}{BW}$$

Thus, if either rise time or bandwidth is known, the other can be calculated. A 10 MHz scope has a rise time of 35 ns and a 15 MHz scope will have a rise time of 24 ns.

A fast rise time is often required when working with digital equipment.

A desirable situation is to have the rise time of the scope 1/5 or less of the rise time of the waveform being measured. A rise time of 10 ns is adequate for most applications. From the preceding information, we can see that this requires a scope bandwidth of 35 MHz. Some digital circuits have very fast rise times. It is fairly common to have rise times of less than 20 ns. A signal rise time of 17.5 ns requires a scope rise time of 3.5 ns, which means a bandwidth of 100 MHz.

This does not mean that you cannot use your scope to measure rise time just because it is slow. It does mean that you cannot read the rise time directly from the scope. You will have to perform some calculations. Suppose you are measuring the rise time of a square wave, using a 15 MHz scope with a rise time (T_{rs}) of 24 ns. You find that the displayed rise time (T_{rd}) is 40 ns. You don't take the difference between the two. Total rise time is a sum of the squares.

$$T_{rd}^2 = T_{rw}^2 + T_{rs}^2$$

Solving for T_{rg}, we get

$$T_{rw}^2 = T_{rd}^2 - T_{rs}^2$$

or

$$T_{rw} = \sqrt{T_{rd}^2 - T_{rs}^2} = \sqrt{40^2 - 24^2} = 32 \text{ ns.}$$

Rise times **do not** simply add together.

Power Supply

In a high quality oscilloscope, the power supply is a very important part of the overall unit. To help you understand some of the reasons this is true, let's look at the incoming AC line voltage. Sometimes, a normal voltage of 115 V will vary from as low as 90 V up to as high as 130 V.

If these changes are allowed to influence the internal scope voltages, several things happen. For one, the length of the sweep is affected by the supply voltage to the sweep generator. If a shorter or longer sweep is generated in the same time as the normal sweep, the TIME/CM will change. This will cause an error in any time or frequency measurement.

Brightness and focus may also be affected. Without proper regulation, the grid voltages on the CRT change with the input voltage, thus causing the trace to grow brighter or dimmer.

Vertical calibration may also change unless the regulation is good. There are two ways that this may be affected. We already know that the acceleration voltage may change, causing a corresponding change in the speed of the electron beam. A slower electron will spend more time in the vicinity of the deflection plates, and thus be deflected more, while a faster electron will pass through the plates faster with less deflection. Thus, the acceleration voltage affects both vertical and horizontal sensitivity.

The supply voltage to the horizontal and vertical amplifiers will affect their amplification and, therefore, the size of the signal applied to the deflection plates.

Any effective method may be used to regulate the power supply voltages for an oscilloscope. However, it is essential in a quality scope that adequate regulation be maintained.

SPECIAL FEATURES

The oscilloscope that we have been discussing is adequate for a large portion of electronic work. However, there are times when we would like to extend this capability. There are many special features available that allow certain parameters of the oscilloscope to be expanded.

The magnifier and the delayed time base allow for more detailed inspection of a selected portion of the trace. The dual beam or dual trace oscilloscope permits the operator to view two signals at the same time. The sampling oscilloscope allows extremely high frequency signals to be displayed while the storage oscilloscope displays very slow signals.

Magnifier

The magnifier or expanded sweep is a very common feature and is found on most moderately-priced and high-priced scopes. The purpose of the magnifier is to enlarge a portion of the waveform for closer observation. The most common method of magnification is to make the sweep longer — usually, by a factor of 5 or 10.

In Figure 33A, we see an oscilloscope presentation as it would normally appear. The sweep is physically ten divisions long. It is obvious that little information can be obtained about the pulse except for pulse width (PW) and pulse recurrence time (PRT). To see more detail of the wave shape, we must expand the waveform.

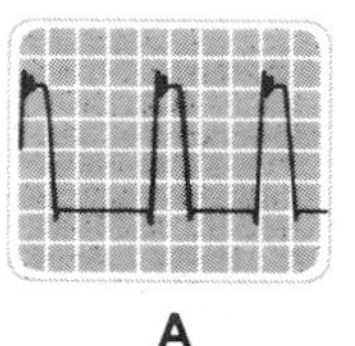

A

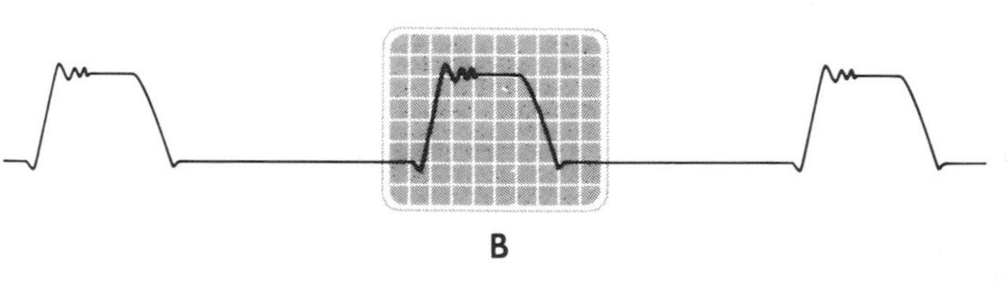

B

Figure 33

Figure 33B shows the same waveform with a ×5 expansion or magnification. The length of the sweep has been increased from ten divisions to fifty divisions. The sweep time remains the same, only the length is changed. If a **TIME/CM** of 2 μS is selected, the presentation in Figure 33A will require 20 μS. The entire waveform in Figure 33B also requires 20 μS; however, the presentation in Figure 33B — the portion visible on the screen — requires 4 μS.

By using the horizontal position control, the waveform can be moved until any 4 μS portion is visible on the CRT. Thus, we have effectively increased the sweep speed of a selected portion of the waveform.

The same results could not be obtained by changing the time-per-division control. That would only allow the first portion of the waveform to be expanded. Therefore, the second and third pulses would be "pushed" off the screen. Also, a portion of the leading edge of the first pulse will be lost. The amount that would be lost depends on the trigger level.

As useful as the magnifier is for viewing a particular section of waveform, it does have disadvantages. Since the magnification is obtained by increasing the gain of the horizontal amplifier, it is limited to about ×10. Even at a ×10 magnification, it is sometimes difficult to locate the exact portion of waveform to be displayed. Also, any errors in the magnification are added to the normal timing errors. Thus, the total timing error, when magnification is used, may approach 10%.

Delayed Sweep

Most of the disadvantages of the magnifier can be overcome by a delayed sweep. The delayed sweep uses a separate time base to generate a faster sweep, which displays a selected portion of the normal waveform. Since both the start and the length of the delayed sweep are controlled by the operator, any given portion of the display may be examined in detail.

Figure 34 shows how a delayed sweep works. Assume that the waveform at (A) is applied to the vertical input of the oscilloscope and that the normal sweep waveform is as shown at (B). Under these conditions, all of the waveform that occurs from T_0 to T_3 will be displayed. The CRT will appear as the one labeled "Normal Presentation."

Suppose that the operator wants to take a closer look at the glitch that appears after pulse number three. The delayed trigger start would be set to a point in time just before the glitch occurs. When the normal sweep

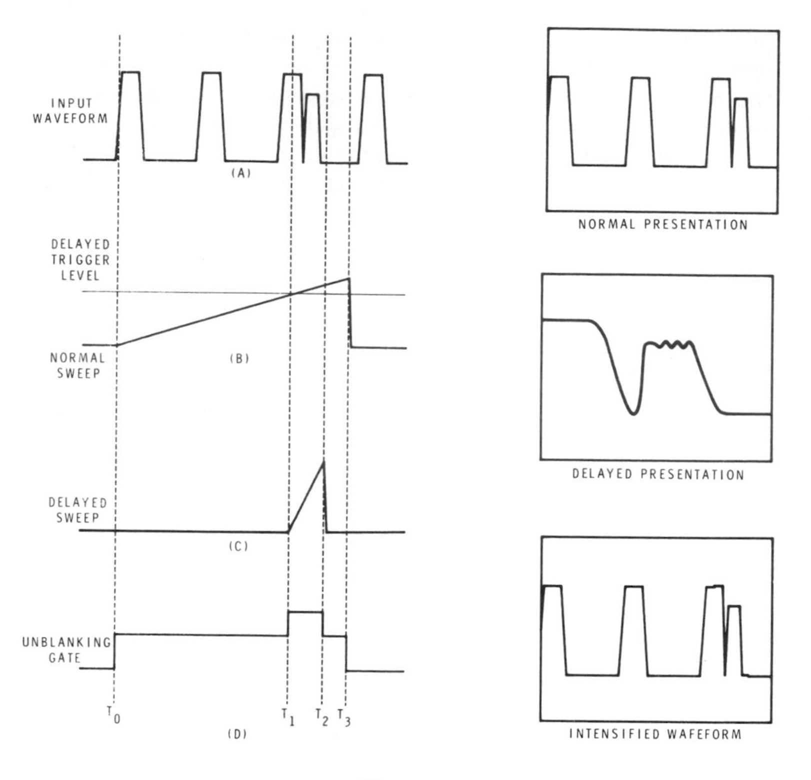

Figure 34

voltage passes through the delayed trigger level, a trigger is generated which starts the delayed sweep. A delayed sweep time is selected that will allow just the desired portion of waveform to be displayed. In this case, the delayed sweep time is between T_1 and T_2. Keep in mind that when delayed sweep is selected, the normal sweep is disconnected from the CRT and the delayed sweep circuits take over. Thus, the delayed sweep display will be similar to the one labeled "Delayed Presentation."

When using delayed sweep, which is often called delayed time base, locating the portion of the normal or main sweep presentation to be studied is quite simple. Provision is made for intensifying the normal presentation.

Remember that the intensity of the trace on the CRT is controlled by the potential difference between G1 and the cathode. If the unblanking gate (waveform D of Figure 34) is increased in amplitude from T_1 to T_2 and applied to the control grid (G1), the trace will be brighter during the time of increased amplitude as shown by the intensified waveform in Figure 34. By varying the delay time and the delayed TIME/DIV, any portion of the input waveform can be expanded.

The operation of the controls varies greatly from scope to scope, so use your handbook.

Two Traces

While the single trace oscilloscope will satisfy a large number of your measurement requirements, having two traces makes troubleshooting easier and faster. You can compare time, phase, amplitude, shape, or pulse width of two waveforms. You can compare output to input or channel A to channel B. If you are serious about digital troubleshooting or design, you will find the dual trace scope to be almost indispensable.

There are basically three ways to obtain dual traces on an oscilloscope. Perhaps the most obvious way would be to use two electron guns and produce two electron beams. If this method is used, the scope will be known as a **dual beam** oscilloscope.

DUAL BEAM

The basic dual beam oscilloscope is essentially two oscilloscopes in one as shown in Figure 35. The CRT contains two electron guns and two sets of deflection plates. Obviously, this is more expensive to build than a single beam CRT. There are also other disadvantages.

Because of the physical separation of the electron guns, the beams will not naturally strike the center of the CRT. Thus, both waveforms will be distorted somewhat. Each waveform will have something like a keystone shape as shown in Figure 36. Of course, the distortion is exaggerated, but the effect is enough to cause inaccuracies in measurement.

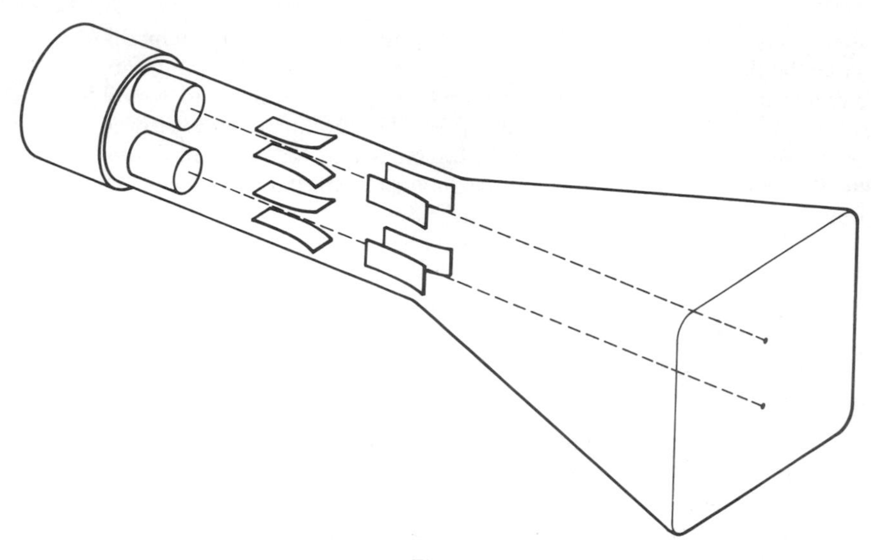

Figure 35

There are other disadvantages of a dual beam scope. For instance, if timing comparisons were to be made, the two horizontal deflection circuits would have to be perfectly in sync. This is impractical, especially at higher frequencies. Therefore, the two sets of horizontal deflection plates, amplifiers, and timing circuits were combined into one set. Then the two electron guns were moved closer together, until the final result was one gun with a split beam. This effectively eliminates the distortion caused by off-center gun placement. Thus, the typical **dual beam** scope has one set of horizontal components and two complete vertical channels.

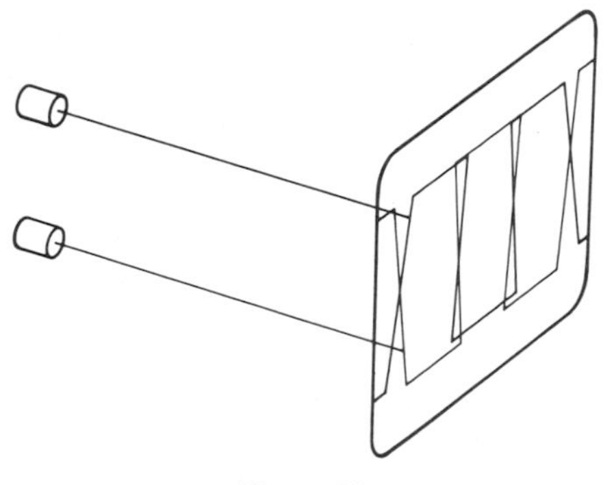

Figure 36

It would appear that the arrangement just discussed would be practically ideal for the presentation of two traces. There are, however, some limitations. As we learned earlier, one of the major features of a dual trace scope is its ability to compare. For accurate comparison, the vertical channels must be identical in gain, phase shift, and frequency response. This is not an impossible task and is, in fact, done on several scopes. However, it does become more difficult and expensive as frequency increases. Thus, the **dual beam** oscilloscope is quite expensive and relatively rare as a maintenance tool. The more common scope uses a single gun, single beam, and a single set of both vertical and horizontal deflection plates. This type of scope is known simply as a **dual trace** oscilloscope.

DUAL TRACE

There are two methods by which a dual trace can be obtained from a single electron beam. They are **chop** and **alternate.**

Chop

In the chop method of dual trace generation, the electron beam is "time shared" between vertical channels at a very rapid rate. With no input to the vertical channels, the CRT presentation will appear as in Figure 37. Here you can see where the name "chop" comes from. Each trace appears to be chopped into sections.

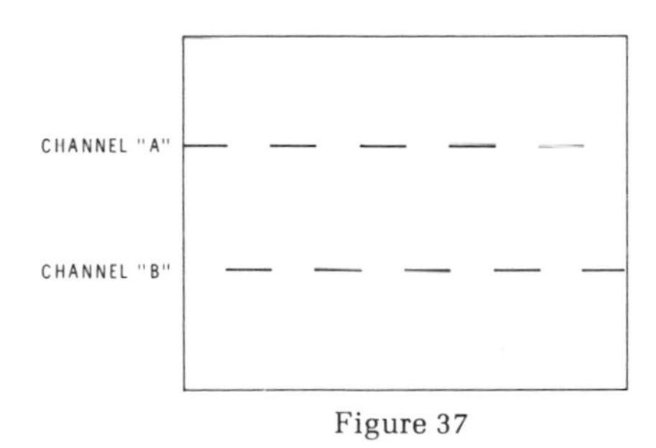

Figure 37

Refer to Figure 38 to see how this is accomplished. The free-running multivibrator constantly changes the switch from channel A to channel B. A typical switching frequency is around 500 kHz. Assume that the switch is in the "A" position to start. The vertical deflection plates are under the control of the channel A vertical amplifier. After a short period of time, the multivibrator will cause the switch to change to channel B. Thus, the sweep is chopped into segments. The first segment is

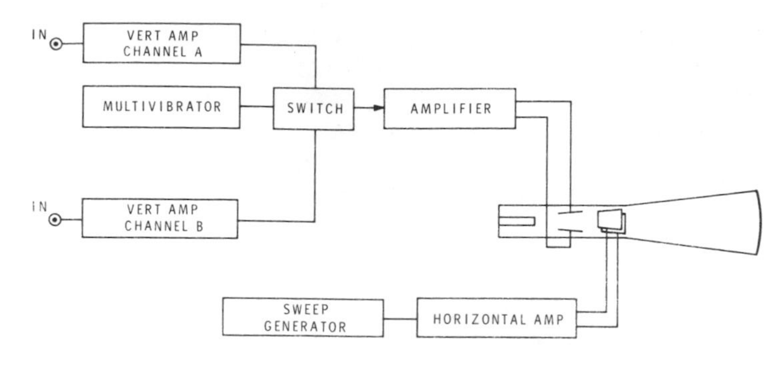

Figure 38

positioned vertically by channel A, the second by channel B, then A, then B, etc.

It might appear that a chopped waveform would cause viewing problems. However, this is rarely the case and only occurs when the chop rate is an exact multiple of the input frequency.

In Figure 39, the input to channel A is a pulse, and the input to channel B is a sine wave. During the high portion of the chop waveform, channel A is connected to the vertical deflection plates, and during the low portion, channel B is connected. The sweep is repetitive and equal to the period of the sine wave.

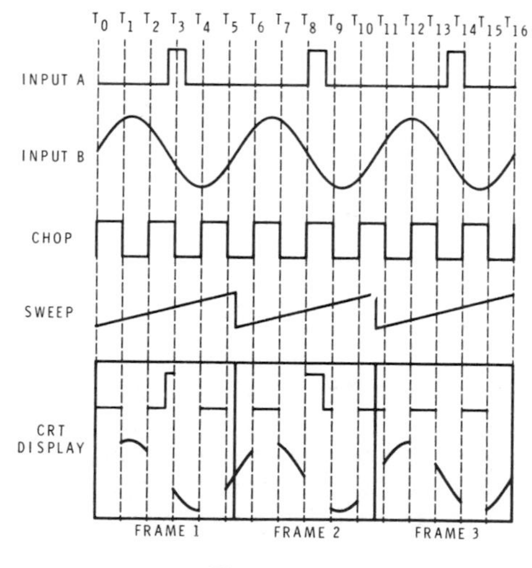

Figure 39

The action starts with the chop waveform high. Thus, from T_0 to T_1, input A has vertical control of the electron beam. During that time, input A is a straight line. Therefore, a straight line segment will be drawn on the CRT as shown in frame 1. From T_1 to T_2 input B has control. During that time, input B is at the peak of the sine wave. So the top of the waveform is drawn as shown in frame 1. The switching continues, and alternate segments are drawn on the scope. However, the first sweep ends shortly after T_5.

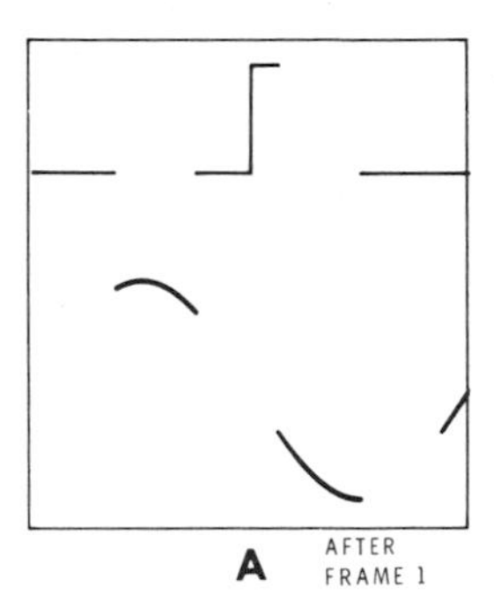

At the end of the sweep, the electron beam will retrace to the left side of the screen. Retrace occurs while the chop waveform is low. Therefore, when the beam returns to the left of the screen, it will continue drawing waveform B as shown at the start of frame 2.

We have shown a recurring sweep with instant retrace. You know that an instant retrace cannot happen. However, if it did, the effect would be the same as a triggered sweep. Therefore, we have shown it this way to simplify the explanation. The action continues as long as the inputs and sweep controls remain unchanged. However, three frames are suitable to show what happens.

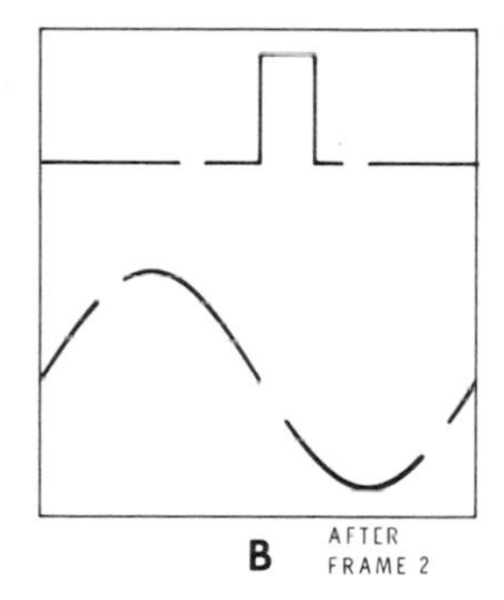

If you could stop the action after one frame or sweep, you would see a display like that in Figure 40A. The persistence of the phosphor is such that what was drawn during frame 1 will still be there a few frames later. So, after two frames, the display will be a combination of frame 1 and frame 2 and will appear as in Figure 40B.

After three frames have been completed, all of the gaps will have been filled in, and the display will be like that in Figure 40C. Because of the high chopping frequency, the persistence of the phosphor and the characteristics of the human eye, the viewer will see two complete waveforms and will not normally be able to discern the different segments.

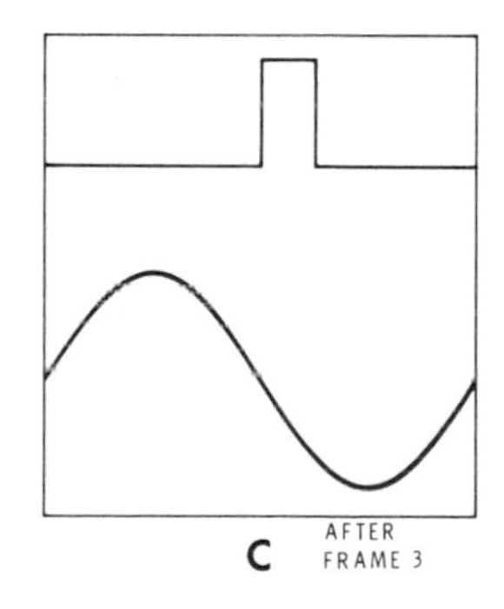

Figure 40

If the chopping frequency is an even multiple of the input frequency, gaps in the waveform may be visible. This could be a problem at the higher sweep rates where the chop interval is a significant portion of the sweep time. This can be overcome by switching to the **alternate** mode.

Alternate

In the alternate mode, the scope makes one complete sweep for channel A, then makes a complete sweep for channel B as shown in Figure 41. Here, when the alternating waveform is high, input A is fed to the vertical deflection plates, and when the alternating waveform is low, input B goes to the plates.

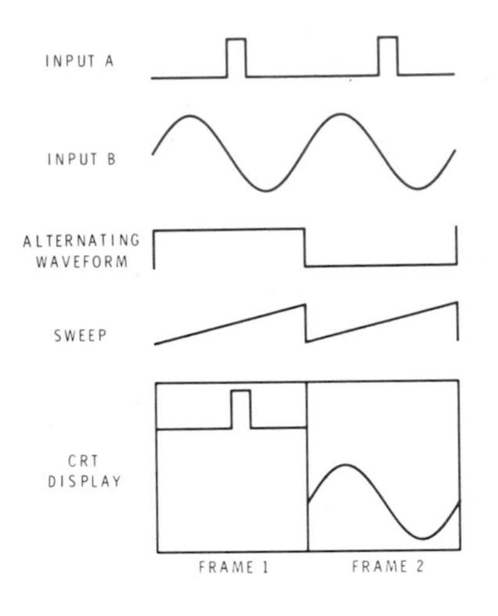

Figure 41

As the action starts, the alternating waveform is high and stays high for one complete sweep. Thus, during that time, only input A is fed to the CRT, and the display will be as shown in frame 1. During retrace time, the alternating waveform goes low so that, during the second sweep, input B is displayed as shown in frame 2. The display will then alternate between A and B continuously. To the operator, the display will appear as in Figure 40C.

If the alternate mode is used at low sweep rates, the operator will be able to discern first one trace, then the other. This can be very disconcerting. Therefore, it is normal operation to use the alternate mode at fast sweep rates and the chop mode at low sweep rates. In fact, some scopes will automatically switch from chopped to alternate at a predetermined point when sweep speed is increased and from alternate to chopped as speed is decreased. If a scope has either an alternate or a chopped mode, it will most likely have both so that the best mode for the measurement being made can be selected.

Summary

There are the three methods of achieving two traces on a scope. The dual beam method provides continuous presentation of both inputs. The chopped mode chops both channels into segments, and the alternate mode changes channels with each trace. As we have seen, each method has its advantages and its limitations. The **dual trace** scope offering both alternate and chopped is the most common.

Storage Scope

It is sometimes necessary to observe information which is not repetitive in nature. Information of this type cannot be viewed on a "normal" oscilloscope because of the relatively short persistence of the phosphor. Long persistence phosphors are available that would overcome this problem; however, they would introduce more serious problems. The long persistence would be unsuitable for the "normal" recurrent signal. Both types of signals can be accommodated by a specially constructed CRT known as a **STORAGE** CRT.

There are several types of storage CRT's available. They are used in radar, data transmission, medical equipment, and test equipment. The type most commonly used in test equipment oscilloscopes is the **variable persistence/storage cathode-ray tube.** This is the only one we will discuss here.

CONSTRUCTION

The basic construction of a storage CRT is shown in Figure 42. Notice that this CRT has all the standard components; electron gun, deflection plates, and phosphor coating. In addition to these, the storage CRT contains a flood gun, collimator, collector mesh, and storage mesh.

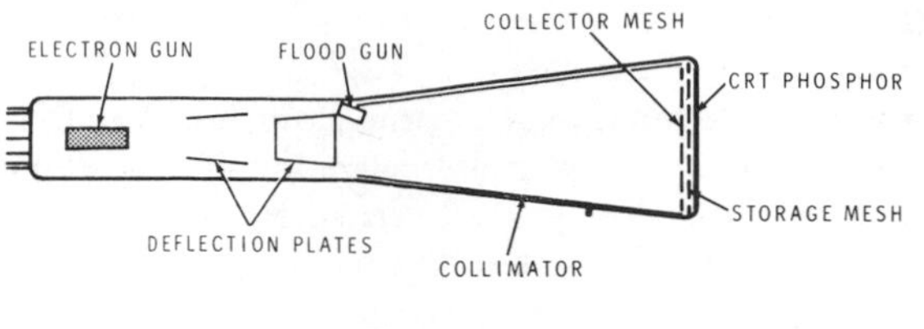

Figure 42

The standard electron gun is called the **write** gun in a storage scope. You will remember that it emits a high velocity stream of electrons which are focused into a very thin beam. In contrast to this, the flood gun emits a low velocity flood of electrons that covers the entire screen. The difference in electron velocity can be better appreciated by remembering that the write gun has a cathode voltage of about −5 kV, while the flood gun has a cathode potential of 0 V.

The collimator is a conductive coating on the inside of the CRT that modifies the path of the electrons from the flood gun so they are parallel to the walls of the tube and each other. Thus, these electrons will strike the screen nearly perpendicular. The collimator potential is approximately 50 V. This voltage will direct the low velocity flood gun electrons but will be too low to affect the high velocity electrons from the write gun.

The collector mesh is a very fine wire screen mounted parallel to the face of the CRT and held at a constant potential that is slightly more positive than the collimator. The correct potential will be around +150 V.

The storage mesh is similar to the collector mesh in construction except that the surface facing the collector mesh is coated with a very thin layer of insulating material. This material is selected to have high secondary emission. That is, when it is struck by high velocity electrons, it gives off electrons and becomes positively charged in that area. This portion of the storage is usually maintained at about −10 V.

The phosphor coating on the CRT is connected to a very high positive potential, usually somewhere between 5 kV and 10 kV.

WRITING

Writing or storage is accomplished by the write gun. As you saw earlier, the relatively low voltages on the collimator and the collector mesh have almost no affect on the write gun electrons. These electrons crash right through with some of them striking the insulated surface of the storage mesh. The rest strike the phosphor screen and cause a trace to appear. The secondary emission from the storage mesh has caused that portion of the mesh to become slightly positive. Once a presentation has been written or stored, the write gun can be turned off and the presentation will remain visible. The electrons emitted by the flood gun are accelerated toward the phosphor coating by the positive potential on the collimator and the collector mesh. These potentials are rather low and the electrons will not have enough energy to pass the −10 V potential on the storage mesh. They can, however, penetrate the areas that have been made positive by secondary emission.

Once an electron penetrates the storage mesh, it will be accelerated by the high positive potential on the phosphor and will have enough energy to cause the phosphor to glow. This charge on the storage mesh will remain for about two minutes. Thus, we see that storage time is limited by the ability of the storage mesh to retain a charge.

Initially, the insulated surface of the storage mesh is charged to −10 V, much as a capacitor is charged. It is the ability to retain this charge that determines storage time. As soon as the tube is placed in the storage mode, that is, the flood gun is turned on, the gas molecules within the tube are ionized by the flood gun electrons. These positive ions are attracted to the negative storage surface, making it less negative.

As the storage surface becomes less negative, it loses its ability to hold back the flood of electrons. Eventually, the entire phosphor screen begins to glow and contrast between the background and the displayed waveform is lost.

Storage time can be increased by turning the flood gun on for a short period and then off again. However, this reduces the brightness of the display. A compromise must therefore be made between storage time and brightness.

ERASING

Often it is desirable to erase the stored presentation before the storage time has elapsed. To accomplish this, storage scopes are provided with an **ERASE** control. The erase control applies collector mesh potential to the storage mesh. The entire storage surface is momentarily charged to a uniform high positive potential after which it is returned to its original voltage of −10 V. The entire erase process requires only a few hundred milliseconds.

VARIABLE PERSISTENCE

Storage scopes are often referred to as variable persistence scopes. Variable persistence means variable storage time. This variable storage time is accomplished by the erase function. In the variable mode, the display is not erased suddenly with a high potential, but slowly with pulses of a lower potential.

Before the display is stored, the potential on the insulated surface of the storage mesh is −10 V. In the area where information is stored by the write gun, the storage potential is near 0 V. Thus, in order to erase the display, the entire storage mesh can be made 0 V. This is accomplished in the variable persistence mode by applying positive pulses of about 10 V to the storage mesh. Each time a pulse is applied, the storage mesh becomes less negative. Therefore, the rate at which the display is erased can be controlled by adjusting the PW and PRT of the pulse. A wide PW and a short PRT will erase the display rapidly, while a narrow PW and a long PRT will allow the display to persist for a relatively long time.

Another term that is used for the variable persistence oscilloscope is "continuous erase scope," because the display is continuously being erased even while it is being displayed.

You have seen that the storage oscilloscope provides a means for displaying very low frequency information. There is also an oscilloscope available for viewing very high frequency signals. It is called a sampling oscilloscope.

Sampling Scopes

The sampling technique extends the frequency range of the oscilloscope far beyond what is possible with a real time scope. Effective bandwidth of up to 18 GHz is possible when using the sampling technique. This is accomplished by converting a high frequency, repetitive signal to a low frequency.

Figure 43 shows how such a conversion is made. The waveform shown in black is the one we wish to measure. The sampling sequence starts at a reference point (R) which is time zero. After one time interval (1T), the incoming signal is sampled and the sample stored in memory. Nothing else is done until the input once more passes the reference. This time, the sampling circuits wait for two time intervals after R to take a sample. The result of this sampling is also stored in memory. The sequence is continued until a predetermined number of time intervals have passed.

When the sampling is complete, all of the samples, in this case, 10, are recalled from memory and plotted on the oscilloscope as shown in Figure 44. Notice that the shape of the waveform is the same as in Figure 43; however, it is expanded in time.

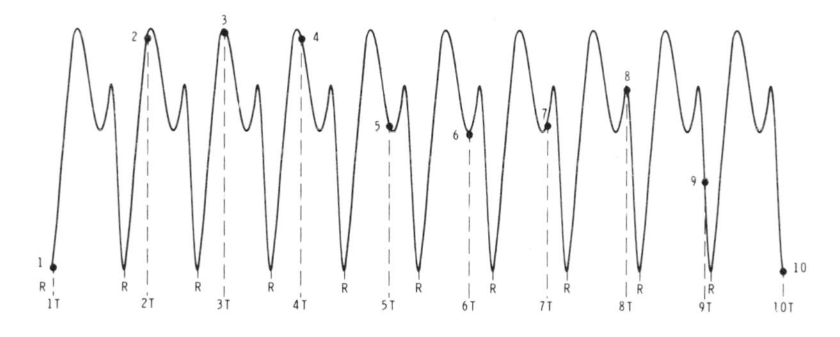

Figure 43

In the example used here, ten samples were taken, one from each of ten waveforms. This gives a 10:1 reduction in frequency and provides a reasonable reproduction of the waveform. However, in actual practice, 100 or more samples would be taken for a greater frequency reduction and much better waveform fidelity.

The major drawback of the sampling oscilloscope is that the input must be repetitive. With that limitation, the sampling can perform any measurement that can be performed with a real time scope. Modern sampling scopes are easy to use, many having only one additional control.

This course will not discuss sampling techniques in detail; however, most manufacturers provide excellent operating manuals which should enable you to use their scopes.

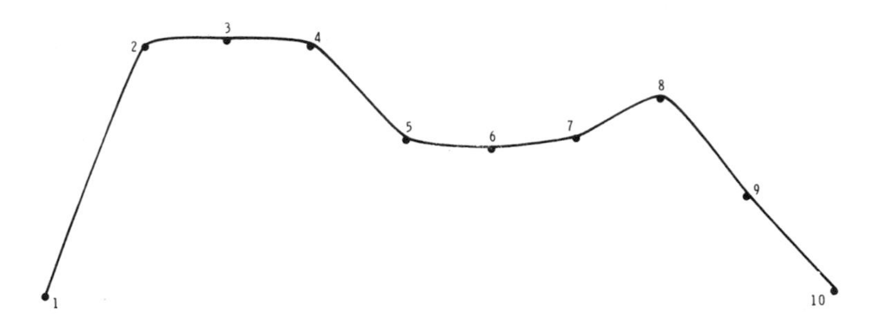

Figure 44

CONTROLS

So far, we have discussed oscilloscope circuits. In the rest of this unit, we will discuss the operation of oscilloscopes. In order to properly operate a scope, you must thoroughly understand the function of all the controls. The following explanation is based on a typical low-cost oscilloscope. It is a relatively simple scope and easy for the beginner to understand.

CRT Controls

Figure 45 shows the front of the oscilloscope with the CRT controls highlighted. A **graticule** is placed on the face of the CRT. This is an 8 × 10 cm grid which allows us to measure the time and amplitude of a signal. With this scope, the graticule is a separate piece of clear plastic mounted in front of the CRT. Since this mounting arrangement places the graticule well in front of the trace on the CRT, the possibility of parallax exists.

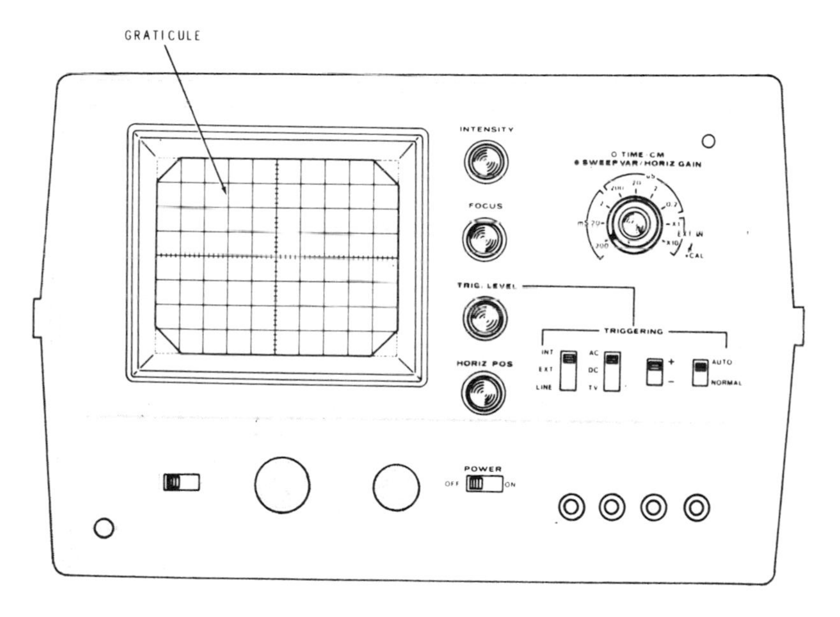

Figure 45

On more expensive scopes, the graticule will be etched on the outside face of the CRT. The graticule is separated from the trace only by the thickness of the CRT face. While possible parallax error is reduced with this arrangement, it is not eliminated. For this reason, the very best scopes have the graticule etched on the inside face of the CRT on the same plane as the phosphor coating. This removes any possibility of parallax error. While the graticule is not actually a control, it is such a necessary part of the scope that meaningful measurement would be very difficult without it.

The control located in the upper center of this scope is the **intensity** control. This control is equivalent to the brightness control on your television and is present on all oscilloscopes. The intensity control varies the intensity or brightness of the trace on the CRT. CAUTION: If the brightness is set too high, there is a danger of burning the phosphor. Once the phosphor has been burned, the damage is permanent. Therefore, the intensity control should only be adjusted for the minimum brightness for comfortable viewing. If a very bright trace is necessary, it should be left at that level only for the time necessary to make the reading, then reduced. Never allow a bright spot to remain in one place on the screen. This is a sure way to burn the phosphor.

The **focus** control adjusts the trace, or the spot on the screen. It, therefore, affects the thickness of the trace. It is usually adjusted for the thinnest trace possible. When high brightness is required, it is sometimes necessary to defocus the trace to prevent burning the phosphor.

The **astigmatism** control changes the shape of the spot on the screen. It is usually adjusted for a round spot. On many scopes, the astigmatism control is not a front panel adjustment. The cover must be removed to make this adjustment. With other scopes, there may be a hole in the case, allowing a screwdriver adjustment.

The **power** switch is included in this section because it is often incorporated into the intensity control. Obviously, it must be on before any other controls are functional.

Sweep and Horizontal Controls

Figure 46 shows the Heathkit IO-4541 Oscilloscope with the sweep controls highlighted. Sweep controls are used to control the start, time, length, and position of the horizontal trace.

The **triggering** controls control the start of the sweep. Since the sweep must start at the same point relative to the input waveform, the trigger source should be a signal which is related in time to the incoming signal. This scope and most other scopes have three trigger sources. They are the incoming signal, the power line voltage, and an external source.

Slide switches, mounted in a row, control the trigger functions of the example scope. In other scopes, the switching arrangement may be different but the functions will be similar. The extreme left-hand switch is the **source** switch, and has three positions:

> The **INT** (Internal) position causes the trigger to be derived from the signal being displayed, and is probably used more than any other source.

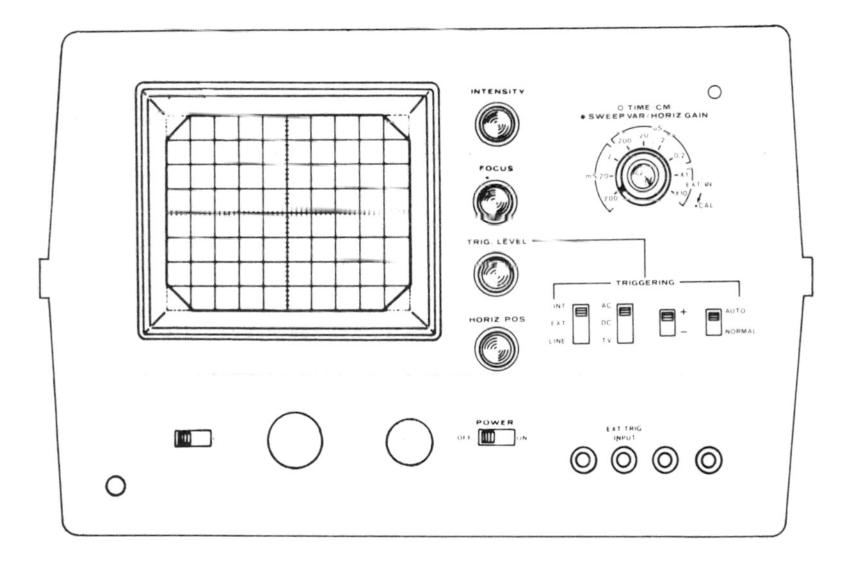

Figure 46

The **EXT** (External) position requires that a signal be fed into the **external trigger input.** This arrangement is used quite often when working with digital circuits.

The **Line** position uses a sample of the power line voltage to trigger the sweep. This position is most useful when working with power supply circuits. Since the ripple frequencies are a multiple of the line frequency, the timing will be correct for a readable display. The sample from which the trigger is derived is constant in amplitude and not dependent on the incoming signal. Therefore, the trigger level does not need to be readjusted for different amplitude signals.

The selected trigger source must be coupled to the trigger circuits. There are typically two types of **trigger coupling;** AC and DC. In the **AC** position, the trigger signal is coupled through a capacitor to the trigger circuits. The capacitor blocks the DC portion of the signal and passes the AC portion.

The **DC** position allows the sweep to trigger on very slow changes in voltages. In this position, the signal upon which we wish to trigger is coupled directly to the trigger circuits.

In addition to the two types of triggering just mentioned, the oscilloscope has a position labeled **TV**. In the TV position, the signal is coupled through C_1 (Figure 47) to a low-pass filter composed of R_1 and C_2. This removes the video from the incoming TV signal. D_1 insures that the output is a positive pulse with a reference of 0 V. This circuit allows the triggering circuits to be synchronized to the vertical frame rate of a video TV signal.

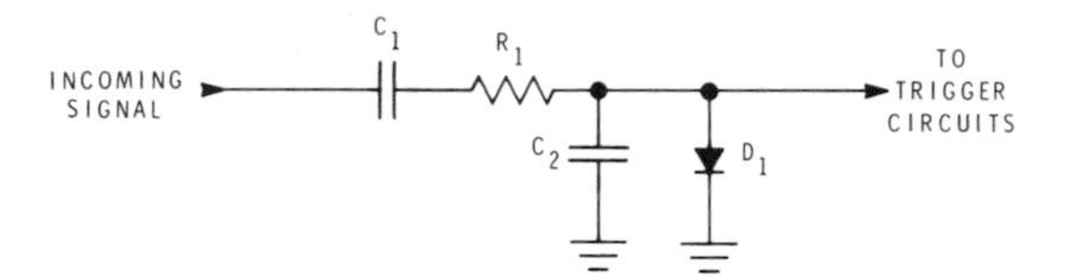

Figure 47

The next switch is the **SLOPE** selector. The sweep may be triggered on the positive-going (+) or the negative-going (−) part of the signal.

The **TRIG. LEVEL** sets the amplitude on the waveform where the trigger occurs. This control, along with the **SLOPE** selector, lets us look at any portion of the waveform we wish.

The **AUTO/NORMAL** or mode switch is next. The difference between these two positions only becomes apparent when the displayed signal is lower in amplitude than the selected trigger level. In the **NORMAL** position, when the incoming signal is too small or the **TRIG. LEVEL** is not set right, there will be no trace on the screen. However, if the switch is placed in the **AUTO** position, an automatic baseline or trace will be generated. Of course, any incoming signal will be superimposed on this trace. The presentation will most likely not be properly synchronized. If the trigger level is adjusted so that the trigger voltage is greater than the trigger level, the display will synchronize. This mode is very useful for locating a waveform when you are unsure of the amplitude.

The final sweep control is actually two controls in one. The outer ring is labeled **TIME/CM** and controls the time it takes the sweep to transverse one centimeter on the screen. As shown in Figure 46, the control is in the 200 mS/CM position. In this position, it will take the trace 200 mS to cover 1 centimeter. Since the screen is 10 centimeters wide, it takes the beam two seconds to complete one sweep. Speeds of 200 mS/CM up to 0.2 μS/CM are available with this scope. The ×1 and ×10 positions have a different purpose and will be discussed later.

The **TIME/CM** which is selected by the outer ring is only accurate if the inner control, the **SWEEP VAR/HORIZ GAIN** control, is turned all the way clockwise to the **CAL** position.

When used as a sweep control, this knob varies the sweep time. It is useful for viewing a waveform, the period of which is not equal to one of the times selected. However, it must be remembered that no accurate time or frequency measurement can be made unless this control is in the **CAL** position.

The entire display can be moved left and right by means of the **HORIZ POS** control. This allows any part of the waveform to be aligned with a point on the graticule; thus, allowing more accurate measurement of time and frequency.

It is also possible to feed external information into the horizontal channel. To accomplish this, the **TIME/CM** control must be turned to one of the **EXT IN** (external in) positions. There are two positions, ×1 and ×10. In the ×10 position, the input is attenuated by a factor of 10.

When in the **EXT IN** position, the **SWEEP VAR/HORIZ GAIN** control is used to control the gain in the horizontal channel. The **HORIZ POS** control works as before and is used to position the presentation left and right on the screen. In this mode of operation, the input to the horizontal channel is connected to the **EXT INPUT** jack in the lower right corner of the scope. The **GND** is a ground connection for the horizontal channel.

Vertical Controls

The remainder of the controls apply to the vertical channel. Vertical signals are fed into the scope through the coaxial connector in the lower left corner labeled **INPUT**. The signal is then coupled to the vertical amplifier. The **INPUT COUPLING** switch selects the method of coupling.

The switch has three positions, **AC**, **GND**, and **DC**. In the **GND** position, chassis ground is placed on the vertical input. This establishes a zero-volt reference from which DC voltages can be measured. The **AC** position of the switch couples the incoming signal through a capacitor which blocks any DC component and allows the AC to pass. This position is most useful for measuring AC which is superimposed on a relatively large DC. In the **DC** position, all components of the incoming signal are coupled directly to the vertical amplifier. In this position, both AC and DC measurements can be made.

The **VOLTS/CM** control sets the vertical sensitivity in volts-per-centimeter when the **VARIABLE** control is in the **CAL** position. Thus, it is possible to obtain the value of either AC or DC voltages. We will discuss later exactly how a measurement is made.

When the **VARIABLE** control is moved from the **CAL** position, the volts-per-centimeter reading is no longer accurate. The **VARIABLE** control is usually used to compare the input with a known value.

The **VERT POS** control is used to position the presentation in an up and down direction.

A **1 V (P-P 60 Hz)** jack is provided on the front of the scope to aid in calibration. On this particular scope, a 60 Hz sine wave with a peak-to-peak amplitude of 1 V is furnished. More expensive scopes will provide a square wave with a frequency of about 1 kHz and a specified amplitude. The 1 V amplitude is common. The square wave allows probe calibration as well as amplitude calibration. However, the probe can be calibrated from any square wave source.

Probes

While the oscilloscope probe isn't actually one of the scope controls, its selection and use is so crucial to accurate measurement that we have included it in this section.

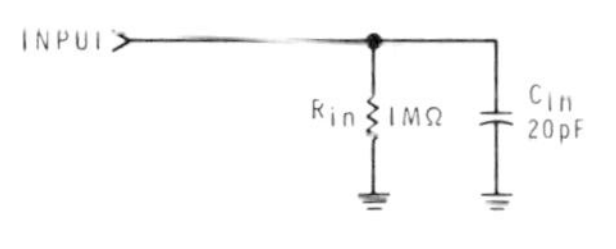

Figure 48

Before starting the discussion of probes, let's look at the characteristics of the oscilloscope input. In the typical scope, the input is a high resistance shunted by a low value of capacitance as shown in Figure 48. When the oscilloscope is connected to the circuit under test, the scope input modifies that circuit and may, depending upon the circuit, cause changes in circuit operation. In any case, the scope loads down the circuit so that the voltage being measured is less than the unmodified circuit voltage. This loading effect depends on the ratio of scope input resistance (R_{in}) to signal source resistance (R_g). You can calculate it with the formula:

$$\text{Percent error} = \frac{R_g}{R_g + R_{in}} \times 100$$

You can see by this that an R_{in} equal to 10 times R_g will have a loading error of less than 10%. For a loading error less than 1%, an R_{in} of at least 100 times R_g is required.

All of this is fairly simple, but it only applies as long as the input capacitance (C_{in}) is not a factor. However, C_{in} is almost always a factor. Therefore, to determine loading effect, we must consider the input impedance (Z_{in}) not just the resistance. Since R_{in} and C_{in} are in parallel, you can calculate the impedance by,

$$Z_{in} = \frac{R_{in} \times X_{cin}}{\sqrt{R_{in}^2 + X_{cin}^2}}$$

where

$$X_{cin} = \frac{1}{2 \pi f C_{in}}$$

This should bring us to the conclusion that Z_{in} changes with frequency, becoming less as frequency increases. To understand the magnitude of this change, compare the X_c of a typical 20 pF C_{in} at the following frequencies.

f	X_c
500 Hz	16 MΩ
1 kHz	8 MΩ
10 kHz	800 kΩ
100 kHz	80 kΩ
1 MHz	8 kΩ
10 MHz	800 Ω
100 MHz	80 Ω

Since Z_{in} can never be greater than the smaller of R_{in} and X_{cin}, but can be considerably less than either, the frequency doesn't get very high before the effect of C_{in} becomes evident. As a matter of fact, it is often of more importance than resistance loading.

It should now be apparent that the percentage of error is a factor of Z not of R. Therefore, you can find the percentage of error by:

$$\% \text{ of error} = \frac{Z_{in}}{Z_{in} + Z_g} \times 100$$

where Z_{in} is the scope input impedance and Z_g is the impedance of the signal source.

Up till now, we haven't even connected the scope to the circuit under test. Obviously, to do so requires the use of some sort of cable or test leads. A simple test lead is subject to a large amount of external interference, such as, 60 Hz line frequency or stray RF fields. This noise often makes it impossible to measure small signals. To prevent this interference, some form of coaxial cable is usually used as a connection. However, this too, has its drawbacks.

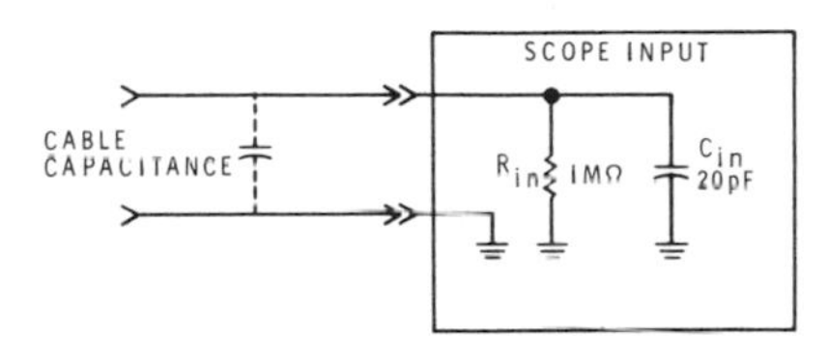

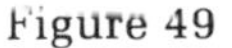
Figure 49

For example, one of the most common coaxial cables, RG-58/U, has 28.5 pF of capacitance per foot. This capacitance is connected in parallel with the input capacitance of the scope as shown in Figure 49. Thus, if a 3 ft. length of cable is used, the cable capacitance will be 85.5 pF. This is added to the scope capacitance making the total capacitance connected to the circuit 105.5 pF.

At 1 MHz, this capacitance has a reactance of only 1509 Ω. If we calculate Z_{in} using the formula discussed earlier, we find it to be only 1508 Ω. Thus, we can see that X_{cin} is so low that R_{in} can be ignored. If this scope lead combination is connected to an R_g of 1 kΩ, the voltage measured will be only 60% of the actual voltage.

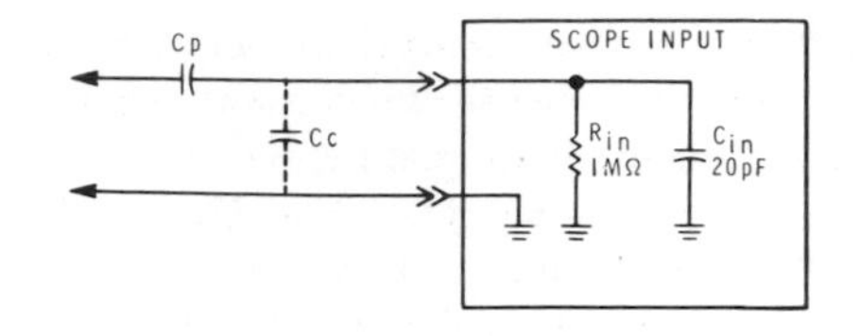

Figure 50

If the frequency increases to 10 MHz, X_c drops to a mere 151 Ω and the measured voltage drops to 13% of the actual value. Obviously, with this amount of error, a meaningful measurement is impossible.

Since the capacitance of the cable and the scope input cannot be eliminated, it must be compensated for. To do this, a capacitor is placed in series with the input, as shown in Figure 50.

We know that when capacitors are connected in series:

$$C_t = \frac{1}{\frac{1}{C_1} + \frac{1}{C_2} + \frac{1}{C_3} + \cdots \frac{1}{C_n}}$$

Therefore, the total capacitance will always be less than the smallest capacitor. We can, thus, adjust the value of probe plus scope capacitance by controlling the value of C_p. If C_p is one ninth the value of C_c and C_{in} combined, the result will be a 10:1 capacitive voltage divider.

For example, using the same scope and cable combination we discussed earlier, C_{in} is 20 pF and C_c is 85.5 pF for a total of 105.5 pF. If an 11.7 pF capacitor is connected at C_p (Figure 50), the total capacitance will be 10.5 pF. At 1 MHz, X_c is 15,157 Ω, which is ten times the previous X_c of 1508 Ω. This, of course, presents a much smaller load to the circuit under test. Assuming the same 1 kΩ source used previously, the percentage of signal loss is:

$$\% \text{ Loss} = \frac{Z_{in}}{Z_{in} + Z_g} \times 100$$

$$= \frac{15157}{15157 + 1000} \times 100$$

$$= 93.8\%$$

which means that only about 6% of the source signal is lost due to loading.

All of this signal, however, is not applied to the scope input. There is a 10:1 division between C_p and $C_{in} + C_c$, allowing only 10% of the available signal to be applied to the scope input. Therefore, the VOLTS/DIV setting must be multiplied times ten to get the actual volts/division.

While the load on the circuit under test changes with frequency, the capacitive voltage divider ratio remains constant. The actual voltage division, however, does not remain constant due to the effect of R_{in}. As the input frequency decreases, the reactance increases until, at some frequency below 1 MHz, R_{in} becomes the determining factor for Z_{in}. Therefore, while X_{cp} continues to increase, Z_{in} will tend to remain constant, and the voltage division ratio will change.

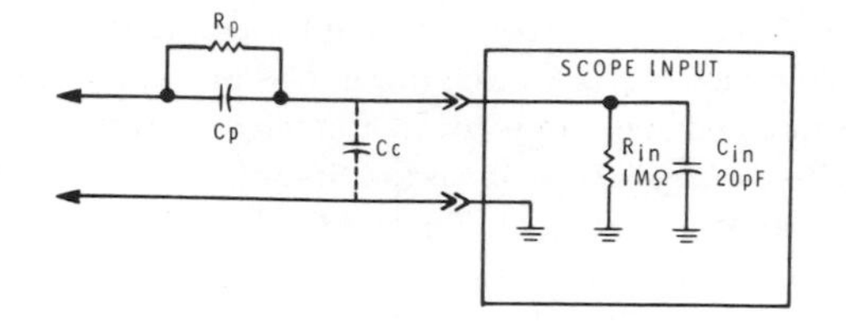

Figure 51

To overcome this tendency, a resistor is added in parallel with C_p, as shown in Figure 51. If this resistor (R_p) is made 9 times the value of R_{in}, the voltage ratio will remain essentially constant at all frequencies. For instance, if R_p is 9 MΩ, the voltage ratio is 10:1 for DC and remains 10:1 as frequency increases. In this manner, we have effectively increased the input impedence of the scope by a factor of 10 and reduced the loading on the circuit under test.

The constant division ratio is only present if the probe is properly compensated. For the probe to be properly compensated, the RC time of the probe must equal the RC time of the scope input, which includes any cable capacitance. For instance, in the scope and probe just discussed, R_p = 9 MΩ and C_p = 11.7 pF for an RC time of 10.5 mS. The scope has an R_{in} of 1 MΩ and a $C_{in} + C_c$ = 105.5 pF for an RC time of 10.55 mS. Thus, the RC time of the probe equals the RC time of the scope, and the probe is properly compensated.

Most probes are designed for use on a variety of scopes. Since C_{in} may vary from scope to scope, C_p is usually a variable capacitor to allow matching of the probe to the scope input. If you know the value of R_{in} and C_{in}, you can calculate the value of C_p as follows:

$$R_p \, C_p = R_{in} \, C_{in}$$

solving for C_p you have:

$$C_p = \frac{R_{in} \, C_{in}}{R_p}$$

If R_{in} is 1 MΩ and C_{in} is 47 pF, let's see what C_p and R_p will be. For a 10:1 probe, $R_p = 9 \, R_{in}$, in this case 9 MΩ.

$$C_p = \frac{1 \text{ M}\Omega \times 47 \text{ pF}}{9 \text{ M}\Omega}$$

$$= 5.2 \text{ pF}$$

Actually, the value of C_p is seldom calculated. Instead, you usually would make the adjustment by observing a square wave and adjusting for the best shape, as shown in Figure 52. Figure 52A shows a square wave with the high frequencies attenuated too much. Figure 52B is the correct waveform with all frequencies handled the same, while Figure 52C does not attenuate the high frequencies enough. Thus, when C_p is adjusted until the display looks like Figure 52B, with a square wave input, the probe is properly compensated. The calibration control can be a screw-driver adjustment, a knurled ring, or some other arrangement.

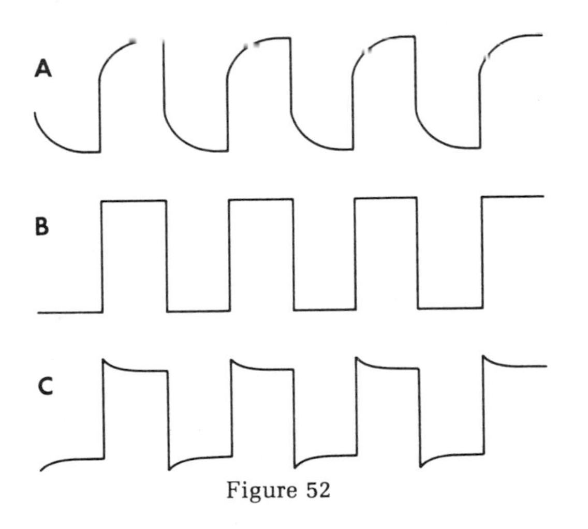

Figure 52

OSCILLOSCOPE MEASUREMENTS

The oscilloscope has often been called the most versatile measuring instrument in the shop or laboratory. Whether this statement is true or not will not be discussed here. However, it is true that the oscilloscope is capable of providing a variety of information. In spite of this claim of versatility, the oscilloscope is capable of measuring only two quantities — voltage and time. In this section, you will see how the two quantities can be used to represent a number of parameters.

Voltage

There are both advantages and disadvantages to measuring voltage with an oscilloscope. The advantages are that shape, phase, and frequency can be viewed while the measurement is being made. Poor resolution is probably the greatest disadvantage, especially at higher voltage levels.

All voltage measurements, and time measurements too, are made with the graticule. The graticule as you learned earlier is a grid, usually 8 × 10 cm superimposed on the face of the CRT. Figure 57 shows how a graticule might appear. Notice that on the center horizontal line (A) and the center vertical line (B), each major division is subdivided into five smaller divisions. Each vertical division has a certain value, depending on the setting of the **VOLTS/CM** control. For instance, if the **VOLTS/CM** is set at 5, then each major division has a value of 5 V, and each subdivision has a value of 1 V. Thus, the resolution is only about ±0.5 V. Depending on the amplitude of the input, resolution varies between 2% and 10%.

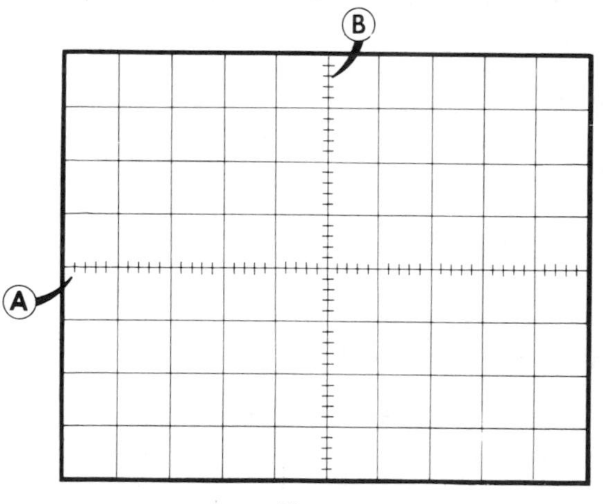

Figure 57

Voltage is measured by counting the number of divisions subtended by the incoming signal and multiplying that by the **VOLTS/CM** setting. For instance, if the display covers four major divisions plus three subdivisions and the **VOLTS/CM** control is set at 2, the voltage at the input is $4.6 \times 2 = 9.2$ V. To determine the actual voltage being measured, this voltage must be multiplied by the probe attenuation. If a $\times 10$ probe were being used, the voltage being measured would be 92 V.

IF the scope being used has DC input coupling available, it is possible to measure DC voltage as well as peak-to-peak voltage and instantaneous voltage.

DC VOLTAGE

As we said, it is possible to measure DC on an oscilloscope if the scope has DC input coupling. Most of the steps in setting up a scope to measure DC are similar to those for measuring any quantity; however, there are differences. Use the following procedure to measure a DC voltage.

1. Turn on the oscilloscope and set up a recurrent trace.

2. Select a deflection sensitivity (Volts/CM) that will give approximately five full divisions of deflection. If you don't know the approximate value of voltage being measured, do as you would with a meter and select the highest value of volts-per-division. This is not to protect the scope. It just keeps the trace on the screen so you can find it. Of course, it is possible to damage the scope with excessive voltage, but this has nothing to do with the sensitivity setting. Make sure that the **VARIABLE** control is in the **CAL** position.

3. Next, establish a reference from which to make the DC measurement. If your scope has a ground **(GND)** position on the input coupling switch, select that. If you have no **GND** position, select **DC** and connect the probe to ground. If you do not have a **DC** position on your scope, you can't measure DC with it. If the voltage to be measured is a positive voltage, use the **VERT/POS** control to position the trace directly on one of the major horizontal grid lines near the bottom of the screen. If the voltage is negative, position the trace near the top of the screen. If the polarity is unknown, just position the trace in the middle of the screen.

4. Using the proper probe, connect the voltage to be measured to the vertical input. If you don't know what probe to use, use a ×10. This will protect the scope from excessive input voltage. If you have the input coupling switch in the **GND** position, now is a good time to move it to the **DC** position.

5. Adjust the **VOLTS/CM** control until the trace moves several divisions from the reference trace. If necessary, change from a ×10 probe to a ×1 probe. A positive voltage will cause deflection toward the top of the screen, and a negative voltage will cause deflection toward the bottom of the screen.

6. Now, interpret the display by counting the number of major and minor divisions subtended by the input. Before going further, it is necessary to convert the number of minor divisions into tenths of major divisions. On Heath scopes and most others, this is easy, as each major division is divided into five minor divisions. Therefore, each minor division will equal 0.2 of a major division.

The unknown voltage can now be determined by multiplying the number of divisions of deflection times the **VOLTS/CM** times the probe attenuation. For instance, if the total deflection is 4.4 cm, the **VOLTS/CM** is set at 0.5, and a ×10 probe is being used, the voltage being measured (V_{in}) can be determined by:

$$V_{in} = 4.4 \times 0.5 \times 10 = 22 \text{ V}$$

Thus, it is quite possible to measure DC voltage with most oscilloscopes. A good meter will do a better job, but the scope can do it.

PEAK-TO-PEAK VOLTAGE

One of the simplest of all voltage measurements to make with the oscilloscope is the peak-to-peak amplitude of an AC voltage. The basic steps required are similar to those for making any type of measurement on an oscilloscope. They are as follows:

1. Turn on the oscilloscope and obtain a free-running trace.

2. Using the proper probe, connect the signal in question to the vertical input. It is usually best to use a ×10 probe unless the signal amplitude is very small.

3. Select the **VOLTS/CM** setting that gives a deflection of about 4 to 6 cm on the screen. Be sure that the **VARIABLE** control is in the **CAL** position. There is nothing magic about the height of the display. Anything that stays on the screen will do.

4. Choose the proper trigger source, coupling, slope, and mode. That usually means **INT, AC,** "**+**", and **AUTO**, respectively. On some older scopes, **AUTO** means a free-running trace that you can't trigger. In this case, use the **NORMAL** mode.

5. Adjust the trigger level for a stable display. You'll find out here which mode you should be in. If you can't get a stable display in **AUTO,** switch to **NORMAL**.

6. Set the **TIME/CM** and the **SWEEP/VAR** controls for a proper display. Unless you are measuring time or frequency, there is no need to have the **SWEEP/VAR** in the **CAL** position. The important thing is that one or more complete waveforms appear on the scope.

7. Using the **VERT POS** control, position the presentation so the bottom of the waveform rests on one of the major horizontal lines. Next, use the **HORIZ POS** control to position one of the peaks exactly on the center vertical line. Figure 58 shows how the presentation might appear.

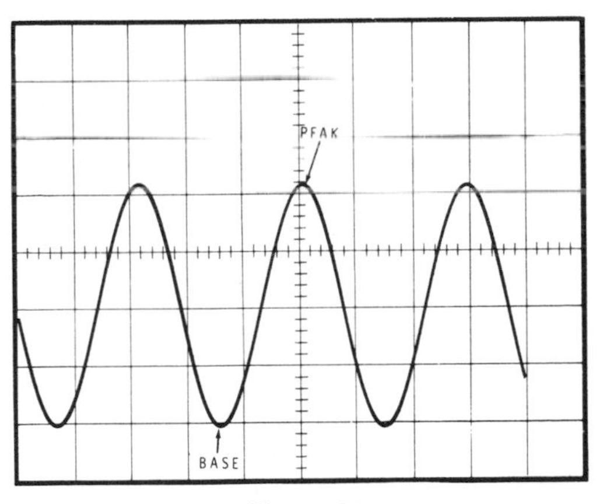

Figure 58

8. Last, interpret the measurement. In Figure 58, if we assume a $\times 10$ probe and a **VOLTS/CM** setting of 0.2, the voltage being measured is $4.4 \times 0.2 \times 10 = 8.8$ V P-P.

If the waveform being measured is a sine wave, the RMS or average value can be calculated by the following formula:

$$\text{RMS} = 0.354 \text{ peak-to-peak}$$

and

$$\text{Average} = 0.318 \text{ peak-to-peak}$$

If the waveform is a square wave, the RMS and average values are the same. The average value may be found by multiplying the peak value times the duty cycle. Duty cycle (DC) is a ratio of pulse width (PW) to pulse-repetition-time (PRT).

Figure 59 shows how a typical square wave might look. Assume that the peak voltage is 10 V, the PW is 10 μS and the PRT is 30 μS. In other words, the voltage is present for 10 μS and absent for 20 μS. The duty cycle is equal to:

$$\text{DC} = \text{PW/PRT}$$

$$= 10/30$$

$$= 0.333$$

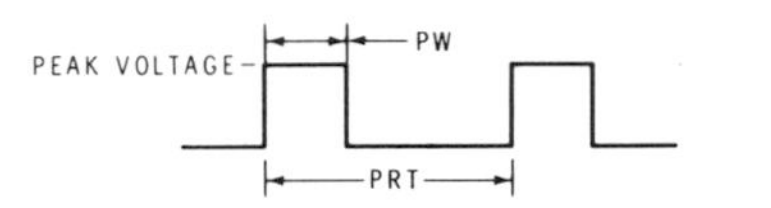

Figure 59

The average or RMS value may then be found by:

$$\text{Average} = \text{PP} \times \text{DC}$$

$$= 10 \times 0.333$$

$$= 3.33 \text{ V}$$

If the waveform is other than a sine wave or a square wave, it is very difficult to find the average or RMS value. It is seldom necessary, and most troubleshooting charts show a pictorial of the waveform with peak values given.

INSTANTANEOUS VOLTAGE

The instantaneous value is the level of voltage at any given instant of time. We could also say that it is the value of any given point on a waveform. To measure instantaneous voltage, your scope must be capable of measuring DC voltage.

The first five steps in this measurement are the same as for DC voltage, so they will not be covered in detail here.

1. Set up a recurrent trace.

2. Select the proper deflection sensitivity.

3. Establish a reference.

4. Connect the signal to the input.

5. Adjust the **VOLT/CM** control for proper amplitude.

6. Set the triggering controls for a stable display.

7. Adjust the **TIME/CM** and **SWEEP/VAR** controls for the proper display. Figure 60 shows how such a display might appear.

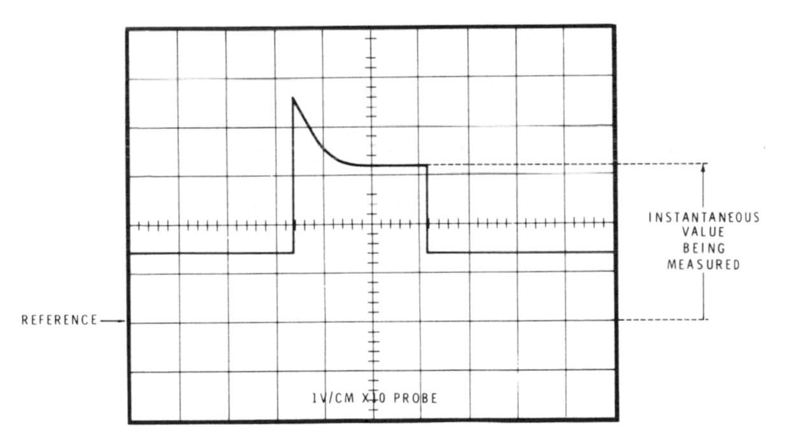

Figure 60

8. Count the number of divisions for the reference line to the instantaneous value being measured. In Figure 60, we are measuring to the plateau of the waveform. However, any other point on the waveform could be selected.

9. Multiply the number of divisions times the **VOLTS/CM** times the probe attenuation.

   For example, assume that the waveform in Figure 60 is being measured with **VOLTS/CM** setting of 1 V/cm and a ×10 probe is being used. To find the instantaneous value, use vertical deflection × V/cm × probe attenuation = instantaneous voltage. Substituting the proper values, we have 3.2 × 1 × 10 = 32 V. Thus, the instantaneous value is 32 volts.

You now have the three basic voltage measurements that can be made with an oscilloscope. There are many applications of these measurements which you will use; however, they are all based on one or more of these techniques.

Time Duration

Before we measure time with the oscilloscope, we will review the time relationships of some waveforms. The first and perhaps the most common is the sine wave as shown in Figure 61. We see here that a cycle is complete when the voltage or current has gone from zero through a positive alternation and a negative alternation back to zero. The time it takes to complete one cycle is the period of the waveform. This period or time can be measured with the oscilloscope.

In your earlier lessons, you learned that frequency = $\frac{1}{\text{time}}$. Therefore, if you know the time, it is a simple mater to compute the frequency.

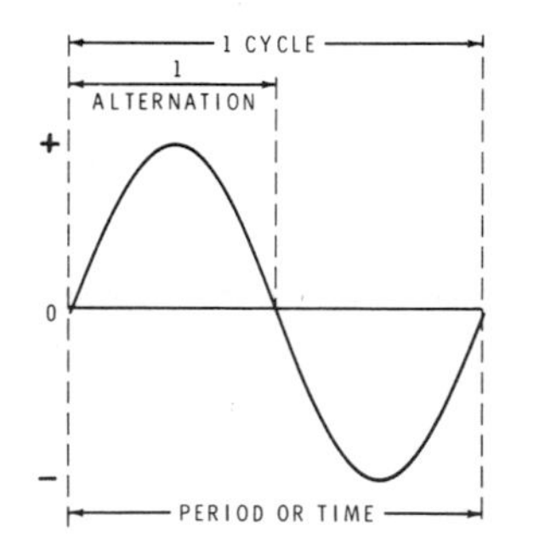

Figure 61

To measure the time or period of a sine wave, use the following procedure:

1. Set up the scope using the procedures mentioned earlier.

2. Connect the signal to the vertical input and adjust the controls for a stable display.

3. Be sure that the **SWEEP/VAR** control is in the **CAL** position.

4. Set the **TIME/CM** control for more than one but less than three – if possible – complete waveforms.

5. Using the vertical position control, position the waveform so it is approximately bisected by the horizontal centerline.

6. Using the horizontal position control, position the waveform so it crosses the horizontal centerline at one of the major vertical lines. The presentation should be similar to that shown in Figure 62.

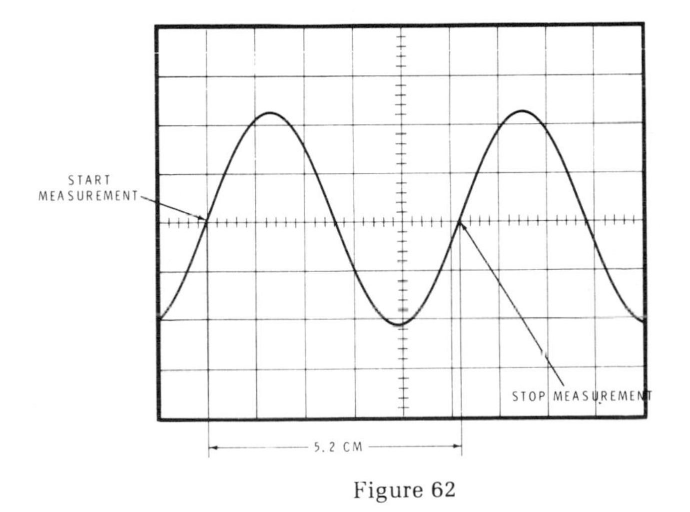

Figure 62

7. Count the number of divisions between the first crossing of the horizontal centerline and the second crossing in the same direction. In Figure 62 the distance is 5.2 cm.

8. Multiply the number of divisions times the setting of the **TIME/ CM** control.

   For example, assume that the TIME/CM control is set at 20 μS/Div and the number of divisions is 5.2 as indicated above. The time or period of one cycle will be:

$$\text{time} = \text{number of divisions} \times \text{sweep setting}$$

or

$$t = 5.2 \times 20\ \mu\text{Sec.}$$

$$t = 104\ \mu\text{Sec.}$$

Frequency may be found by:

$$f = \frac{1}{t}$$

$$= \frac{1}{104\ \mu\text{Sec.}}$$

$$= 9.6\ \text{kHz}$$

We have consistently used cm as a unit of distance when measuring both voltage and frequency because it is one of the most common units. You should remember, however, that any unit of measure can be used. The only thing that matters is the calibration in units/division.

Pulse Measurements

A pulse is a voltage that rises sharply from some value, usually zero, to another value, remains at that value for a period of time, then returns to the original value. If this excursion is repeated periodically, a pulse string exists. In today's world of computers and other digital electronics, the pulse is one of the most common waveforms requiring measurement. The oscilloscope is probably the best instrument for measuring pulse waveforms. However, before we discuss pulse measurement, we will briefly review the characteristics of the pulse waveform. Figure 63 shows the various times involved in a pulse waveform and should be referred to during this explanation.

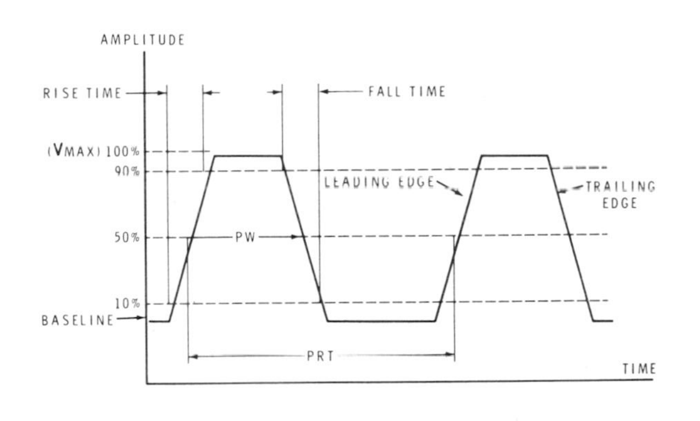

Figure 63

Each pulse will have a certain amplitude or voltage level (V_{max}). Since nothing can happen instantaneously, some time is required for the voltage to rise from the baseline voltage to V_{max}. This is known as rise time (Tr) and is defined as the time required for the voltage to change from 10% of V_{max} to 90% of V_{max}. The 10% and 90% figures have been selected because most pulses have a certain amount of distortion at the base line and the peak, which makes it difficult to obtain a precise measurement at those points. Figure 64 shows some of the more common types of distortion. This distortion must be compensated for when making pulse measurements.

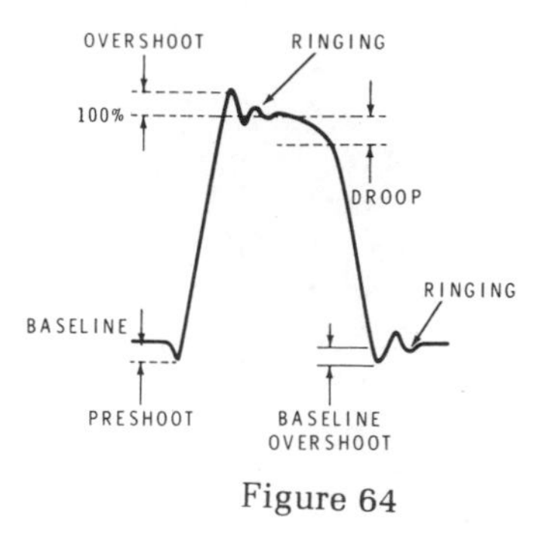

Figure 64

Once the pulse has risen to its maximum value, it will remain there for a period of time, then rapidly drop to the baseline voltage. The time required for a pulse to decrease from 90% of V_{max} to 10% of V_{max} is known as fall time (Tf) as shown in Figure 63.

The duration of the pulse, or how long it remains above the 50% point is known as the pulse width (PW). Thus, pulse width is measured at the 50% point on the waveform.

The time from the 50% point on the leading edge of one pulse to the 50% point on the leading edge of the next pulse is called the pulse recurrence time (PRT). The PRT of a pulse waveform is analogous to the period of a sine wave. Thus, the number of pulses-per-second, which is known as the pulse recurrence frequency (PRF), can be calculated by $PRF = \frac{1}{PRT}$.

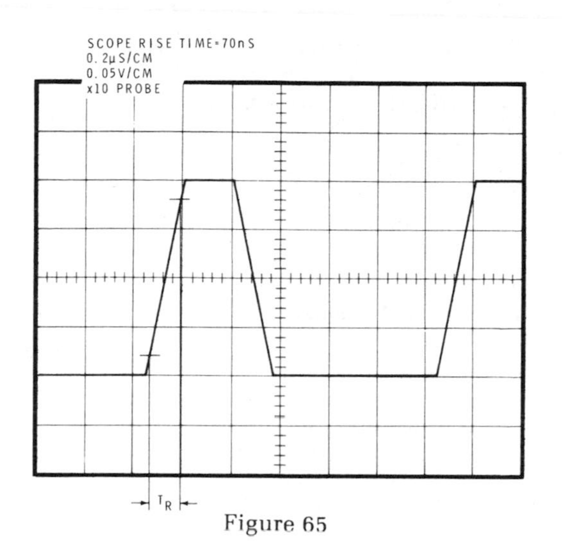

Figure 65

For example, assume that the CRT display and the control settings are as indicated in Figure 65. The waveform is positioned properly for measuring PW and PRT.

$$PW = 1.8\ CM \times 0.2\ \mu S/CM$$

$$= 0.36\ \mu S$$

$$PRT = 6\ CM \times 0.2\ \mu S/CM$$

$$= 1.2\ \mu S$$

$$PRF = \frac{1}{1.2\ \mu S}$$

$$= 833\ kHz$$

The rise time displayed on the CRT (T_{rd}) is approximately 0.6 CM.

$$T_{rd} = 0.6 \text{ CM} \times 0.2\ \mu\text{S/CM}$$

$$= 0.12\ \mu\text{S}$$

$$= 120 \text{ nS}$$

This is the displayed rise time and does not accurately represent the rise time of the square wave being measured. The rise time of the scope is a determining factor in the accuracy of the measurement. The relationship may be shown by the following formula:

$$T_{rd} = \sqrt{T_{rs}^2 + T_{rw}^2}$$

Where T_{rd} equals the displayed rise time, T_{rs} equals the rise time of the scope and T_{rw} is the rise time of the waveform being measured. Thus, to find the rise time of the waveform being measured use:

$$T_{rw} = \sqrt{T_{rd}^2 - T_{rs}^2}$$

which in this case is:

$$T_{rw} = \sqrt{120 \text{ ns}^2 - 70 \text{ ns}^2}$$

$$= 97 \quad \text{nS}$$

If the displayed rise time is at least three times the scope rise time, the rise time of the scope may be ignored and the accuracy of the measurement will be within 5%. If the displayed rise time is five times the scope rise time, the accuracy will be within 2%. Of course, the above inaccuracies are in addition to any inaccuracies in the sweep timing circuits.

When the displayed rise time approaches the rise time of the scope, it is impossible to obtain an accurate measurement. Therefore, to measure fast rise time, the scope must have a very fast rise time also. However, the very fast rise time may not be required to measure other waveform parameters such as PW and PRT.

X-Y Measurements

We mentioned earlier in our discussion of the basic oscilloscope block diagram (Figure 19, Page 3-30) that there were two possible inputs to the horizontal amplifier, one from the sweep generator and the other from an external source. At this time, we will study the action of the oscilloscope with other than a sweep voltage applied to the horizontal deflection plates.

Assume that the inputs to the vertical and horizontal amplifiers are two sine waves of the same frequency and amplitude. Also, assume that the two signals are in phase. Figure 66 shows how the presentation will appear on the CRT. AT T_0, both inputs are at zero volts. Therefore, no deflection occurs and the electron beam will be positioned at the center of the CRT. At T_1, both signals are positive and the beam is deflected upward and to the right, placing the spot in the upper right corner of the CRT. At T_2, the beam has returned to the center and at T_3, both signals are negative, deflecting the beam to the lower left corner of the CRT. At T_4, both signals have returned to zero and the beam is back in the center. Thus, we can see that the result of two sine waves in phase is a straight line. If the two signals are of equal amplitude, the line on the CRT will be at a 45° angle as shown. However, if the vertical signal is larger in amplitude, the angle will be greater than 45°. If the horizontal signal is larger, the angle will be less than 45°.

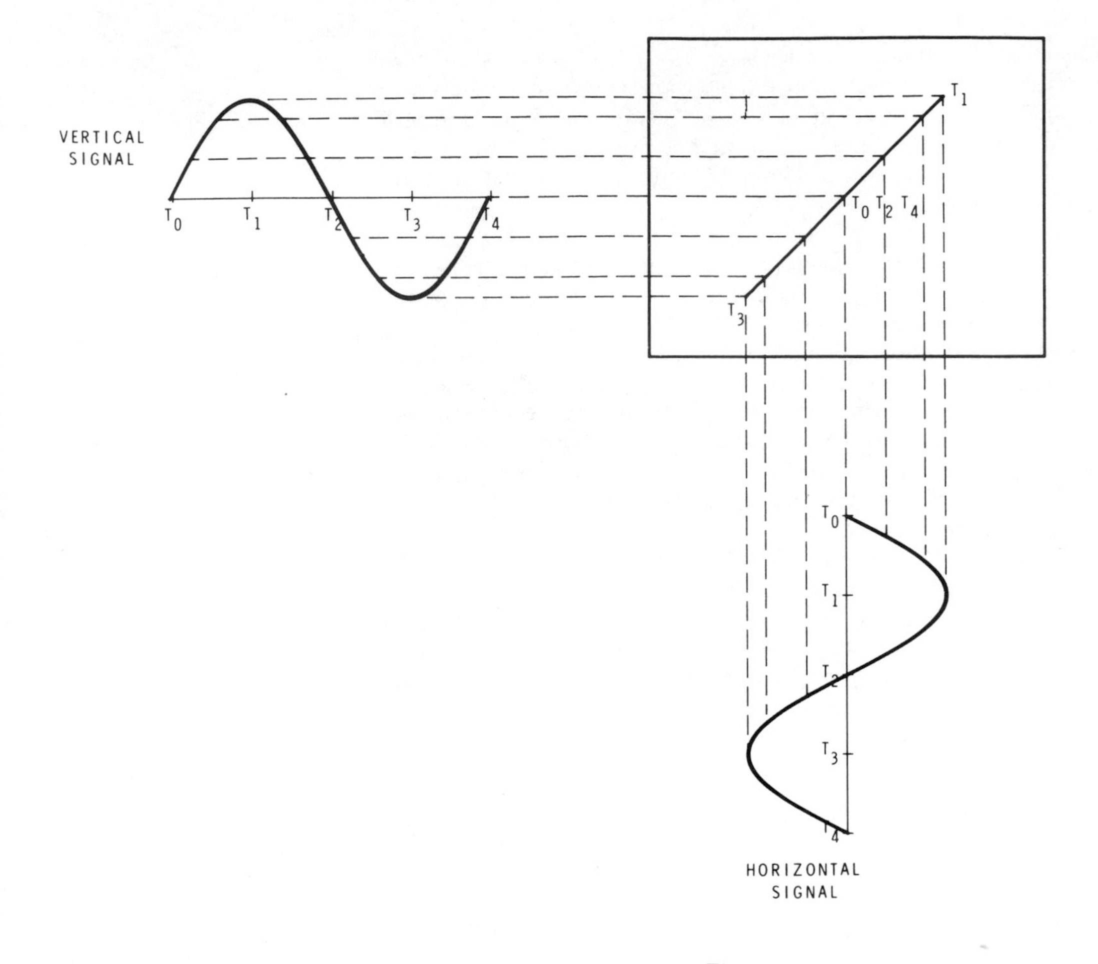
VERTICAL
SIGNAL
T_0
T_1
T_2
T_3
T_4
T_1
T_0 T_2 T_4
T_3
T_0
T_1
T_2
T_3
T_4
HORIZONTAL
SIGNAL

Figure 66

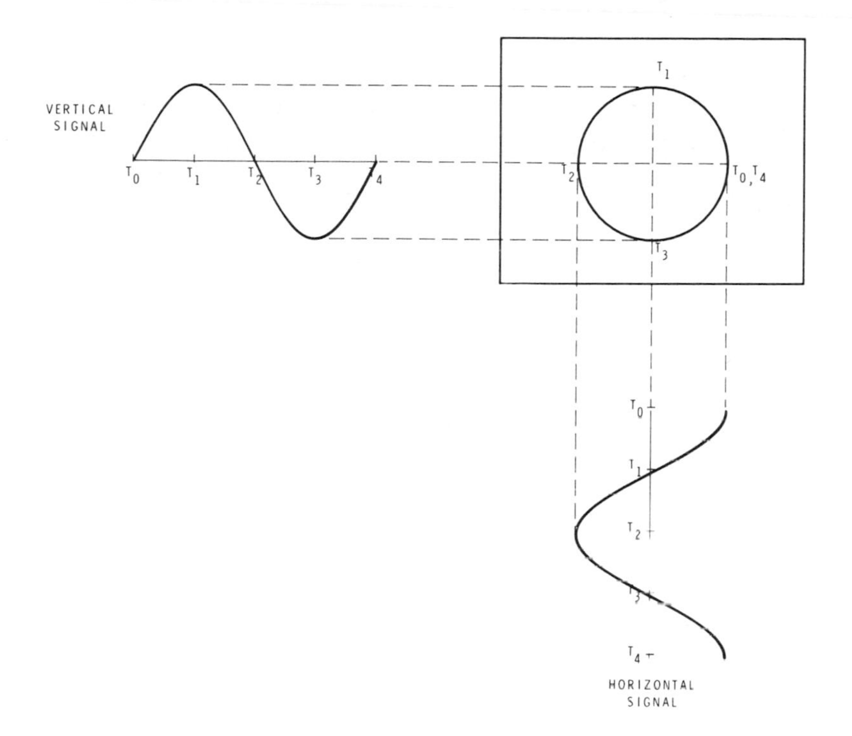

Figure 67

As long as the two inputs are in phase, the result will be a straight line. When the two signals are of the same frequency and amplitude but out of phase, the result will not be a straight line. Figure 67 shows the inputs 90° out of phase. The result, as you can see, is a circle on the CRT. If the phase difference is other than 90°, the result will be an elliptical presentation. An elliptical pattern will also be produced if the inputs are of unequal amplitude.

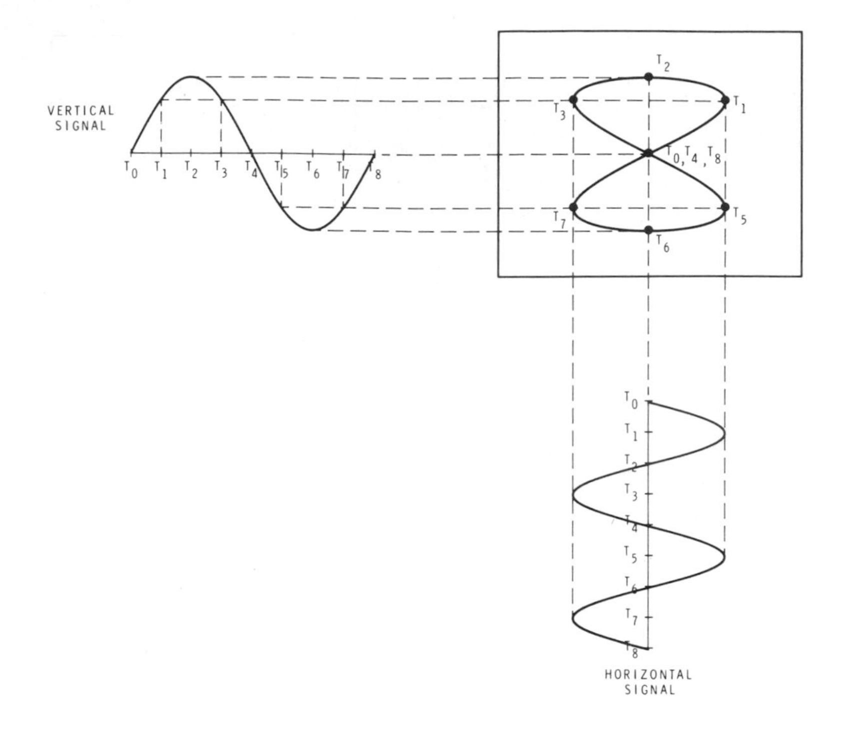

Figure 68

When the two inputs are of different frequencies, a more complicated presentation is produced. Figure 68 shows what happens when the horizontal is exactly twice the frequency of the vertical input. Notice that there is a direct relationship between the number of loops and the frequency difference. One frequency is twice the other and there are two loops. We will use this ratio later to determine an unknown frequency. The presentations just illustrated are called **Lissajous** figures. They can be used to compare the phase and frequency difference between two waveforms. The standard procedure is to apply the reference signal to the horizontal input and the unknown frequency or phase to the vertical input.

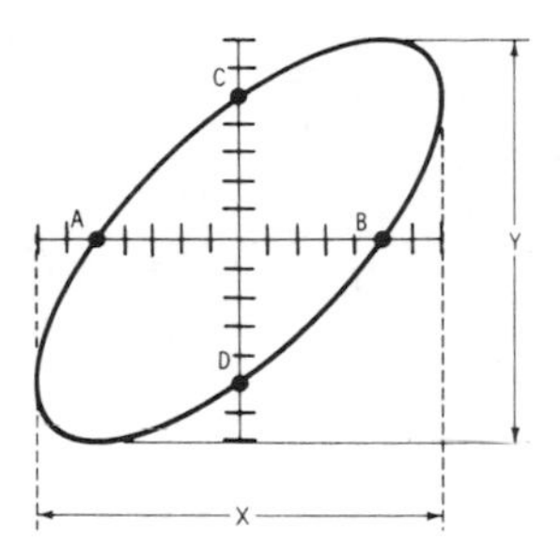

Figure 69

PHASE MEASUREMENTS

You have already learned how to determine if signals are in phase or 90° out of phase. Interim angles can be calculated from two simple measurements. See Figure 69.

$$\text{Sin} \angle \emptyset = \frac{AB}{X} = \frac{CD}{Y}$$

In Figure 69, AB equals about 10 divisions and X equals about 14 divisions. Therefore,

$$\text{Sin} \angle \emptyset = \frac{10}{14}$$

$$= 0.714$$

Looking at the table of trigonometric functions in Appendix A, we find that the angle corresponding to a sine of 0.714 is nearest to 46°. Thus, the phase difference between the vertical and horizontal inputs appears to be 46°. Actually, it may be 46°, 136°, 226°, or 316°, as this procedure does not tell us which quadrant the angle is in.

Part of the ambiguity is easily resolved. Figure 70 shows how this can be done. In this illustration, A is a 0° or 360° difference, B is 45° or 315°, C is 90° or 270°, D is 135° or 225° and E is 180°. The remaining 180° ambiguity cannot be resolved using this procedure without additional equipment.

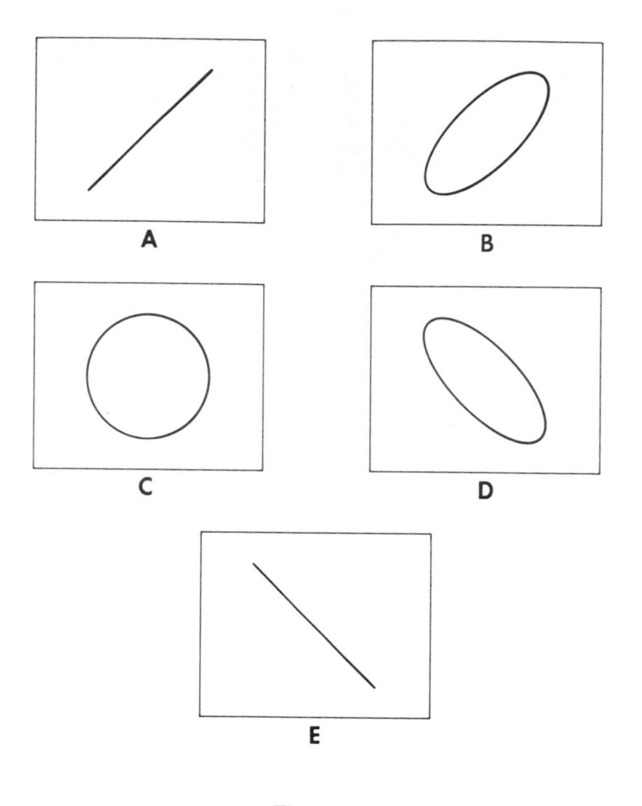

Figure 70

Phase differences as small as 1° can be detected using this procedure. However, as the difference approaches 90°, the measurement becomes more difficult. You should remember that the phase difference displayed on the CRT includes any phase shift induced by the oscilloscope amplifiers, leads, and probes. Therefore, a single signal should be fed to both inputs in parallel to detect any equipment induced difference. A good modern scope should induce no appreciable phase difference; however, if necessary, continuously variable delay lines can be connected in series with the leads to compensate for the shift.

Another potential source of error is the presence of harmonics on either of the inputs. Even a low level of harmonics can cause a significant error.

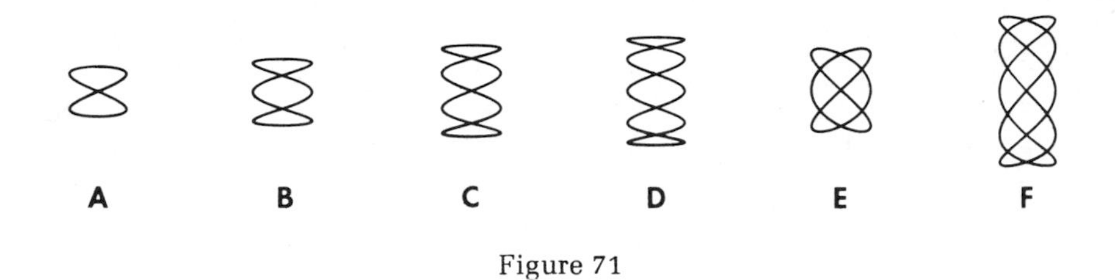

Figure 71

FREQUENCY MEASUREMENTS

The Lissajous pattern can also be used for frequency measurement. In Figure 68 we saw that a figure eight was formed on the screen by placing a frequency on the horizontal input that was twice that on the vertical input. This principle can be used to compare a known frequency with an unknown.

Figure 71 shows some typical patterns that would be obtained if the frequency on the vertical plates is less than that on the horizontal plates. If the reference frequency is applied to the horizontal plates, the unknown frequency can be found by dividing the reference frequency (f_r) by the number of horizontal loops (N_h) then multiplying by the number of vertical (N_v) loops. For instance, Figure 71C has four horizontal loops and one vertical loop. If the reference frequency is 1000 Hz, the unknown frequency will be

$$f_u = \frac{f_r}{N_h} \times N_v$$

$$= \frac{1000}{4} \times 1$$

$$= 250 \text{ Hz}$$

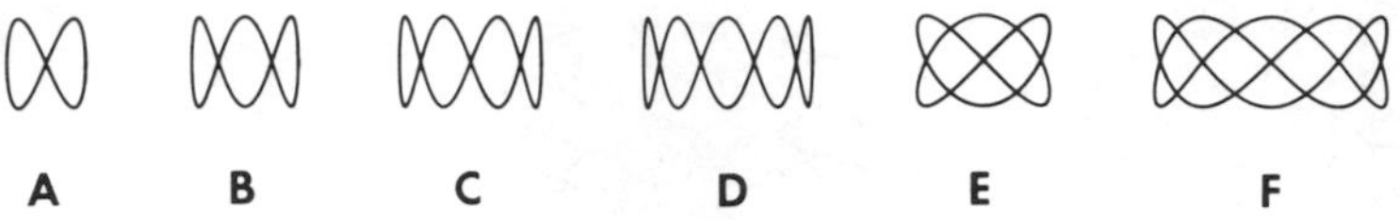

A B C D E F

Figure 72

Figure 72 shows some patterns where the reference frequency is less than the unknown. The procedures for finding the unknown frequency is the same as before.

$$f_u = \frac{f_r}{N_h} \times N_v$$

Thus, if we assume the same 1000 Hz reference and calculate the unknown frequency for Figure 72E, we have:

$$f_u = \frac{1000}{2} \times 3$$

$$= 1500 \text{ Hz}$$

Thus, we can see that it is not necessary for the unknown frequency to be an exact multiple or sub-multiple of the reference. However, there is a practical limit to the number of loops that can be accurately counted.

It may appear that frequency measurement by means of a Lissajous pattern is easy. It is not. Accurate frequency measurement requires a considerable amount of skill. In most cases, it is difficult to obtain a stationary pattern which is required for accurate measurement. However, if an accurately calibrated reference source is used, very accurate measurements can be made.

Phase measurement is not as difficult, and the oscilloscope is one of the best phase detecting devices available.

Dual Trace Measurement

As you might expect, the major application of the dual trace scope is comparing two signals. Comparisons of phase, shape, amplitude, timing, etc. can easily be made.

PHASE MEASUREMENT

When measuring phase with the dual trace scope, start by setting the scope for dual trace operation using your handbook as a reference. Feed one of the signals to the A channel and the other to the B channel. It is not necessary that the two signals be of the same amplitude, frequency, or shape. Do not, however, overdrive the vertical amplifiers.

Adjust the TIME/DIV and HORIZONTAL GAIN CONTROLS so that either one alteration or one complete cycle of the reference waveform requires exactly nine major horizontal divisions. Either channel can be used for a reference, however, the scope should be triggered off the reference waveform. It is also normal practice to use the lower frequency signal for reference when the inputs are of a different frequency. The horizontal sweep does not need to be calibrated for this measurement so the horizontal gain control can be adjusted as needed for proper display.

Figure 73 shows a typical display with one alteration equal to nine divisions. Since one alteration is equal to 180 electrical degrees, each division will equal $\frac{180}{9}$ or 20 degrees. Thus, in the waveform shown, waveform A leads wave B by three divisions or 60 degrees. The dual trace method of phase measurement is very accurate. However, you must be sure that both waveform A and waveform B are vertically centered.

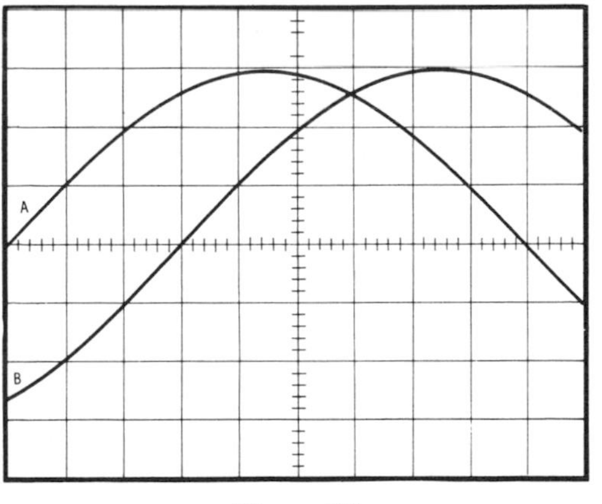

Figure 73

TIME DIFFERENCE

Time difference measurements are a snap with the dual trace scope. Set up is similar to that used for phase measurements. However, for time measurement, the horizontal time base control must be in the CALIBRATE position. Adjust the TIME/DIV until both pulses appear on the CRT as shown in Figure 74. The vertical control should be adjusted so the waveforms are bisected by the major horizontal line as shown. The time difference between the pulses can then be found by counting the number of divisions from the leading edge of the first pulse to the leading edge of the next pulse and multiplying that times the μS/DIV setting. For instance, in Figure 74, it is three divisions from the leading edge of A to the leading edge of B. Three divisions times 2 μS per division equals a delay of 6 μS.

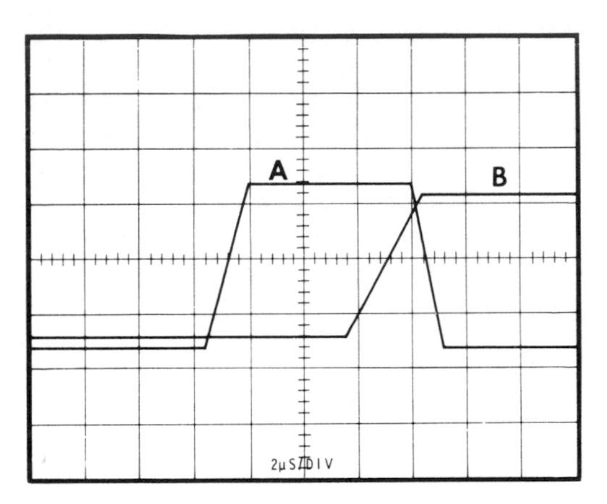

Figure 74

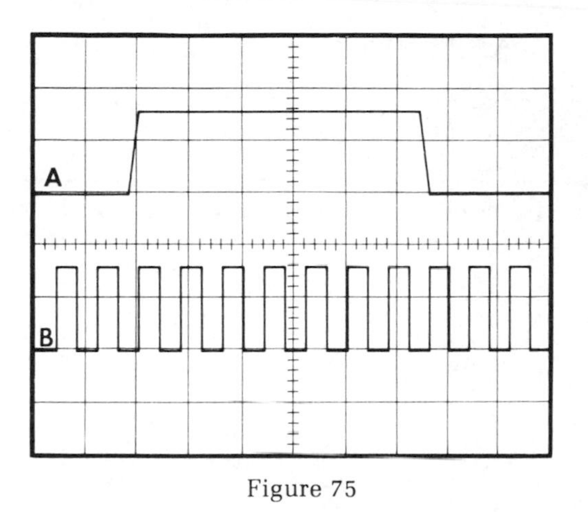

Figure 75

Another useful timing measurement which can be made with a dual trace scope is determining the number of pulses that occur during a given gate length. Simply adjust the horizontal controls so that the gate is expanded as much as possible, without running off the screen. There is no need to maintain time calibration if all we are interested in is relative numbers. However, if we wish to know the gate length or the pulse width, calibration must be maintained. Set the vertical controls for any comfortable viewing amplitude. The display might appear like the one in Figure 75.

Once the proper display has been established, count the number of pulses that occur from the leading edge of the gate (waveform A) to the trailing edge of the gate. In Figure 75, seven pulses occur during gate time.

OTHER USES

There are many more ways that the dual trace scope can make your troubleshooting easier. The output of an amplifier can be compared to the input. Channel A of a stereo amplifier can be compared with channel B. Simultaneous comparisons of amplitude, phase, frequency, time, distortion, and other factors can be made in this manner. Of course, many of these factors can be checked with a single trace scope, but the dual trace does a better job. In many digital applications, the dual trace scope becomes essential.

APPENDIX A
Table of Trigonometric Functions

Degrees	Sine	Cosine	Tangent	Degrees	Sine	Cosine	Tangent
0	0.000	1.000	0.000				
1	0.017	1.000	0.017	46	0.719	0.695	1.036
2	0.035	0.999	0.035	47	0.731	0.682	1.072
3	0.052	0.999	0.052	48	0.743	0.669	1.111
4	0.070	0.998	0.070	49	0.755	0.656	1.150
5	0.087	0.996	0.087	50	0.766	0.643	1.192
6	0.105	0.995	0.105	51	0.777	0.629	1.235
7	0.122	0.993	0.123	52	0.788	0.616	1.280
8	0.139	0.990	0.141	53	0.799	0.602	1.327
9	0.156	0.988	0.158	54	0.809	0.588	1.376
10	0.174	0.985	0.176	55	0.819	0.574	1.428
11	0.191	0.982	0.194	56	0.829	0.559	1.483
12	0.208	0.978	0.213	57	0.839	0.545	1.540
13	0.225	0.974	0.231	58	0.848	0.530	1.600
14	0.242	0.970	0.249	59	0.857	0.515	1.664
15	0.259	0.966	0.268	60	0.866	0.500	1.732
16	0.276	0.961	0.287	61	0.875	0.485	1.804
17	0.292	0.956	0.306	62	0.883	0.469	1.881
18	0.309	0.951	0.325	63	0.891	0.454	1.963
19	0.326	0.946	0.344	64	0.899	0.438	2.050
20	0.342	0.940	0.364	65	0.906	0.423	2.145
21	0.358	0.934	0.384	66	0.914	0.407	2.246
22	0.375	0.927	0.404	67	0.921	0.391	2.356
23	0.391	0.921	0.424	68	0.927	0.375	2.475
24	0.407	0.914	0.445	69	0.934	0.358	2.605
25	0.423	0.906	0.466	70	0.940	0.342	2.748
26	0.438	0.899	0.488	71	0.946	0.326	2.904
27	0.454	0.891	0.510	72	0.951	0.309	3.078
28	0.469	0.883	0.532	73	0.956	0.292	3.271
29	0.485	0.875	0.554	74	0.961	0.276	3.487
30	0.500	0.866	0.577	75	0.966	0.259	3.732
31	0.515	0.857	0.601	76	0.970	0.242	4.011
32	0.530	0.848	0.625	77	0.974	0.225	4.332
33	0.545	0.839	0.649	78	0.978	0.208	4.705
34	0.559	0.829	0.675	79	0.982	0.191	5.145
35	0.574	0.819	0.700	80	0.985	0.174	5.671
36	0.588	0.809	0.727	81	0.988	0.156	6.314
37	0.602	0.799	0.754	82	0.990	0.139	7.115
38	0.616	0.788	0.781	83	0.993	0.122	8.144
39	0.629	0.777	0.810	84	0.995	0.105	9.514
40	0.643	0.766	0.839	85	0.996	0.087	11.43
41	0.656	0.755	0.869	86	0.998	0.070	14.30
42	0.669	0.743	0.900	87	0.999	0.052	19.08
43	0.682	0.731	0.933	88	0.999	0.035	28.61
44	0.695	0.719	0.966	89	1.000	0.017	57.29
45	0.707	0.707	1.000	90	1.000	0.000	

Unit 4

FREQUENCY MEASUREMENT

INTRODUCTION

Why should you be interested in measuring frequency? Because without the ability to accurately measure frequency there could be no space program, radio communications, electronic home entertainment systems, ship and aircraft all-weather navigation, or a multitude of other things we take for granted. Anyone engaged in the design, manufacture, servicing and, in some cases, operation of today's electronic equipment should be aware of the techniques and equipment used to obtain accurate frequency measurements.

The following pages will show you that two distinct concepts are used in the instruments you can use to measure frequency. One type detects frequency either by resonating with the measured signal, or by using a mixing action. The second type of instrument compares the number of events (or cycles) being measured to a very accurate time interval. Each of these devices fills a specific need, so you will study several different types.

In the first part of this Unit, we will examine the vibrating reed meter, the wavemeter, and the dip meter to demonstrate the use of resonance in frequency measurement. Then the heterodyne meter and the transfer oscillator will show you how frequency is measured by a mixing action.

The remainder of this Unit will be devoted to the electronic counter, which compares the input signal to a standard interval of time.

In this Unit we will discuss:

- the nature of time and frequency;
- the standards upon which all frequency measurements are based;
- the types of test equipment available in the field;
- the capabilities and limitations of each type of test equipment;
- and the theory and operation of this equipment, as well as its expected accuracy and calibration requirements.

BASIC CONCEPTS

Before we begin discussing the various methods of measuring frequency, let's take a few moments to review some basic concepts.

Time and Frequency

The fundamental unit of frequency measurement, as you know, is the hertz, abbreviated Hz. This is the number of times an alternating current goes through its complete cycle in one second. These "cycles" can come in many different shapes, such as sinusoidal (sine waves), square, triangular, or sawtooth waves.

In this section we will use a wide range of frequency values. You probably also recall that a series of prefixes have been assigned to various magnitudes raised to some power of ten (10^x). You should already be familiar with most of these, but Figure 1 will provide you with a quick review and a convenient reference source.

	Prefix	Symbol	Definition	10^x	Example
Fractional	PICO	p	one trillionth	10^{-12}	picofarad = pF = $\frac{1}{1\ 000\ 000\ 000\ 000}$ of one farad
	NANO	n	one billionth	10^{-9}	nanosecond = ns = $\frac{1}{1\ 000\ 000\ 000}$ of one second
	MICRO	μ	one millionth	10^{-6}	microhenry = μH = $\frac{1}{1\ 000\ 000}$ of one henry
	MILLI	m	one thousandth	10^{-3}	milliamp = mA = $\frac{1}{1000}$ of one ampere
Greater than one	KILO	k	one thousand	10^{3}	kilovolt = KV = 1000 volts
	MEGA	M	one million	10^{6}	megohm = MΩ = 1 000 000 ohms
	GIGA	G	one billion	10^{9}	gigahertz = GHz = 1 000 000 000 hertz
	TERA	T	one trillion	10^{12}	terahertz = THz = 1 000 000 000 000 hertz

Figure 1

To further simplify matters, groups of frequencies, or "bands," have been established to identify different frequency ranges. The *audio frequencies* (AF) are generally considered to be from 15 Hz to 20 kHz. Frequencies from 20 kHz to 300 GHz are loosely defined as *radio frequencies* (RF). As the electronics field has grown and expanded, various terms have been adopted to identify certain frequency bands within the RF spectrum. Figure 2 lists these bands.

The right-hand column in Figure 2 lists the wavelength of the signals in these RF bands. In certain applications, wavelength calculations are more convenient than frequency. This is especially true when you want to determine antenna lengths. Early radio receivers had tuning dials marked in wavelengths. Radio amateurs still refer to their frequency allocations in terms of the 2, 10, or 20-meter bands. It is also more convenient to refer to a radar system as operating at 3.2 cm rather than 9.375 GHz.

You can make an approximate conversion of wavelength to frequency by using the simple equation

$$\lambda = \frac{300,000,000}{F}$$

where λ (Lambda, the symbol for wavelength) is the wavelength in meters, 300,000,000 meters per second is the approximate speed of light, and F is the frequency in hertz. From this relationship, you can see that the 10 meter band is about 30 MHz, and that 100 kHz on the AM broadcast band is 300 meters.

Band	Frequency	Wavelength (λ) in meters
VLF (Very Low Frequency)	3 kHz to 30 kHz	100 km to 10 km
LF (Low Frequency)	30 kHz to 300 kHz	10 km to 1 km
MF (Medium Frequency)	300 kHz to 3 MHz	1 km to 100 m
HF (High Frequency)	3 MHz to 30 MHz	100 m to 10 m
VHF (Very High Frequency)	30 MHz to 300 MHz	10 m to 1 m
UHF (Ultra High Frequency)	300 MHz to 3 GHz	1 m to 10 cm
SHF (Super High Frequency)	3 GHz to 30 GHz	10 cm to 1 cm
EHF (Extremely High Frequency)	30 GHz to 300 GHz	1 cm to 10 mm

Figure 2
Frequency Bands.

In certain applications, it is also convenient to express frequency as a function of time. To measure frequency with an oscilloscope, we measure the elapsed time between two successive points on a recurring waveshape. Next, we compute the frequency by the formula $F = \frac{1}{t}$, where F is the frequency in hertz and t is the elapsed time in seconds. As you will learn later in this unit, elapsed time measurements can also provide more precise readings at low frequencies by increasing the resolution.

One final factor to consider is the manner in which manufacturers state the frequency accuracy or stability of their equipment. You may see these accuracy figures stated as, for example, 3 parts per million, .001%, or 5×10^{-9}. Exactly what do these values mean and how can you compare them? Figure 3 shows the relationship between the various terms.

PARTS PER			PERCENT
1 PART PER HUNDRED	$= 1 \times 10^{-2} =$	$\frac{1}{100}$	1.0%
1 PART PER THOUSAND	$= 1 \times 10^{-3} =$	$\frac{1}{1{,}000}$	0.1%
1 PART PER 10 THOUSAND	$= 1 \times 10^{-4} =$	$\frac{1}{10{,}000}$	0.01%
1 PART PER 100 THOUSAND	$= 1 \times 10^{-5} =$	$\frac{1}{100{,}000}$	0.001%
1 PART PER MILLION	$= 1 \times 10^{-6} =$	$\frac{1}{1{,}000{,}000}$	0.0001%
1 PART PER 10 MILLION	$= 1 \times 10^{-7} =$	$\frac{1}{10{,}000{,}000}$	0.00001%
1 PART PER 100 MILLION	$= 1 \times 10^{-8} =$	$\frac{1}{100{,}000{,}000}$	0.000001%
1 PART PER BILLION	$= 1 \times 10^{-9} =$	$\frac{1}{1{,}000{,}000{,}000}$	0.0000001%
1 PART PER 10 BILLION	$= 1 \times 10^{-10} =$	$\frac{1}{10{,}000{,}000{,}000}$	0.00000001%
1 PART PER 100 BILLION	$= 1 \times 10^{-11} =$	$\frac{1}{100{,}000{,}000{,}000}$	0.000000001%
1 PART PER 1,000 BILLION (1 PART PER TRILLION)	$= 1 \times 10^{-12} =$	$\frac{1}{1{,}000{,}000{,}000{,}000}$	0.0000000001%
1 PART PER 10,000 BILLION	$= 1 \times 10^{-13} =$	$\frac{1}{10{,}000{,}000{,}000{,}000}$	0.00000000001%

Figure 3

For example, Figure 3 shows that one part per million or 1×10^{-6} can also be expressed as 0.0001%, but what is three parts per million? This can be converted in the same way. Three parts per million is 3×10^{-6} or 0.0003%.

Frequency Standards

Frequency is nothing more than the number of events which occur in a given span of time. If two people count the number of cars which pass a given point in one hour, they should both count the same number. Any difference between the two results would likely be caused by an inability to determine an accurate one-hour time interval. Therefore, the key to accurate frequency measurement is an accurate standard of time.

Man first marked the passage of time by the changing of the seasons. This served his purpose nicely and a tolerance of a few days was quite acceptable. But as man's knowledge increased, so did his need to accurately measure time. This need has dramatically increased since the beginning of this century. In 1920, the National Bureau of Standards (NBS) began maintaining a national time and frequency standard. At that time it was possible to determine elapsed time to within 1 part in 10^4 or $\pm$ 100 μsec.

In 1967 the "second" was redefined, by international agreement, to provide a more precise time reference. This reference is based upon the radiation period of the cesium-133 atom. By using cesium-beam oscillators under carefully controlled conditions, current NBS standards are capable of measuring time with an uncertainty of only 1 part in 10^{14}, or 0.000000000001%. Research is constantly being conducted to improve the accuracy and stability of this standard.

In addition to the National Standard, you may also hear reference made to Primary and Secondary Standards.

A primary standard is one which does not require calibration against any other reference. Until recently, the National Bureau of Standards maintained the only primary standard in the U.S., but cesium-beam frequency standards are now available commercially. However, their high cost limits their use to large calibration laboratories and manufacturers with extremely critical time and frequency requirements.

Secondary standards are oscillators which are used locally for reference and calibration purposes. These standards must be periodically calibrated against a primary standard to insure their accuracy.

One of the most popular sources of standard time and frequency information is high frequency (HF), shortwave, radio broadcasts. HF signals from stations such as WWV (Ft. Collins, Colorado), WWVH (Kauai, Hawaii), and CHU (Ottawa, Canada) are readily available and provide essentially worldwide coverage. In addition, these signals can be picked up with relatively inexpensive receiving equipment.

Types of Frequency Measuring Instruments

Many types of frequency measuring instruments have been used over the years. A large number of these have become obsolete and are seldom found in the field today. This course will include only those types of instruments you might reasonably expect to find in common use. By learning their principles of operation, you should be able to apply this knowledge to any older instruments you might encounter.

Two very common types of frequency measuring instruments, as we stated in the "Introduction," are the electronic counter and the oscilloscope. Both of these instruments compare an input signal to a standard unit of time. You will study the counter in detail beginning on Page 4-43, and the oscilloscope is described in Unit 3.

A frequency meter, by definition, is any instrument that is used to measure frequency. Therefore, do not assume that all "meters" employ a scale and pointer. Some instruments are self-indicating while others rely on external indications. Frequency meters are divided into two basic groups, *passive meters* and *active meters*.

Passive meters are not powered and do not process (amplify, divide, shape, etc.) the measured signal in any way. As a general rule, they are less expensive than active meters and require less frequent calibration. They are, however, usually less accurate than active meters.

Active meters require a power source. This source can be AC line voltage or a battery. Some types (electronic counters and heterodyne converters, for example) are able to process the incoming signal by dividing, shaping, amplifying, or attenuating it. This allows them to detect and measure a wider range of signal shapes and amplitudes.

Active meters are generally more accurate than passive instruments, but are also more expensive and require more frequent calibration. You should keep in mind that the comparisons of cost and accuracy are generalizations. A top quality passive instrument could cost much more and be far more accurate than an "economy" model active instrument.

PASSIVE FREQUENCY METERS

Passive frequency meters are not powered and do not process the input signal in any way. All of the passive frequency meters we will discuss operate on the principle of resonance. The vibrating reed meter makes use of mechanical (physical) resonance, while wave meters employ electrical resonance. Therefore, before discussing meter operation, a quick review of resonance is in order.

Resonance

As you know, electrical resonance occurs in LC circuits at a frequency where X_c is equal to X_L. What, then, is mechanical resonance?

This term refers to the physical property of a material which causes it to vibrate easily (resonate) at one specific frequency. If brought close to a source of vibration, an object will vibrate at the same frequency. When the driving frequency is the same as the object's natural, resonant frequency, the amplitude of vibration will be maximum.

All passive frequency meters react, in some manner, when the frequency of the applied signal is equal to the resonant frequency of the meter.

Vibrating Reed Meters

The vibrating reed meter is one of the simplest types of frequency meters. It is used to monitor line power frequency and is commonly found on motor driven power units such as mobile generators. It is designed to operate at a specific fixed frequency and is commonly available in 50 Hz, 60 Hz, 400 Hz and 1000 Hz models.

This meter consists of a series of thin metal strips. These strips (or reeds) are all of equal length, but each has a slightly different weight attached to one end. The amount of weight placed on the reed mechanically "tunes" it (makes it resonant) to a specific frequency.

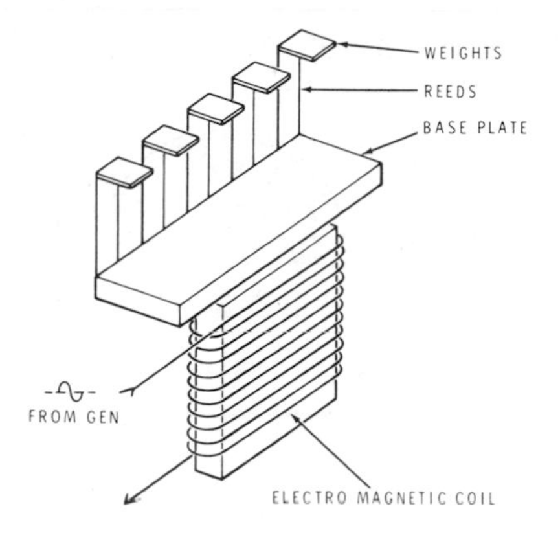

Figure 4
Vibrating reed frequency meter – internal construction.

These reeds are connected, at their unweighted ends, to a common baseplate, as in Figure 4. The reed/baseplate assembly is mounted close to an electromagnetic coil. Alternating current from the generator output is fed through the coil, causing the baseplate to vibrate at the generator frequency. This vibration is felt by the attached reeds. All of the reeds will vibrate slightly, but the reed which is tuned to that frequency will display a large amplitude vibration as shown in Figure 5A.

If the induced frequency is midway between the tuned frequency of two adjacent reeds, both reeds will vibrate at a reduced but equal amplitude (Figure 5B). If the induced frequency lies between two adjacent reeds but not midway, both reeds will vibrate, but the reed closest to the induced frequency will vibrate at a greater amplitude (Figure 5C). Figure 5D shows how the meter appears with no frequency applied. As you might guess, mechanical inertia limits the use of these meters to very low frequencies.

The weighted ends of the reeds are the only portions of the mechanism visible from the outside of the meter. These are usually painted white and displayed against a dark background, making them appear as a series of white dots. This type of meter is usually mounted permanently on a control panel. By observing one of these meters while adjusting the motor speed, the frequency of a generator can be set to within about 1/4 Hz of its desired value.

A 60 Hz INPUT

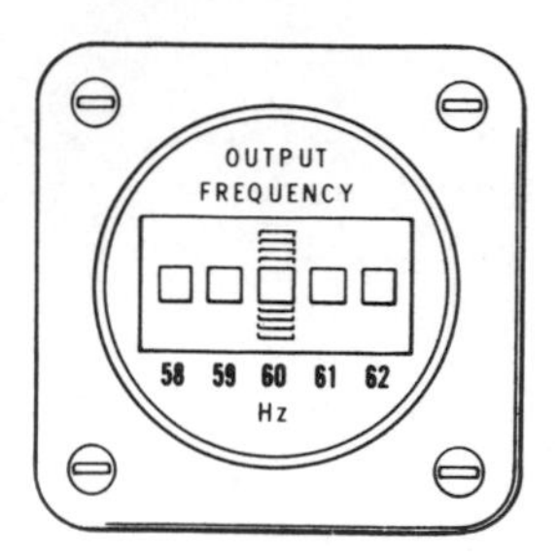

B 59-1/2 Hz INPUT

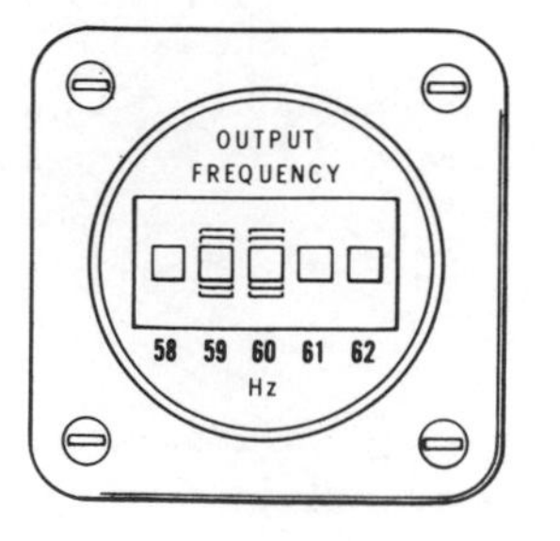

C 59-3/4 Hz INPUT

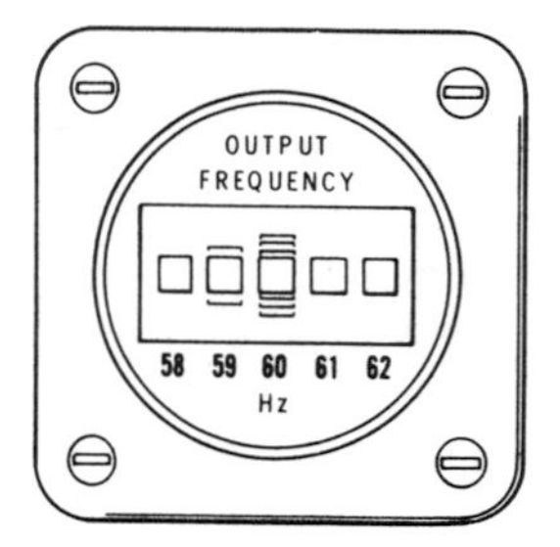

D NO FREQUENCY

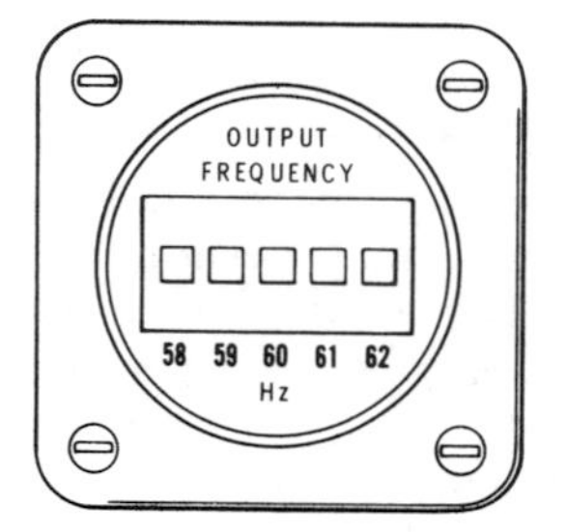

Figure 5

Wavemeters

Resonant wavemeters are portable and are tunable by the operator. This makes them much more versatile than vibrating reed meters. They operate on the principle of electrical resonance, and therefore are not limited in frequency response by mechanical inertia as is the vibrating reed meter.

Wavemeters absorb a portion of the output power from the circuit being tested. This is done by inductively coupling an LC tank circuit, in the wavemeter, to the circuit being tested. The wavemeter tank circuit is then tuned to resonance and the frequency is read from the wavemeter tuning dial. Before going deeper into wavemeter operation, let's review a few basic principles.

As you remember from your studies of AC electronics, a coil has no reactance (X_L) to DC current. As the frequency increases, so does the X_L. This is shown by the formula

$$X_L = 2\pi \times \text{frequency} \times \text{inductance} = 2\pi FL$$

Conversely, a capacitor shows infinite reactance (X_C) to DC current and, as frequency increases, reactance decreases. This is shown by the formula

$$X_C = \frac{1}{2\pi \times \text{frequency} \times \text{capacitance}} = \frac{1}{2\pi FC}$$

Figure 6A shows a simple circuit, containing a coil and a capacitor, coupled to a source of AC voltage. Let's assume the AC source to be a very low frequency. In this case, X_L will be very low and X_C will be quite high. The exact values will depend on the values of L, C, and the applied frequency.

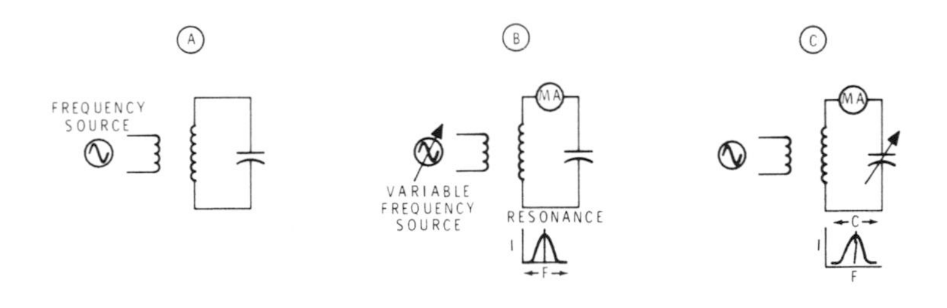

Figure 6
Evolution of the wavemeter circuit.

Now assume that the frequency source is adjustable. If you gradually increase the frequency, the X_L will gradually increase while the X_C gradually decreases. As you increase the frequency further, a point is eventually reached at which the value of X_L is the same as the value of X_C. The exact frequency at which this occurs depends upon the values of the coil and capacitor. However, for any combination of capacitance and inductance, one frequency exists where $X_L = X_C$. This is the resonant frequency of the circuit.

If you induce a signal into the tank circuit, you will find that the tank circuit has very little reaction to any frequency other than its resonant frequency. At resonance, however, it breaks into oscillation, causing current to flow within the tank. By placing a current meter in the tank, you could obtain an indication of resonance. Referring to Figure 6B, if you were to vary the frequency of the input signal and plot the results, the frequency versus current plot would appear very much as indicated. The peak current reading would correspond to the resonant frequency of the tank circuit.

This would work well if we had a precision frequency source and wanted to determine the tank's resonant frequency. However, when using a wavemeter, we usually have known values of tank circuit components and are looking for an unknown input frequency. By carrying this circuit one step further, you can substitute a precision variable capacitor for the fixed capacitor, as in Figure 6C. Now you can apply an unknown frequency to the input and slowly vary the capacitor value until the meter peaks at resonance. As the value of L and C are known, the input frequency could now be computed using the formula,

$$F_O \text{ (resonant frequency)} = \frac{1}{2\pi \sqrt{LC}}$$

The instrument manufacturers have done this for us, however, by calibrating the tuning dial for direct frequency readout.

Another factor to consider is the figure of merit, or "Q" of the wavemeter tank circuit, which is inversely proportional to the resistance within the circuit. If you had a perfect coil and capacitor connected by perfect conductors, the circuit would respond to only one exact frequency, and current within the tank would be very high. In practice, however, all coils, capacitors, and conductors have some resistance, which effectively lowers the Q of the circuit.

Figure 7 shows the effect of Q on amplitude and frequency response. As you can see the higher the Q, the greater the circuit's selectivity will be. Therefore, to obtain a sharp, well-defined peak on the current meter, a relatively high Q circuit is necessary.

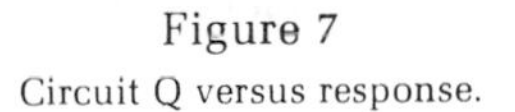

Figure 7
Circuit Q versus response.

Another factor affecting the selectivity of this circuit is the *degree of coupling* between the wavemeter coil and the signal source. As you might assume, the closer the wavemeter is to the signal source, the stronger the indication will be. However, it is undesirable to hold the wavemeter too close since looser coupling will produce a narrower bandwidth. Figure 8 shows how the degree of coupling affects bandwidth and amplitude. To pinpoint a specific frequency, the most selective (loosest) coupling will give you the most accurate reading.

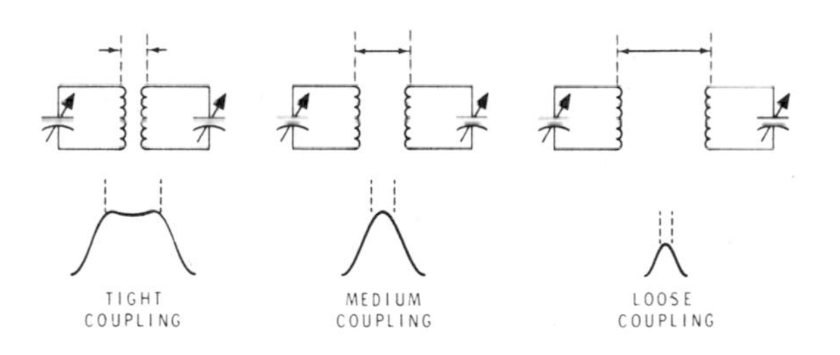

Figure 8
Coupling effect on selectivity.

Coupling too tightly can also, in some cases, have an adverse effect on the circuit being tested. Not only does the wavemeter respond to the frequency of an oscillator, but the oscillator circuit will also respond to the reactive components of the wavemeter. This places an inductively coupled load on the oscillator circuit and results in a shift (change) in oscillator frequency which is called *frequency pulling*. The result is much the same as voltmeter loading, in that what is measured with the test instrument in the circuit is not exactly what is present when the test instrument is not in the circuit.

The looser the coupling, the less loading effect is felt by the oscillator circuit. Therefore, when you use this type of frequency meter, hold it as far away as possible from the circuit being measured. The most accurate readings will occur when you are obtaining the smallest usable indication.

If you cannot get an indication of resonance, it might be due to the *angle of coupling*. Figure 9 shows the effects of angular difference, between primary and secondary coils, on induced signal strength.

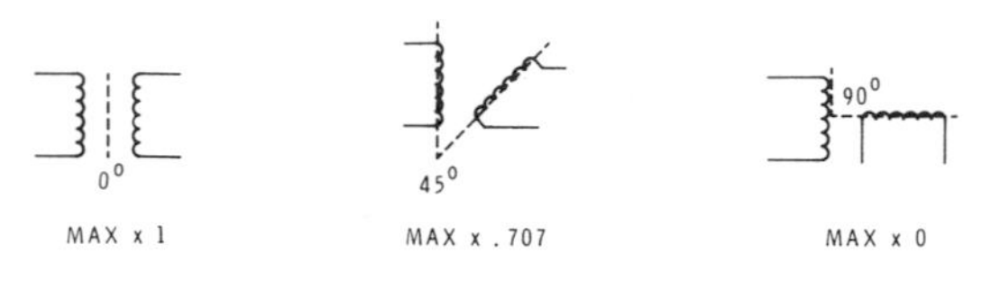

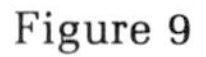

Figure 9

Simply stated, the maximum induced signal for any given degree of coupling will occur when the primary coil and the secondary coil are physically parallel. At any other angle, you can determine the induced amplitude by multiplying the maximum induced signal by the cosine of the angular difference. Thus, if 5 volts were induced with the coils parallel, and then the secondary was rotated to 45° with respect to the primary, the induced voltage would drop to 5 volts × .7071 (cosine of 45°) or 3.536 volts. With the secondary at 90° to the primary, no voltage would be induced (cosine 90° = 0).

Two types of resonant wavemeters are available; the reaction type and the absorption type.

The *reaction wavemeter* (described below), while not as accurate as the absorption type, absorbs very little power. It is, therefore, perferred when you are measuring frequencies in low-power equipment. As shown in Figure 10, this wavemeter has no indicating device. The indication of resonance is supplied by the unit under test. This is usually in the form of an ammeter, of suitable range, inserted in the oscillator circuit.

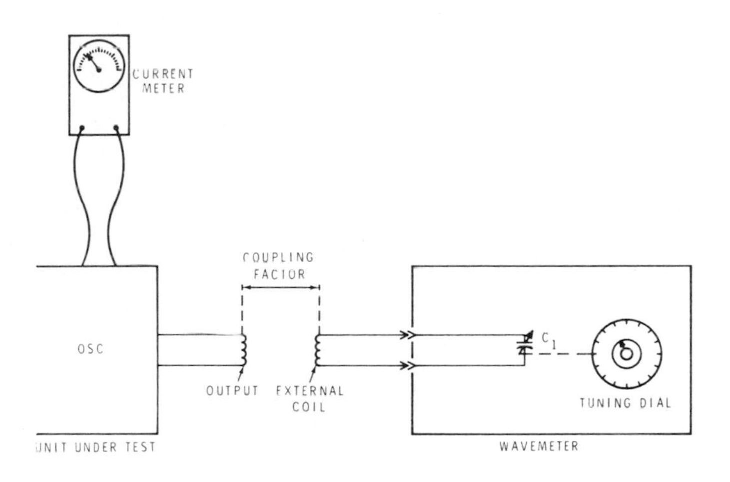

Figure 10
Reaction wavemeter – simplified circuit.

CAUTION: If the wavemeter coil or case should come in contact with high voltages in units under test, a severe or fatal shock could result.

To obtain a frequency measurement you should hold the wavemeter close to the LC tank circuit of an oscillator or low-power amplifier. Vary the wavemeter tuning dial until you see an indication on the ammeter. This may be either a peak or a dip, depending upon the placement of the meter in the circuit. Then move the wavemeter as far from the circuit as you can while still obtaining a positive indication (minimum coupling); readjust the tuning dial as necessary. When you have obtained the lowest usable reading and adjusted the tuning for optimum peak, you can read the unknown frequency from the dial.

The *absorption wavemeter* is shown in Figure 11. Note that a meter has been added to the wavemeter, making it a self-indicating instrument. The fixed capacitor across the meter permits a voltage to be developed at resonance.

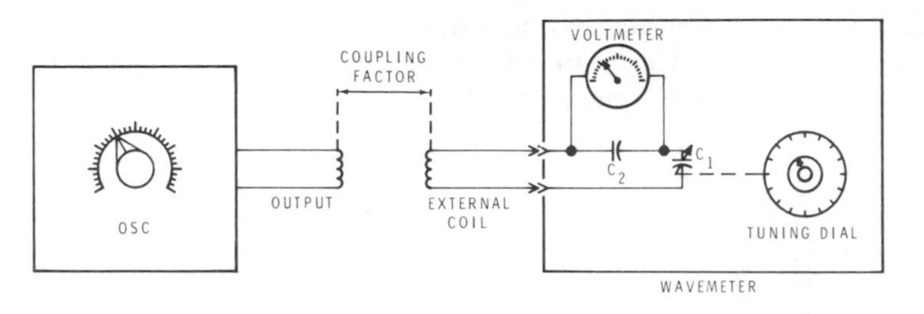

Figure 11

This unit operates much the same as the reaction wavemeter. However, more care is required as you approach resonance, as it is possible to exceed the current range of the indicating meter. Therefore, you should always adjust the tuning dial slowly until the meter starts to move up scale. Then reduce the coupling by moving the wavemeter further away from the oscillator circuit, while continuing to adjust the tuning dial. As with the reaction wavemeter, you should take the final reading when the meter is tuned to a well-defined peak with the minimum usable meter deflection.

The frequency range of these instruments is limited by component size and availability. For this reason their use is confined to radio frequencies (RF) between approximately 1 MHz and 1 GHz. Below 1 MHz the coils and capacitors become bulky and cumbersome, while above 1 GHz they become too small to be economically manufactured.

At the higher frequencies, stray capacitance and inductance from leads, wires, and connections also become increasingly critical. A few picofarads of stray capacitance at 1 MHz have little effect on circuit operation. At 1 GHz, however, this may be a dominant factor. Therefore, if wavemeter type instruments are needed at microwave frequencies, resonant cavities must be used. While a detailed study of microwave devices is beyond the scope of this course, the following simplified explanation of resonant cavity wavemeters should enable you to understand their principles of operation.

At higher frequencies, L and C must be reduced to achieve resonance. Thus, as frequency increases, the number of required coil-turns decreases, as does the required capacitance. This is shown by Figure 12. In the UHF band (above 300 MHz), the number of coil turns can be reduced to one turn and the required circuit capacitance is only the distributed capacitance between the two ends of the coil wires. This is known as a "hairpin" tank. If we place two of these hairpin tanks in parallel, "L" will be reduced but "C" will be increased resulting in the same resonant frequency. The only change will be an increase in Q, making the circuit more selective.

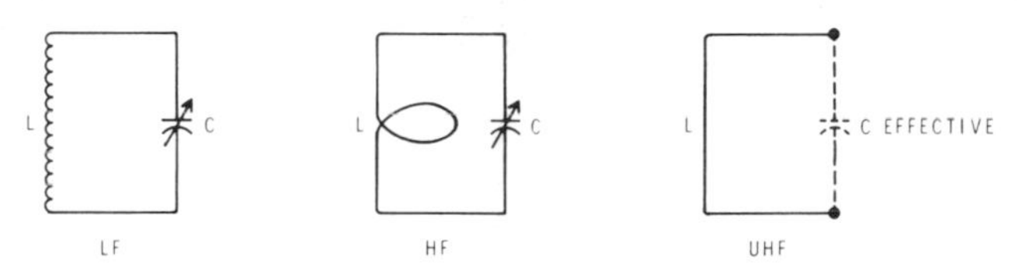

Figure 12
Evolution of the hairpin tank circuit.

If you add an infinite number of these circuits in parallel, as shown in Figure 13, you will have a device that resembles a round can. The natural frequency of this device is dependent upon its physical size, the length of L (Figure 13A) being equal to 1/4 wavelength ($\lambda 1/4$) of the resonant frequency. Thus, if we assume L to be 1 cm per side (3 cm overall), its fundamental frequency would have a quarter wavelength of 3 cm (1 cm times 3 sides) for a wavelength of 12 cm or .12 meters. By applying this to the formula $F\ (\text{MHz}) = \dfrac{300}{\lambda\ (\text{meters})}$

we find the cavity will resonate at $\dfrac{300}{.12}$ or 2500 MHz.

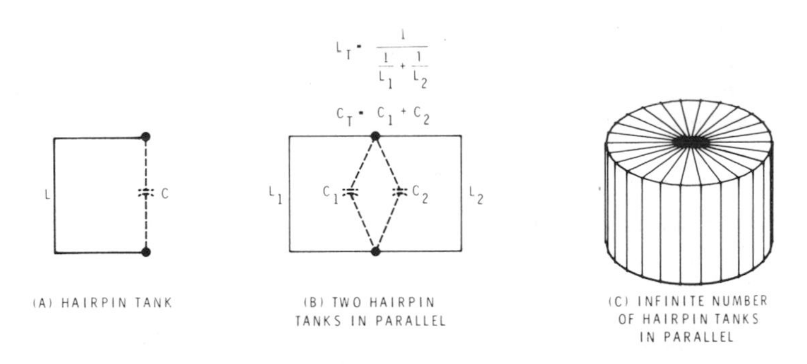

Figure 13
Evolution of the Microwave resonant cavity.

As you can see from this example, cavity size is inversely proportional to its frequency. To change the resonant frequency of a cavity, it is only necessary to change its physical dimensions. A typical method of doing this is shown in Figure 14. Turning micrometer adjustment "A" causes large changes in the size of "C" by moving piston "D" either up or down. This is a coarse frequency adjustment. Fine tuning is accomplished by adjustment "B", which flexes the cavity walls slightly.

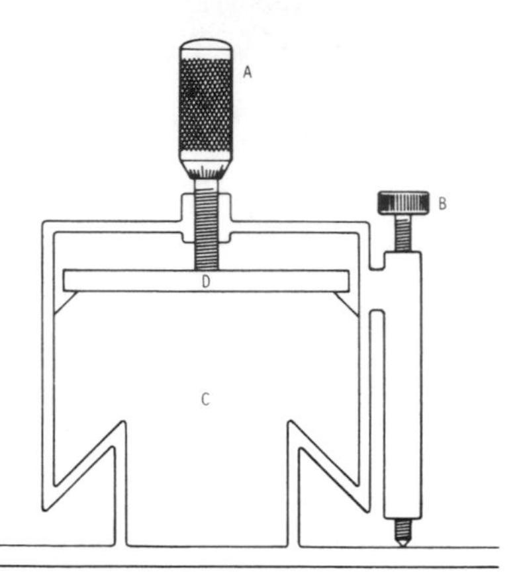

Figure 14

In use, these cavity wavemeters are inserted into the transmission line between the output and load. An output power indicator is connected to the load and monitored while the wavemeter is being tuned. At resonance, the wavemeter will absorb a small amount of power, resulting in a dip in the output power reading, as the wavemeter is tuned across the resonant frequency.

These meters are available for frequencies from 1 to 40 GHz, with an accuracy of about 0.1%. They may be designed for use with either coaxial or waveguide transmission lines, with 5 to 10 GHz being the dividing line between the two types. Below 5 GHz, the physical size of waveguide transmission line becomes quite large, while above 10 GHz coaxial cable losses become significant.

This completes our discussion of passive frequency meters. The remainder of this unit will be devoted to active meters.

As a group, active frequency meters have the potential for much greater accuracy and versatility than passive meters. They also tend to require more frequent calibration to maintain this accuracy and are more susceptible to malfunctions and damage from improper operating techniques.

DIP METERS

The first and simplest active meter we will discuss is the dip meter. It is also the least accurate of the group; but properly used, it can be a very useful device. Figure 15 shows a typical Solid-State Dip Meter.

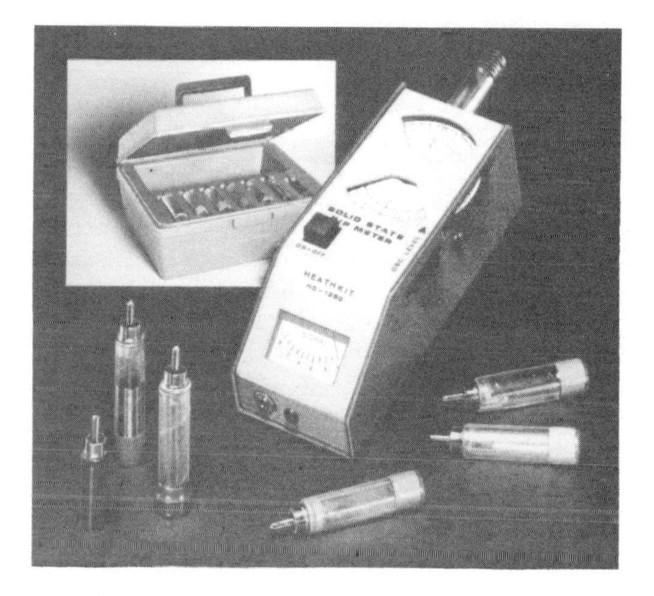

Figure 15

Principles of Operation

As you can see from the simplified circuit in Figure 16, a dip meter is basically an absorption wavemeter with a built-in oscillator circuit. When a vacuum tube circuit is used, as in Figure 17, it is called a grid dip meter; however, the operation is much the same for both instruments. A series of plug-in coils are usually supplied with each meter in order to provide a wide operating range.

These units can be used in one of two modes. With the oscillator power level reduced to the point where oscillations stop (zero indication on the meter), you can use them to measure the frequency of active circuits. This is called the **wavemeter mode**, and operation is exactly the same as the absorption wavemeters you just studied.

In the **injection mode**, the oscillator level is increased until the meter reads upscale. The dip meter then acts as a miniature transmitter and can be used to determine the resonant frequency of de-energized circuits. It is coupled into the circuit in the same manner as a wavemeter. As it is tuned to the resonant frequency of the test circuit, current will begin to circulate in the test circuit. This reaction will cause some power to be absorbed from the dip meter, which loads the dip meter's oscillator, causing a decrease or "dip" in the meter indication. After you reduce the coupling to minimum, while carefully tuning the meter for the exact center of the dip, you can read the resonant frequency directly from the dip meter frequency dial.

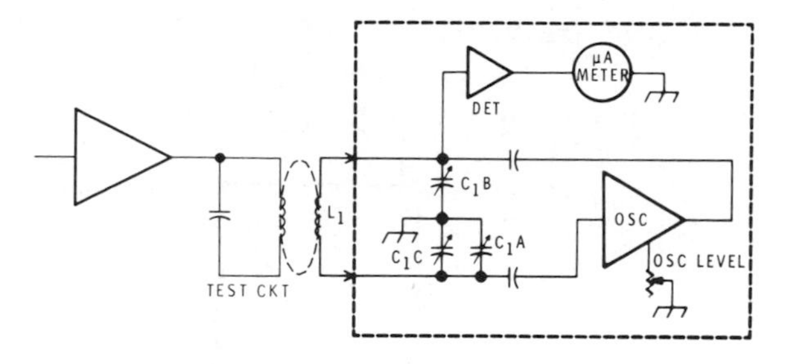

Figure 16

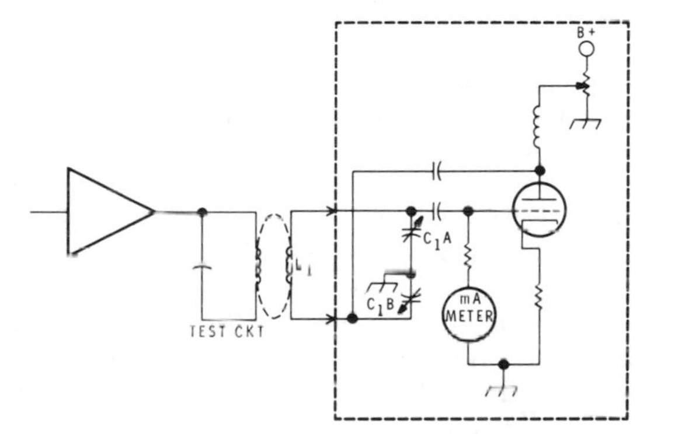

Figure 17

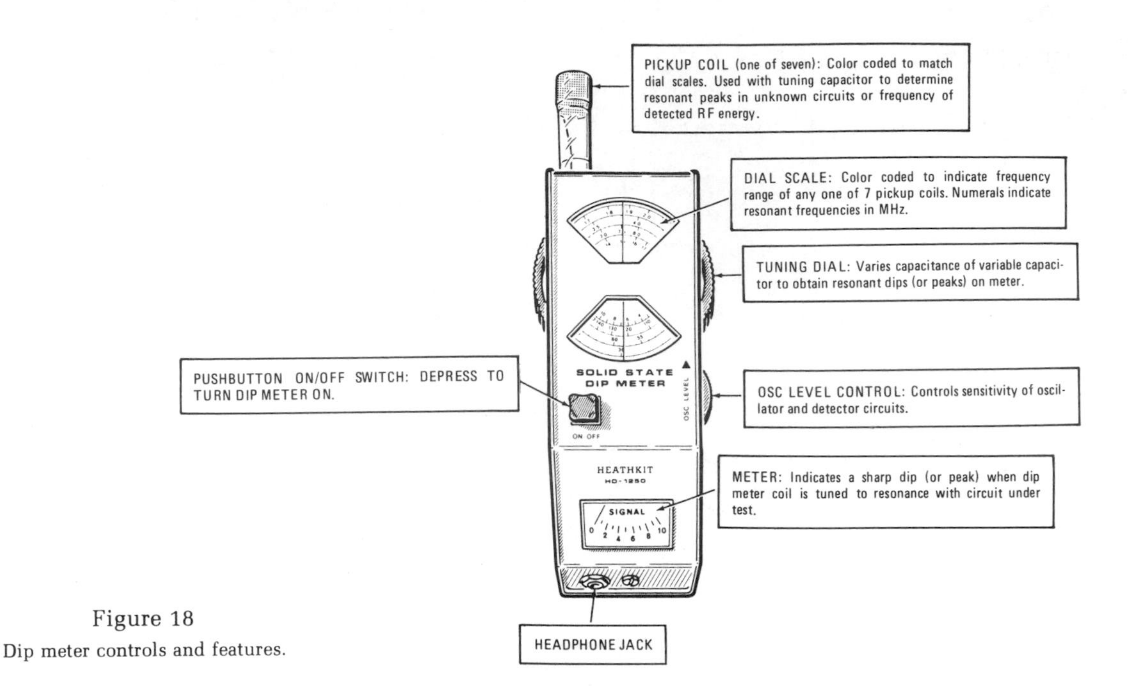

Figure 18
Dip meter controls and features.

Most meters now have a multiscale tuning dial with one scale for each plug-in coil. (See Figure 18.) Some older meters, however, used a single scale, with divisions graduated in arbitrary units (0 - 100 for example), and provided a calibration chart for each coil. An example of this type dial and chart is shown in Figure 19. In this example, with the dial set at 50 divisions and L-1 installed, the dip meter would be tuned to approximately 5.4 MHz. Notice that the calibration line is actually a curve. This is caused by nonlinearity of the tuning capacitor. By using more expensive, linear variable capacitors, a nearly straight line could be achieved. This is not a major drawback, however, as it has little effect on accuracy after initial calibration.

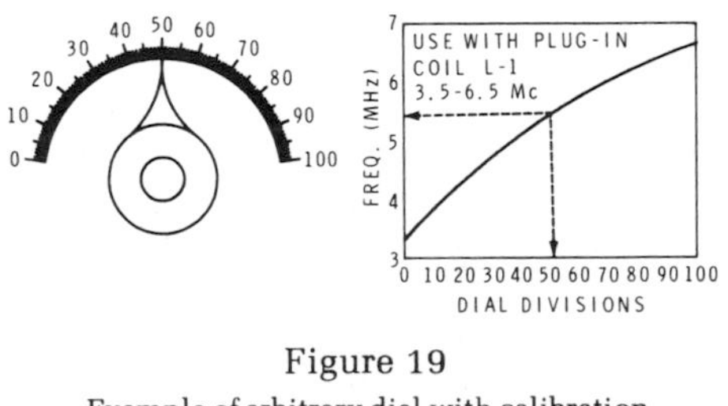

Figure 19
Example of arbitrary dial with calibration chart.

Operating Precautions

Observe the following precautions when you are using a dip meter.

1. **Always be aware of shock hazards.** The case is not grounded on these units. Therefore, if you come in contact with any voltage, either with the instrument or your hand, the only current path to ground is through your body.

2. Do not expose these units to strong RF fields. These can damage electronic components in the instrument, even when it is turned off.

3. Do not force the tuning dial past the mechanical stop. This can ruin the calibration and possibly damage the tuning mechanism.

4. Protect plug-in coils from dust, dirt and damage. A dirty or bent connector can cause intermittent or false indications.

5. If it is battery powered, always turn the unit off when it is not in use. This will conserve your batteries. Never store a unit for long periods of time with batteries installed. They can corrode and cause serious internal damage to the instrument.

General Applications

You can use dip meters to determine the approximate resonant frequency of either energized or de-energized circuits in the frequency range of 1 to 300 MHz. This can be very useful when you are building or repairing receivers, transmitters, signal generators, or any other equipment operating in this frequency range. Keep in mind, however, that readings and adjustments, using this method, should only be considered preliminary. They will be close enough to insure that the system will operate; but dip meters do not have the accuracy or resolution to be used for calibration or final tuning operations.

In addition to the obvious applications of this instrument, you can also use it to measure the bandwidth, bandpass, and Q of tuned circuits, and to determine the unknown values of capacitors and inductors.

Bandwidth and Q Measurement

You can estimate the relative Q of a resonant circuit by observing the dip meter indication when it is used in the injection mode. As you tune the meter very slowly and steadily across resonance, a sharp, pronounced dip and rise indicates a circuit with a relatively high Q. A broad shallow dip indicates a relatively low Q.

If you need to find a more definite value of Q, you can use a high input impedance voltmeter and an RF probe in conjunction with the dip meter. This is accomplished in the following manner.

As you will remember from your studies of resonant circuits, these instruments do not respond to one exact frequency but rather to a group, or band, of frequencies, centered on the resonant frequency. The amplitude of this response depends upon how close the frequency is to the instrument's resonant frequency.

If you were to plot the instrument's response to a series of frequency inputs, the resulting graph would appear as a bell-shaped curve, as shown in Figure 20A. The two frequencies at which the voltage response equals 70.7% of the peak response are called the half-power points. These points are indicated as F_1 and F_2 in Figure 20B.

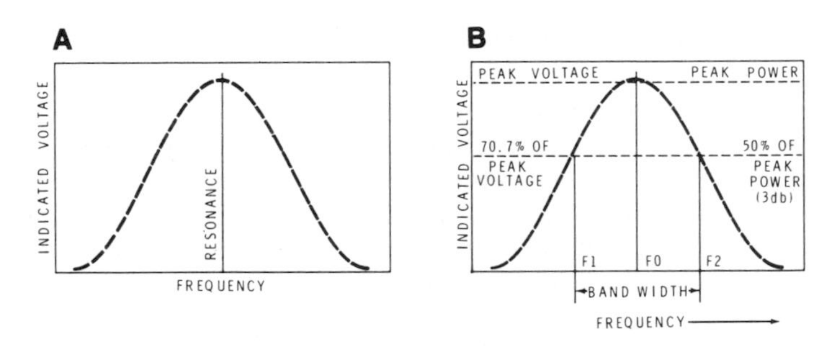

Figure 20

Resonance curve – frequency versus voltage.

The bandpass of this circuit is F_1 to F_2, while the bandwidth is $(F_2 - F_1)$. To calculate the Q, it is only necessary to divide the resonant frequency (F_0) by the bandwidth $(F_2 - F_1)$ or $Q = \frac{F_0}{BW}$.

To measure these characteristics of a resonant test circuit, you would:

A. Connect a high input impedance DC voltmeter that is equipped with an RF probe across the capacitor of the deactivated circuit.

B. Loosely couple a dip meter, operating in the injection mode, to the circuit's inductor, as shown in Figure 21.

C. Carefully tune the dip meter to the circuit's resonant frequency, as indicated by a peak voltmeter reading.

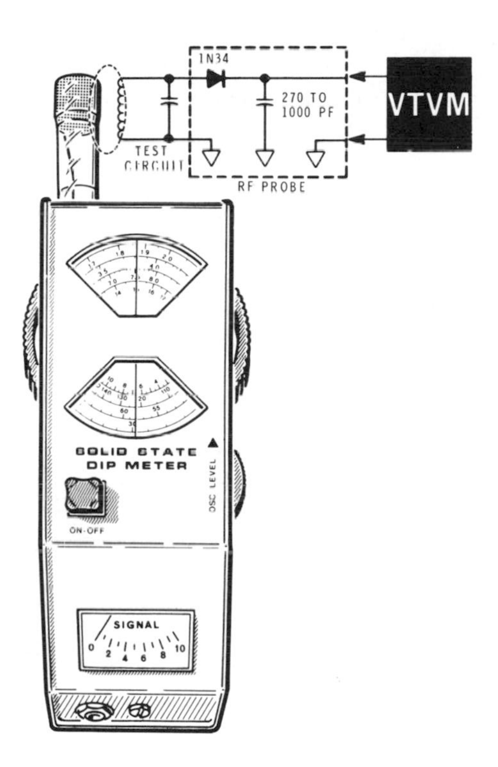

Figure 21

D. Vary the coupling until the peak voltmeter reading is at some convenient value. Avoid too close a coupling — this can result in significant errors. Also avoid moving the dip meter or changing its oscillator level after this step, as this combination of signal strength and coupling establish the reference level for all subsequent readings.

E. Record the dip meter frequency as F_0. Then decrease the frequency until the voltmeter reads 70.7% of its peak value. Record this frequency as F_1.

F. Increase the dip meter frequency until the voltmeter passes through its peak and returns to the 70.7% level. Record the frequency at this point as F_2.

By assuming some typical voltage and frequency values, you can now calculate the bandpass, bandwidth, and Q of this circuit. Assume that:

$$F_0 = 1 \text{ VDC} = 2.50 \text{ MHz}$$

$$F_1 = .707 \text{ VDC} = 2.45 \text{ MHz}$$

$$F_2 = .707 \text{ VDC} = 2.55 \text{ MHz}$$

Thus:

Bandpass = 2.45 MHz to 2.55 MHz

$$\text{Bandwidth} = (F_2 - F_1) = 2.55\text{MHz} - 2.45 \text{ MHz} = 0.1 \text{ MHz}$$

$$Q = \frac{F_0}{BW} = \frac{2.5 \text{ MHz}}{0.1 \text{ MHz}} = 25$$

Once again, keep in mind these are only approximations. The basic accuracy of the dip meter, its limited resolution, and the method of coupling all contribute to uncertainties in the final readings. It is, however, a fairly quick method of arriving at rough values and could prove to be of value when more precise equipment is not available.

HETERODYNE FREQUENCY METERS

The heterodyne frequency meter offers much greater accuracy and precision than the dip meter. While its operation is also based upon interaction with the input signal, it produces a measurement indication through mixing action rather than mutual resonance.

Principles of Operation

The heterodyne frequency meter measures the frequency of an unknown radio frequency signal by matching the unknown frequency with a locally generated signal of the same frequency. This is obtained from a calibrated, high precision, variable oscillator. Figure 23 shows the block diagram of a basic heterodyne frequency meter.

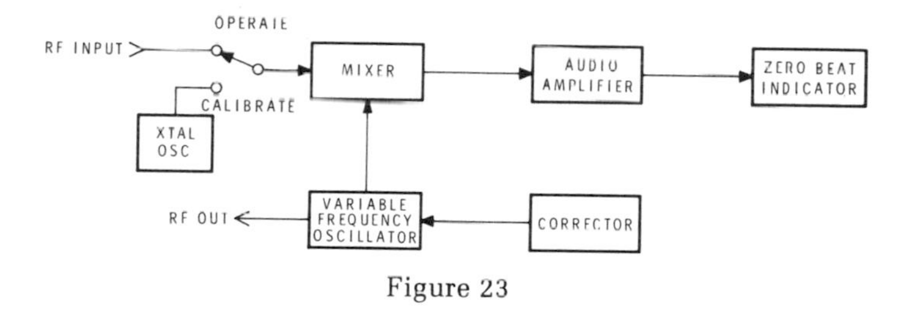

Figure 23

The unknown signal is applied, through the input connection, to the mixer stage where it is mixed, or "heterodyned", with the reference signal from the variable frequency oscillator. As you will remember from your previous studies, this mixing action results in the presence of four basic frequencies: the unknown input signal, the oscillator signal, the sum of the two, and the difference between the two. Harmonics of both primary signals are also present, but we will consider these later. Thus, if the RF input were at 3 MHz and the precision oscillator were tuned to 2.999 MHz, both these frequencies would be present at the mixer output, as well as 5.999 MHz (the sum) and 1 kHz (the difference).

The mixer stage feeds an audio amplifier stage which responds only to relatively low frequency signals. Therefore, the 2.999 MHz, 3 MHz, and 5.999 MHz signals will be blocked but the 1 kHz difference signal, which is called the beat frequency, will be amplified and passed on to an indicating device. This device may be a loudspeaker, headsets, an oscilloscope, or a null meter. For purposes of explanation, we will assume the device to be headsets, however, the results will be much the same in any case. Only the presentation will differ.

Operation

Under the preceding signal conditions, you would hear a 1 kHz tone in the headsets. As you increased the oscillator frequency to 2.999500 MHz, the difference frequency would decrease to 500 Hz, resulting in a lower pitched tone in the headsets. By slowly tuning the oscillator upward, the detected tone would become lower and lower. When the oscillator frequency matched the input frequency, you would not hear any sound at all. This condition is known as a "zero beat". If you continued to increase the oscillator frequency beyond this zero beat point, you would again hear a tone that would start at a very low pitch and increase with the oscillator frequency until it became inaudible.

Figure 24 gives a graphic representation of this effect. Notice that as you approach the zero beat point you enter a region where the tone is below the audible range. Therefore, what you would hear would not be a tone as such, but may be a fluctuating, rushing sound, or a clicking noise, depending upon the type of circuits used. In either case, when you reach this region, you should tune the oscillator very carefully to achieve a point where the rushing sound does not fluctuate or the clicks stop. This will be the exact zero beat, where the variable frequency oscillator output exactly matches the frequency of the RF input. You can then read the oscillator frequency directly from a calibrated dial which is coupled to the oscillator tuning control. In order to increase the resolution of the tuning dial, a vernier dial is often used. Examples of this are shown in Figure 25.

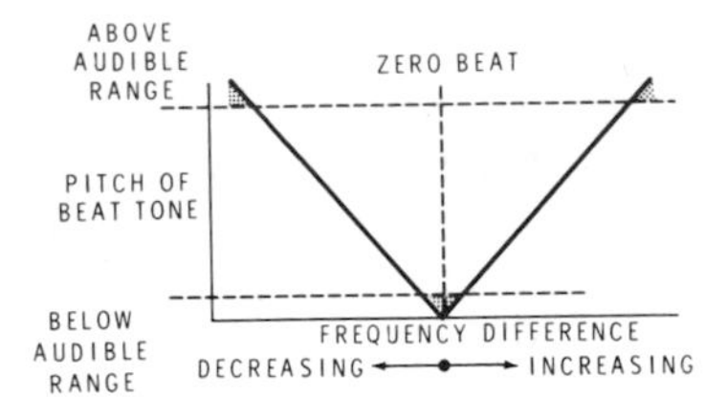

Figure 24

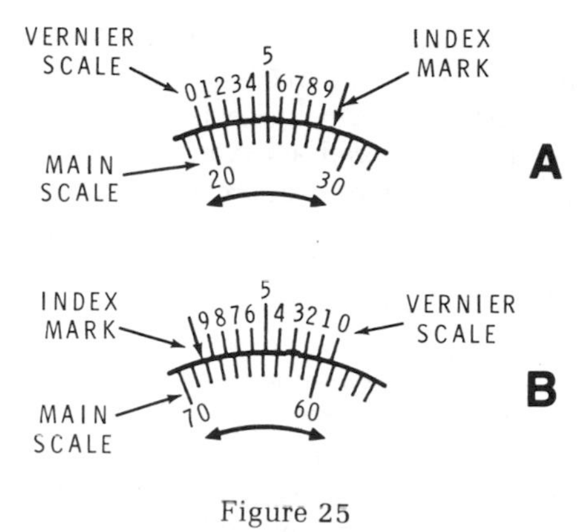

Figure 25
Vernier scale examples.

In these two examples, the main scale moves with the tuning control while the vernier scale is permanently fixed above it at the index point. This effectively divides each division on the main scale into ten equal parts. To read the dial indication in Figure 25A, notice that the main scale reads greater than 28 but less than 29. Our reading, then, is 28, plus some fractional value. To read the venier scale, look for the vernier division mark which most nearly aligns with a main scale division mark. In this case, vernier mark 5 is almost perfectly aligned. Adding this to the main dialing reading, we find our final reading is 28.5.

In Figure 25B, the main dial reads 68 plus. On the vernier dial, the 3 is most closely aligned with a division mark; therefore, the reading is rounded off to 68.3. Some verniers provide for readings up to 1/100 of a major division, but these become so difficult to read that a magnifying glass is often provided.

We should note at this point that, due to the precise tuning required, the tuning control is not connected directly to the variable frequency oscillator; it is connected through a reduction gear assembly. Although great care is taken during design and manufacture to reduce "back lash" in this gear assembly, it cannot be entirely eliminated. This results in slightly different dial readings at zero beat, depending upon whether the point is approached from below or above. Therefore, when you make a series of precise readings, always approach the zero beat from the same direction. A good habit to get into is to approach from below, as this is compatible with other instruments you will be using (notably the counter plug-in heterodyne converter). Most calibration laboratories also calibrate the units in this manner.

The variable frequency oscillator, while very stable, is nevertheless subject to some drift from its calibrated frequencies. This is caused by the sum of a number of factors such as changes in temperature and humidity, imperfections in moving parts, and aging of components. Therefore, most heterodyne frequency meters incorporate a crystal calibrator circuit. Referring back to Figure 23 for a moment, notice that one mixer input goes to a two-position switch. This allows the unit to be operated in either the "operate" mode or the "calibrate" mode.

In the "operate" position the mixer is connected to the RF input connector; but in the "calibrate" position it receives its signal from a highly stable, temperature controlled, quartz crystal oscillator. The operating frequency of this oscillator is chosen to be compatible with the operating range of the frequency meter. For instance, if the frequency meter had an operating range of 100 kHz to 1 MHz, a crystal frequency of 100 kHz might be chosen. By switching the unit to the "calibrate" mode, a stable 100 kHz signal would be coupled into the mixer. In this mode, if you tuned the variable frequency oscillator through its operating range, you would receive a zero-beat every 100 kHz. These "crystal check points" occur at the crystal oscillator's fundamental frequency and at each harmonic of this frequency (200 kHz, 300 kHz, 400 kHz, etc.). A control called the "corrector" is used in conjunction with the crystal oscillator. This control adjusts a small trimmer capacitor in the variable frequency oscillator circuit which will vary its frequency a small amount. Adjustment of this control will not cause the tuning dial to move. To use this circuit, the frequency meter is connected to an RF signal source we wish to measure, the mode switch is placed in the "operate" mode, and an approximate zero beat attained. This gives us a rough idea of the input frequency. If, for instance, a zero beat was received at about 720 kHz and our instrument had 100 kHz crystal check points, we would next set and lock the tuning dial at the exact frequency of the nearest check point; in this case 700 kHz. Now, by switching to the "calibrate" mode and adjusting the corrector control for an exact zero beat, we have transferred the accuracy of the crystal oscillator to this portion of the meter's frequency dial by fine tuning the variable frequency oscillator. Next, by switching back to the "operate" mode and readjusting the main tuning dial for an exact zero-beat with the input signal, we can obtain a very accurate frequency reading.

Some heterodyne frequency meters are equipped so a whip antenna can be connected to the variable frequency oscillator circuit. This allows the oscillator signal to be radiated for short distances. If you are checking transmitters or other high power signals which cannot be safely coupled directly into the meter, an arrangement like the one in Figure 26 can be set up. The radio receiver is tuned to the transmitter frequency. The frequency meter is then placed near the receiver and adjusted until a zero beat is heard through the receiver's loud speaker. With this method, the transmitter frequency can be measured as if it were coupled directly to the frequency meter.

Figure 26

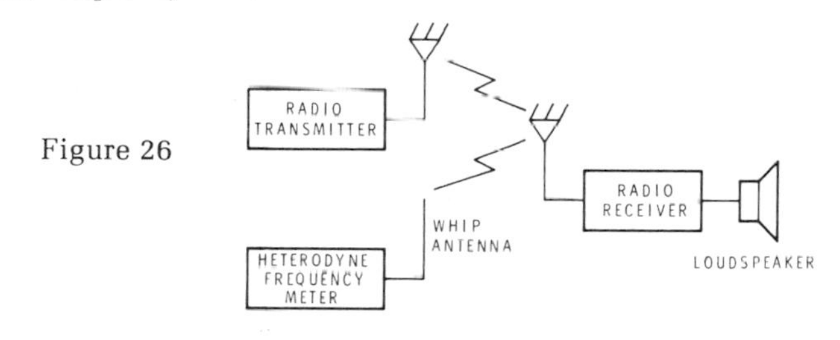

Heterodyne frequency meters are available in various fundamental frequency ranges from 100 kHz to 200 MHz. Above 200 MHz, harmonics of the variable frequency oscillator may be used for zero beats, much as the harmonics of the crystal oscillator are used for check points. By this means, the frequency range may be extended up to 3 GHz (3,000 MHz). But while excellent accuracy is maintained using harmonics, it is sometimes difficult to determine which harmonic you are tuned to. It may, therefore, be helpful to use an absorption wavemeter to determine the approximate frequency and then use a heterodyne frequency meter to provide a more exact measurement. For example, suppose an absorption wavemeter indicated that an oscillator was operating at approximately 900 MHz. When this signal was coupled into a heterodyne frequency meter, a zero beat was obtained at 178.55 MHz on the frequency meter tuning dial. A check of the harmonic frequencies of 178.55 MHz showed that:

- the 4th Harmonic was 714.20 MHz
- the 5th Harmonic was 892.75 MHz
- the 6th Harmonic was 1071.30 MHz

By comparing the wavemeter reading with the harmonic values, you now know that the oscillator under test is operating at a frequency equal to the 5th harmonic of the heterodyne meter, or 892.75 MHz.

Accuracy

The general accuracy of these instruments range from 0.01% to 0.05% when you are using the audible zero beat, the inability to obtain an exact zero beat being the limiting factor. By using sensitive electronic nulling indicators, modern heterodyne meters are capable of accuracies of 0.00025% or better.

TRANSFER OSCILLATORS

The transfer oscillator operates on much the same principle as the heterodyne frequency meter. The unknown frequency is measured by mixing it with a known frequency and observing the resulting interaction. The two methods have approximately the same frequency coverage.

One major advantage of the transfer oscillator over the heterodyne meter is its very broad bandwidth. A single transfer oscillator can measure frequencies from 50 MHz to 18 GHz. As many as four separate heterodyne meters might be needed to cover the same range of frequencies. This results in considerable savings in equipment costs. Transfer oscillators also have greater sensitivity than heterodyne meters. They are able to measure weaker signals, a great advantage in certain applications.

The major disadvantage is the computation required to determine the unknown frequency value. Also, the accuracy and resolution of a transfer oscillator is not as good as the heterodyne meter. In fact, to even approach this accuracy, the transfer oscillator must be used in conjunction with an electronic counter.

Principles of Operation

Figure 27 shows the simplified block diagram of a typical transfer oscillator. It consists of a very stable variable frequency oscillator (VFO), a harmonic generator, a mixer, and a null detecting device.

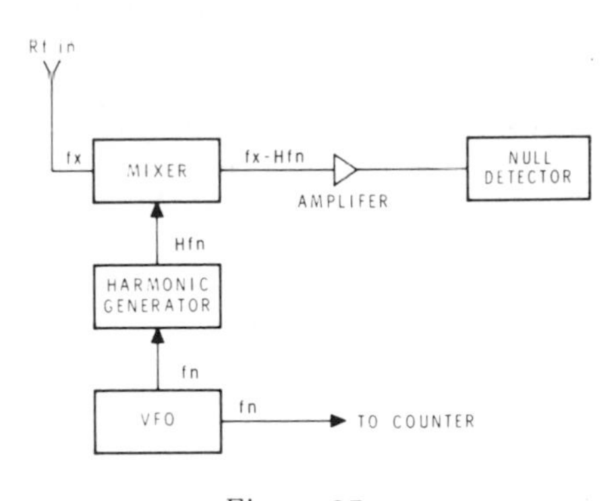

Figure 27
Transfer oscillator – simplified block diagram.

The typical VFO can be tuned over a frequency range of 100 to 220 MHz. This signal, which we will label Fn, is fed to a harmonic generator and also to an electronic counter. Because it is being monitored by a counter, the frequency dial does not have to be as elaborate or precise as the tuning mechanism for the heterodyne meter. The main consideration for the VFO is a high degree of stability.

The harmonic generator accepts the frequency Fn from the VFO and produces a large number of harmonic frequencies (Hfn), each spaced Fn apart. For example, if the VFO were tuned to 150 MHz, the harmonic generator would put out signals at 150 MHz, 300 MHz, 450 MHz, 600 MHz, etc., on up to greater than the 200th harmonic of Fn. Of course, as the VFO frequency is changed, the harmonic frequencies also change. (Remember, a harmonic is a whole-number multiple of the fundamental frequency.)

The mixer stage beats these harmonics (HFn) with the incoming unknown frequency (Fx) to produce a beat frequency which is amplified and applied to a null detector.

Operational Techniques

In operation, the VFO is carefully tuned until one of its harmonics matches the unknown frequency. This frequency match is indicated by a zero beat on the null detector. When the best possible zero beat is obtained, you should read and record the counter reading. *Label this reading* F_1.

The VFO is then slowly tuned upward or downward until the next adjacent harmonic produces a null. Once again, obtain the best possible zero beat. Read and record the counter reading. *Label this reading* F_2.

> Caution: When you are tuning for the second zero beat, be careful not to miss a null point by tuning over it too quickly. If the two nulls (F_1 and F_2) are not adjacent harmonics, gross error in the computation of F_x will result.

These two counter readings are all you need to compute the value of the unknown frequency.

Let's examine exactly what you would have. You know that some multiple of F_1 equals F_x. You also know that F_2 is the next frequency that has a multiple equal to F_x. To find the value of F_x, you must now determine the harmonic number of either F_1 or F_2. You can calculate the harmonic number applicable to F_1 with the formula

$$H_1 = \frac{F_2}{|F_1 - F_2|}$$

(The vertical bars mean a negative sign is ignored. The difference between F_1 and F_2 is always a positive number.)

This will give you the harmonic of F_1 which is equal to F_x. Once you determine this, it is only a matter of multiplying to determine the value of F_x.

$$F_x = F_1 \times H_1.$$

You could also have used F_2 to find F_x by finding its harmonic number. In this case you would use the formula

$$H_2 = \frac{F_1}{|F_1 - F_2|}, \text{ and } F_x = F_2 \times H_2$$

To illustrate this, let's assume that the first null occurs when F_1 equals 88.0 MHz. As the VFO is slowly increased in frequency, the next null is found at 90.0 MHz. You can now find the harmonic number applicable to F_1.

$$H_1 = \frac{90.0}{88.0 - 90.0} = \frac{90.0}{2.0} = 45^{th} \text{ harmonic}$$

Thus, the unknown frequency is equal to the 45th harmonic of 88.0 MHz, or $F_x = 45 \times 88.0 = 3960$ MHz.

If you had to use F_2 to find F_x, you would have computed it in this manner:

$$H_2 = \frac{88.0}{88.0 - 90.0} = \frac{88.0}{2.0} = 44\text{th harmonic} \quad F_x = 44 \text{ x } 90.0 = 3960 \text{ MH}$$

In theory, either method is equally satisfactory. However, in practice, one reading may yield more accurate results than the other. For any given F_x, you may find many null points as you tune the VFO through its range. Using the same F_x of 3960 MHz as an example, you will find an adjacent null pair at VFO settings of $F_1 = 82.50$ MHz and $F_2 = 84.255319194 \ldots$ MHz. Because the fractional part of F_2 does not end, the counter cannot display the whole number. This would not keep you from determining H_2 to be the 47th harmonic because, as harmonics are always even multiples, you can round the answer off to the nearest whole number. However, you cannot calculate F_x with accuracy from the relationship $F_x = H_2 \times F_2$ because there is no way of determining the missing part of F_2. In this case it would be better to calculate F_x from F_1, since F_1 is an exact number.

As with the heterodyne frequency meter, the greatest source of error is the ability of the operator to obtain an exact null or zero beat. The resolution of the null detector and the care taken by the operator to determine an exact null are all important to obtaining accurate measurements. The total measurement error is the sum of the counter error plus the error caused by imperfect null detection. Assuming good quality calibrated equipment, you can make frequency readings by this method with an uncertainty of a few parts in 10^7.

Although the transfer oscillator, as a separate test instrument, can still be found in the field, the major application of this principle is currently in high frequency electronic counters. Units of this type, using advanced circuits, are employed to extend the frequency range of counters to nearly 20 GHz. These will be discussed in greater detail later.

ELECTRONIC COUNTERS, GENERAL PRINCIPLES

The electronic counter, which a few years ago was found only in laboratories and highly specialized organizations, can today be found in repair shops, engineering departments, ham shacks and on production lines throughout the country.

The reason for this vast increase in the use of electronic counters can be traced directly to the development of the integrated circuit. Through the use of these IC chips, the physical size has been reduced from that of a large television receiver, weighing over one hundred pounds, to an instrument that can easily be carried in a briefcase.

The same technical advances which allowed this dramatic reduction in size have also brought about equally dramatic reductions in cost, as well as increased accuracy, stability reliability and frequency range. While a 10 MHz vacuum tube counter cost $2,600 in 1960, you can now purchase a 250 MHz, multi-function counter for less than $350, and basic "no-frills" counters are available at under $100.

Electronic counters have always had the advantages of speed, accuracy, and ease of operation. Now that they have become portable and affordable, their use has become universal. Figures 28 and 29 show the Heathkit 110 MHz counter Model IM-4110 which is typical of the type of general-purpose counter you may encounter in the field.

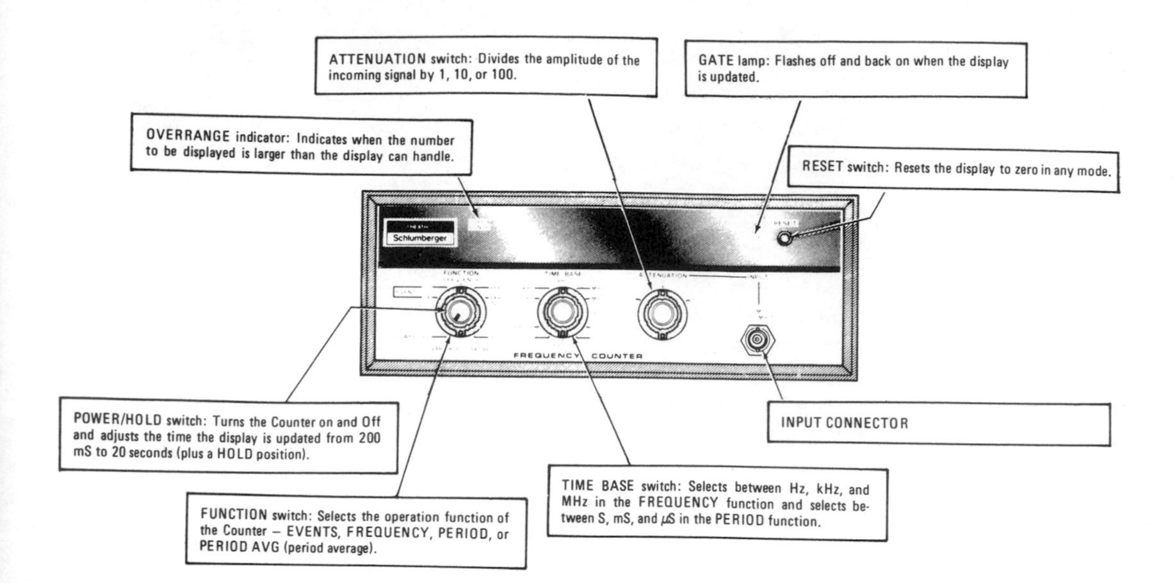

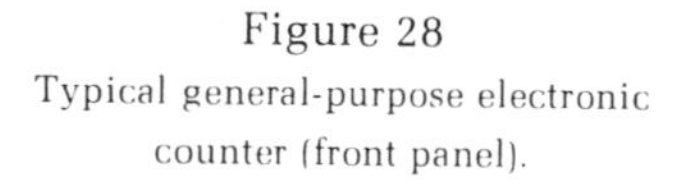

Figure 28
Typical general-purpose electronic counter (front panel).

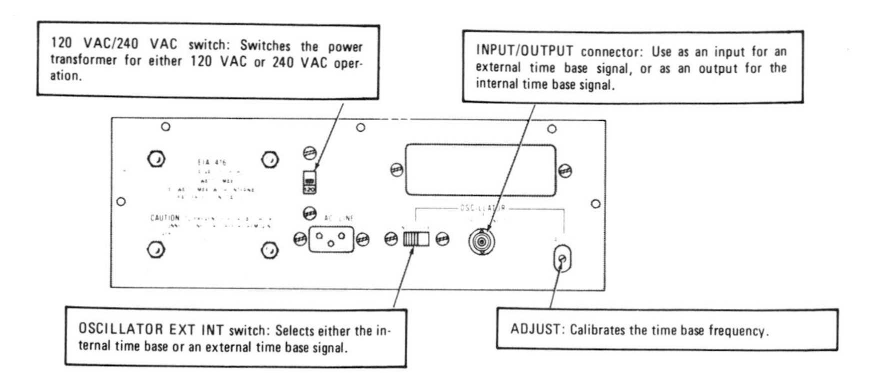

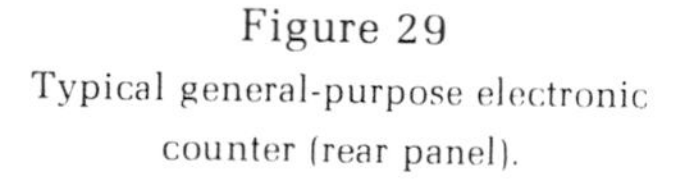

Figure 29
Typical general-purpose electronic counter (rear panel).

Operational Overview

All counters, regardless of type or complexity, measure frequency by comparing the frequency of the input signal with a known frequency or time period. In addition to direct frequency measurement, many modern general-purpose counters have provisions for measuring the time period of the input signal and totalizing the number of events which occur during an externally determined time period. Some of the more expensive counters also have provisions for the insertion of plug-in modules which adapt the counter to an even greater range of applications, such as digital voltage measurements, measurement of time intervals, and range extension to nearly 20 GHz.

Although counters vary greatly in specifications, design, and price, all have certain major functional areas in common. The basic parts of a counter are the *time base, input signal conditioner, gate control, main gate, and decade counter and display*. We will discuss these functional areas in more detail later but for the moment let's establish just what each one does and how it relates to the overall function of the counter. Refer to Figure 30 for the following discussion.

Signal Conditioner

The purpose of the signal conditioner section, or front end, is to convert the input signal into a waveshape and amplitude compatible with the internal circuitry of the counter. This usually entails the use of an amplifier to increase the amplitude of low-level signals and attenuators to reduce the amplitude of strong signals. Trigger level and slope selec tion circuits may also be included. A Schmitt trigger or other pulse shaping stage is used to convert the amplified input signal into a fast-rise, rectangular wave. No matter how a particular manufacturer elects to process the signal, the output of the signal conditioner section consists of a "digital" pulse train in which each pulse represents one cycle of the input signal.

Main Gate

The conditioned signal is applied as one of the inputs to a gate circuit (input A). If the gate is enabled (B high), the signal will pass through to a series of decade counter stages. When the gate is disabled (B low) the input pulses are prevented from reaching the counter stages. Since the counter stages count all pulses which pass through the gate, it follows that the basic accuracy of the instrument hinges upon precise control of the gate. This is a function of the time base section.

Time Base

The vast majority of counters use a crystal-controlled time base oscillator operating at a frequency between 1 MHz and 10 MHz. This frequency is divided down, through a selectable number of decade divider circuits, to more practical values and used to drive the gate control circuit. The front panel "time base" switch allows the operator to select a time base (gate time) compatible with the signal being measured. Typical gate times are 10 seconds, 1 second, 100 ms, 10 ms, etc. The position of this switch also controls the decimal point in some counters.

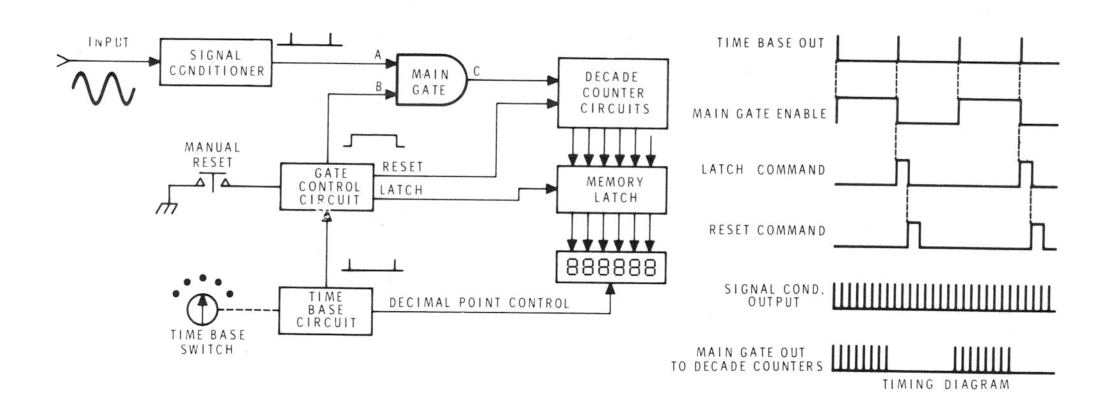

Figure 30
Electronic counter – functional block diagram.

Gate Control

The gate control circuit acts as the synchronization center of the instrument. Not only does it directly control the opening and closing of the main gate, but it also provides signals to latch the completed count into the memory circuit at the end of the counting period and to reset the decade divider circuits in preparation for the next counting period. A manual reset switch is usually included which allows the operator to initiate a new measurement cycle at any time. It may also provide a hold-off function which allows the measurement reading to be displayed for variable lengths of time.

Decade Counter and Display

The decade counters keep a running tally of all the pulses that are permitted to pass through the main gate. The number of decade counter stages determine the number of digits which can be displayed. One counter stage is required for each digit.

Various systems have been used in the past for transferring this count information to a visual display. One system, which is currently popular, employs a memory latch circuit which is monitored by the display unit. At the end of each counting period, the information in the decade counter circuit is transferred to the memory latches. This updates the stored information from the preceding count. Thus, the operator sees a steady display which is updated after each counting period.

Many types of visual display units can be used to provide a readout. These include such devices as gas-discharge tubes, light-emitting diode arrays, and liquid-crystal displays.

Summary

Let's briefly summarize what you have learned so far. Referring to Figure 30, assume for a moment that no input signal is applied. That leaves the time base oscillator as the only signal source in the instrument. Let's also assume the time base switch is set for a one-second time base. This will cause the decimal point to appear in its proper place in the visual readout. It will also result in pulses being routed to the gate control circuit at a one-second rate.

The first time base pulse will cause a bistable circuit (flip-flop) to change state. Assuming it was low, it will now go high which will enable, or open, the main gate. The gate will remain open for exactly one second, at which time the second pulse from the time base reaches the gate control flip-flop, causing it to again change state. Its output now goes low, which inhibits (closes) the gate.

The falling edge of the gate control signal triggers a single-shot multivibrator which generates a short duration pulse. This pulse is routed to the memory latch circuit, causing any data contained in the counter stages to be transferred (or latched) into memory.

The falling edge of this latch pulse, in turn, triggers a second single-shot, which causes the decade counter stages to be reset to zero.

This process, with the gate alternately opening and closing at a one second rate, will be continuously repeated.

If we now apply a signal of 10 kHz to the input of the instrument, 10,000 pulses (10,000 Hz times 1 second) would be counted by the decade counter stages each time the gate is open. When the gate closes, this count is transferred into storage and the decade counters are reset to zero. If the input frequency drifts to 10.005 kHz, the display will be updated at the end of the next one second counting period. This will cause the units digit to change from 0 to 5. All other digits will continue to glow steadily, however, as no change is required.

COUNTER TIME BASE SECTION, TYPICAL CIRCUITS

The single most important section of a digital counter, in terms of system accuracy, is the time base section. Without an accurate, stable time base, the other counter specifications are meaningless. The heart of the time base section is the time base oscillator.

Time Base Oscillator Circuits

Three types of time base oscillators are found in digital counters. These are the room temperature crystal oscillator, the temperature-compensated crystal oscillator, and the oven-controlled crystal oscillator. The main differences between these three types are the steps taken to reduce frequency drift due to temperature change. The design of the oscillator circuit is extremely critical regardless of the type employed.

The room temperature crystal oscillators are usually found in the least expensive counter. Such oscillators will usually vary by about $\pm 5 \times 10^{-6}$ (5 ppm) from center frequency over a temperature range of 0°C to 50°C.

The temperature-compensated crystal oscillator, often referred to as TCXO, employs special compensating components and circuits which tend to counteract changes in temperature. A well-designed TCXO can provide five times more frequency stability than the room temperature type. Thus, stability specifications of $\pm 1 \times 10^{-6}$ (1 ppm) from 0°C to 50°C are quite common. Don't be misled by the type oscillator used in any given instrument, however. Check the entire specification. A well-designed room temperature oscillator may be more stable than a poorly designed temperature-compensated type.

The oven-controlled crystal oscillator is used in the more expensive line of counters. These house the crystal in a thermostatically-controlled oven to reduce temperature changes.

Several types of oven-controlled oscillators are available. The simplest of these is the on-off type, where the heating element is always either fully on or fully off. More sophisticated counters use a proportional oven. This system provides heating which is proportional to the difference between the oven's inside temperature and the outside temperature. This method is superior to the on-off method because it reduces the cycling action of the oven, thus minimizing changes in oscillator frequency. The highest quality oscillators carry this one step further and employ double proportional ovens. These contain one proportional oven inside another which can maintain a constant crystal temperature, within 0.01%. These minimize changes in oscillator frequency due to sudden, short duration changes in room temperature.

In addition to temperature effects on oscillator frequency, a few other factors must also be considered.

As you know from previous studies, changes in power supply voltage will cause changes in oscillator frequency. A well-designed regulated power supply is, therefore, essential in minimizing the effects of varying line voltage. Of course, high stability oven oscillators require more elaborate power regulation than room temperature oscillators, since minute frequency changes due to line voltage fluctuations tend to become lost in the larger frequency variations caused by changes in ambient temperature. A typical oven oscillator power supply will provide frequency stability of about 5×10^{-10} for a 10% change in line voltage. With a room temperature oscillator, on the other hand, stability better than 1×10^{-7}, for the same voltage change, is unnecessary due to the lower inherent stability of the oscillator circuit.

Another specification of interest is "long term stability," or "crystal aging rate." All crystals display a slow, cumulative drift in oscillator frequency over a period of time. This does not mean that the crystal wears out, only that its natural frequency changes slightly. Therefore, periodic calibration is required to insure that the oscillator circuit is operating at its specified frequency.

This crystal aging rate, or speed of frequency drift, is dependant upon the quality of the manufactured crystal. The best crystals are used in double-oven oscillators and have an aging rate in the order of 5×10^{-10} per day or 1.5×10^{-8} per month. Crystals used with room temperature oscillators are typically about 3×10^{-7} per month.

The above figures are misleading, however, as aging only occurs while the oscillator is operating. Room temperature oscillators only operate while the counter is actually turned on, while oven oscillators usually remain on as long as the power cord is plugged in. This is because an oven oscillator requires 24 hours after turn-on to reach its specified stability. Therefore, over a year's time, an oven oscillator may display more aging than a room temperature oscillator which was used very little.

The final oscillator specification you should be concerned with is its "short term stability." This refers to the minor fluctuations of frequency above and below the mean (or long term) value. These are caused by imperfections in the crystal and instability in the oscillator circuit. Short term stability specifications are usually not stated for room temperature oscillators, since these minor fluctuations (typically 5×10^{-11}) are masked by the relative oscillator instability.

Most counter oscillators operate at either 1 MHz or 10 MHz, as these frequencies fall within the frequency range where crystals display their greatest stability. These frequencies are also easily divided down to values which are convenient for counter use.

Time Base Divider Circuits

The time base divider circuit makes up the remainder of the time base section. This circuit is comprised of a series of decade counter stages as shown in Figure 31.

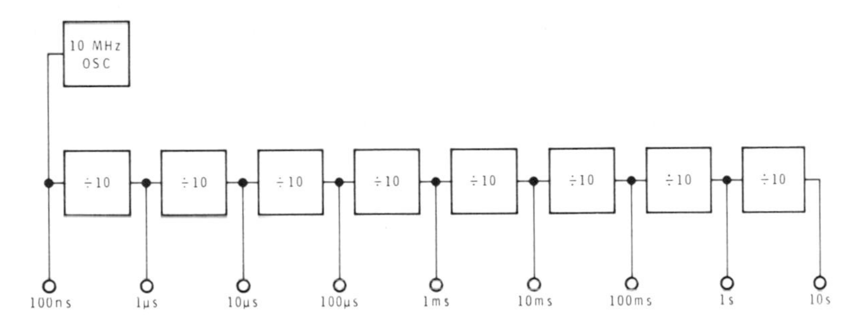

Figure 31
Crystal oscillator and time base divider – functional block diagram.

Each decade counter stage divides its input by ten; or in other words, for every ten pulses of input it provides one pulse of output. Thus, using a 10 MHz oscillator and eight decade divider stages, pulse spacings are available in decade increments from 10 seconds to 100 nanoseconds. But few counters will allow the operator to select all nine of these time base frequencies. Some use only the 1-second gate, which provides 1 Hz resolution and, with eight digit readout, a maximum reading of 99,999,999 Hz. Most general-purpose counters, however, provide operator selection of from four to seven time base frequencies. These are selected by use of the **time base** switch. (Some manufacturers label this switch **gate time**.)

The time base switch selects the submultiple of the crystal oscillator frequency that will be used as the reference for counting. This switch is also wired to control the position of the decimal point in the readout display. For instance, with a 10-second gate time selected, the switch would indicate a "Hz" unit of count and the decimal point positioned preceding the last digit to the right or 000000.0. By moving the time base switch to a 1-second gate time, the units indicator may remain "Hz" and the decimal point moved one position to the right (0000000.); or it may indicate kHz, in which case, the decimal point will move two places to the left (0000.000). Both methods provide a correct display and it is a matter of designer's choice as to which is used.

The selected time base frequency is now routed to the gate control circuit.

COUNTER GATE CONTROL SECTION, TYPICAL CIRCUITS

This circuit is the "brain" of the counter. It determines the operating mode and issues commands to other sections of the instrument. In addition to controlling the main gate, it also controls updating the memory latch circuit, resetting the decade counters, and determining the display time.

Figure 32 shows a simplified block diagram of this circuit and its associated timing diagram. Although the actual circuit is much more complex, and will vary from one manufacturer to another, this will serve to show you the operating principles involved.

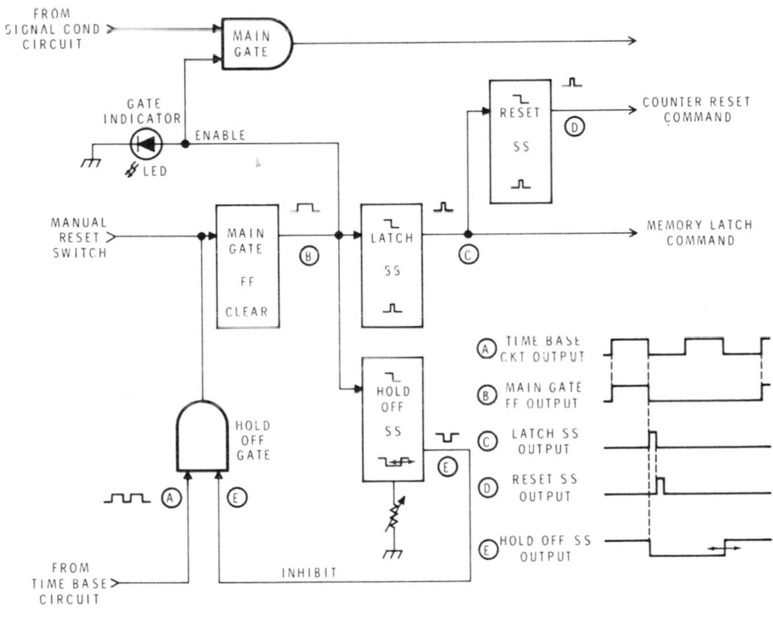

Figure 32

Gate Control

Let's start by assuming that all circuits are in their quiescent (resting) state. In practice, you can do this by depressing and holding the manual reset switch. This will result in the output of the main gate flip-flop being held low, closing the main gate. At this same time, all the single-shot multivibrators are reset, as are the decade counters and memory latch circuits. Releasing the reset switch will allow all of these circuits to resume their normal operations.

The first pulse from the time base circuit is applied to the hold-off gate. Because the output of the hold-off single-shot is high, this pulse will pass through to the input of the main gate flip-flop. This will cause the flip-flop to change states. Its output, which was low in the reset condition, goes high. The main gate is now open; the gate indicating LED is lit; and all pulses felt at the signal input of the main gate will be passed on to the decade counters. The latch and hold off single-shots both trigger on a high-to-low transition. As their inputs are now high, they can be considered "armed," but they will not trigger until the input drops low.

The second pulse from the time base circuit will again pass through the hold-off gate and toggle the main gate flip-flop. Its output will go low, closing the main gate. The latch and hold-off single shots, which were armed by the flip-flop's high output, will feel this high-to-low transition and will trigger.

Data Storage Control

The output of the latch single-shot will go high for a short period of time, during which the count information contained in the decade counters will be transferred into memory. This high will also arm the reset single-shot. Once the count information has been stored in memory, it is no longer of any use in the counters. Therefore, at the end of the latch pulse, its high-to-low transition will cause the reset single-shot to trigger. Its output pulse is routed to the clear inputs of all decade counter stages, resetting them to zero in preparation for the next counting cycle.

Hold-off

At the same time the latch single-shot started the data storage process, the hold-off single-short was also triggered. Its output, taken from the normally high ($\overline{Q}$) output, will go. This low is felt on the control input of the hold-off gate, which will inhibit (close) the hold-off gate. The output duration of the hold-off single-shot is usually adjustable by a front panel control. Thus, the operator can adjust the sampling rate from a few milliseconds to many seconds. At the end of this delay period, the output of the hold-off single-shot will once more go high. The next time base pulse will once again toggle the main gate flip-flop, causing the entire cycle to repeat itself.

Most counters incorporate a display time "hold" feature which causes the hold-off circuit to latch-up each time the hold-off single shot triggers. Thus, in the hold mode, the counter will only sample its input whenever the reset switch is depressed and released.

COUNTER SIGNAL CONDITIONER, TYPICAL CIRCUITS

Now that we have discussed the generation of the time base signal and seen how it is used to synchronize the counter circuits, let's look at the counter's front end for a moment and discuss some of the circuits used to condition the incoming signal.

As we pointed out in the general discussion (Page 4-56), the purpose of the signal conditioner is to accept a wide variety of input amplitudes and waveshapes and convert them into digital "1"s and "0"s which can be processed by the counter's logic circuitry.

Figure 33 shows the simplified circuit of a typical signal conditioning section. Not all counters will provide all of these features, such as selectable coupling and adjustable trigger level, however, you should be aware of their functions.

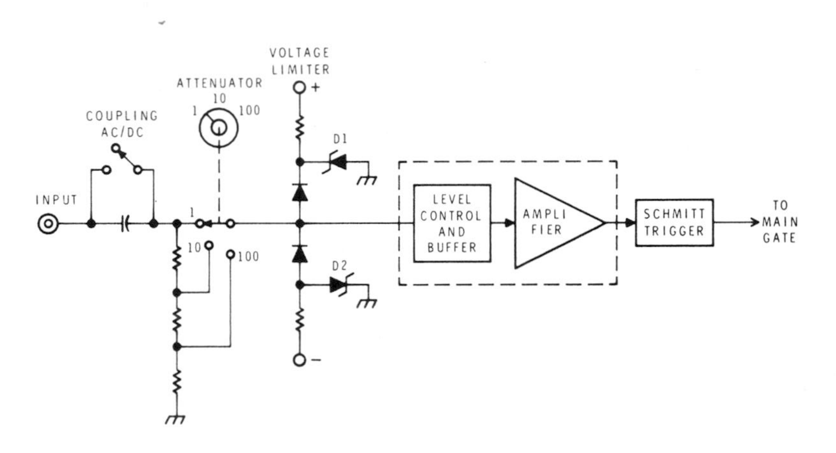

Figure 33
Signal conditioner circuit – simplified diagram.

Input Coupling

The signal applied to the input connector is fed to a coupling selector switch. In the AC position, as shown, the DC component of the signal is blocked and the AC component is passed on to the attenuator switch. In the DC position, the signal, along with its DC component, is passed directly to the attenuator.

Many counters use fixed DC coupling and, as long as the subsequent circuits are well designed for this type coupling, perform quite well. Selectable coupling does provide an extra measure of versatility, however. In some cases, it is advantageous to consider only a small AC signal while disregarding its DC reference level. Under other circumstances, the input coupling capacitor may distort the input signal to such an extent that erroneous readings result.

Attenuator

From the coupling circuit, the signal passes through an input attenuator. In the "1" position, as shown, the signal is fed directly through the switch contacts with no attenuation. Thus, the stage has an input/output amplitude ratio of 1:1. In the "10" position, the signal is reduced, by means of a precision voltage divider, to one-tenth of its original amplitude or a 10:1 attenuation ratio. In the "100" position, this ratio is increased to 100:1.

In addition to its obvious function of preventing strong signals from overdriving the amplifier, it is often effective in reducing noise amplitude to a point where it will not cause false triggering.

Voltage Limiter

A voltage limiting stage is provided to protect the sensitive electronic components from excessively high voltage levels. Should the input signal level at this point exceed the Zener breakdown voltage of either D1 or D2, the appropriate diode would conduct. This would limit the input to the Zener voltage value. D1 protects the circuit from positive voltage overloads, while D2 protects from negative overloads.

Level Control and Amplifier

These stages may be separate discrete circuits or may be combined on one IC chip.

The level control and buffer stage usually consists of two field-effect transistors (FETs) which serve a dual purpose. First, they convert the high input impedance, required of the counter, to a low output impedance to drive the amplifier. Secondly, they establish the DC reference level of the amplifier output. The importance of this will become apparent when we look at the operation of the Schmitt trigger stage. This DC reference may be preset using fixed bias levels, may be an internal calibration adjustment, or may be brought out to the front panel as an operator control, depending upon the counter design.

Schmitt Trigger

The output of the amplifier is still an analog signal. The function of the Schmitt trigger is to convert this to digital form.

As you will remember from your studies of circuits, a Schmitt trigger circuit is a bistable device similar to a flip-flop. Its hysteresis window, however, gives it a certain degree of noise immunity, which makes it ideal for this application.

Due to its critical importance to proper counter operation, let's quickly review the operation of this circuit. Figure 34 shows a discrete component Schmitt trigger circuit and its associated wave forms. The majority of recent applications use an integrated circuit version; however, its basic operation is the same. The input shown is a sine wave but any wave form will produce the same results as long as it passes through both the upper and lower limits of the hysteresis window. The positioning of this signal, in relationship to the hysteresis window, is a function of the amplifiers DC reference level.

With Q1 cut off, its collector voltage is a relatively high positive value. This makes the voltage on the base of Q2 high enough to cause Q2 to saturate. As Q2 conducts, it forces a heavy current through R3. This develops a positive voltage (E_{R3}) at the emitter of Q1 which helps to hold Q1 cut off.

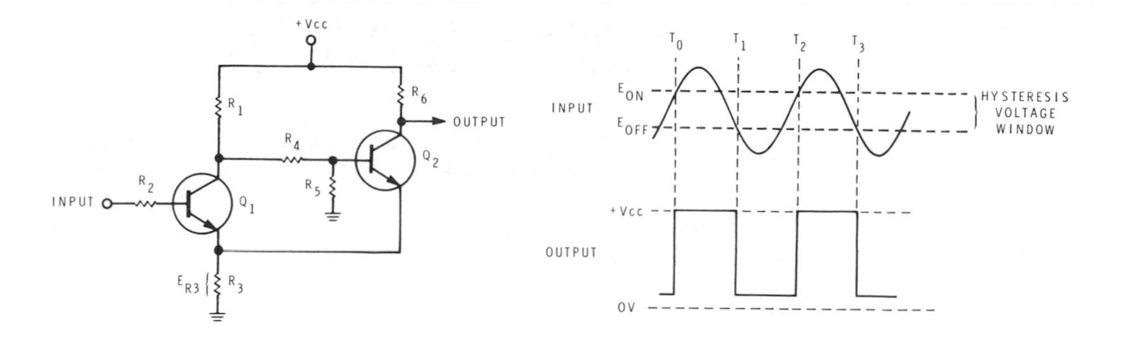

Figure 34

Q1 cannot conduct until the input voltage rises about 0.7 volts above E_{R3}. The voltage at which Q1 conducts is shown in Figure 34, as E_{on}. When Q1 conducts, the collector voltage decreases. In turn, the base voltage of Q2 decreases. Therefore, Q2 conducts less and E_{R3} decreases. The decrease in E_{R3} makes Q1 conduct harder, causing the collector voltage to decrease further. This action is regenerative and continues until Q1 is saturated and Q2 is cut off. As shown in Figure 34, the output voltage suddenly steps to $+V_{cc}$ as Q2 is quickly cut off at time T_0. The circuit remains in this state until time T_1.

It might seem that Q1 would cut off again as soon as the input voltage falls below E_{on}. However, in practical circuits, the reset occurs at a lower voltage called E_{off}. When the input voltage falls below E_{off}, Q1 comes out of saturation and its collector voltage increases. This causes Q2 to conduct forcing additional current through R3. Thus, E_{R3} increases causing Q1 to conduct less. Again, the action is regenerative and Q1 is quickly driven to cut off while Q2 is quickly driven to saturation. Thus, at time T_1, the output voltage falls to a low value.

The difference between E_{on} and E_{off} is called the hysteresis voltage or hysteresis window.

The output of the Schmitt trigger stage is now applied to one input of the main gate as the signal to be counted. Before moving on to the remaining counter circuits, however, let's examine some of the sources of error within the signal conditioning section.

SIGNAL CONDITIONER INPUT CONSIDERATIONS

Up to this point, we've generally considered the input signal to be a pure sine wave, or other idealized waveform. In the real world, however, it's seldom this simple. Complications such as RF noise, switching transients, amplitude modulation, impedance mismatches, harmonic distortion, and many other factors can combine to provide input signals which are far from ideal. The signal conditioner section is the interface between the counting circuits and the outside world. How well a counter performs under adverse signal conditions is directly related to how well the front end is designed and how well the operator understands his equipment. While factors internal to the counter can be controlled to limit errors to a few parts per million, errors due to triggering problems may exceed 100%.

You, as an operator, must be able to recognize these gross errors and reduce or eliminate their causes.

Input Impedance

The first characteristic to consider is the instrument's input impedance. For most low frequency measurements, a high input impedance is considered desirable because it has less tendency to load the circuit under test. But since impedance is both resistive and capacitive, you must consider both of these factors.

At high frequencies (above 100 MHz) capacitive loading becomes the major factor. For instance, a typical specification for this type input is 1 megohm shunted by 25 pF. At 500 MHz this is actually a 25 ohm load, which is closer to a short than it is to a high impedance. Thus, for measurements above 100 MHz, a separate input connector is usually provided which displays a nominal impedance of 50 ohms. As most pieces of equipment which operate at these frequencies are designed to operate into a 50 ohm load, no serious impedance mismatches are likely to occur. However, if your counter has only a "high impedance" input and you wish to measure a 70 MHz oscillator with a 50 ohm output, you may have an impedance matching problem. The shunt capacitance plus the capacitance of the coaxial cable (RG-58 C/U has approximately 30 pF per foot) will tend to load down the oscillator. If it oscillates at all, it will be pulled considerably off its actual frequency. In this case, a 50 ohm feedthrough termination should be used at the counter's input connector. This will provide the proper termination for 50 ohm coaxial cable and provide the oscillator with a load close to 50 ohms.

Even at low frequencies, the shunt capacitance may be too high for some circuits. So you can also reduce capacitive loading by using a 10× oscilloscope probe. These are usually 10 megohms shunted by 3 to 15 pF. Of course, they also reduce the signal amplitude by a factor of 10, so this will not work at very low signal levels.

Sensitivity and Input Signal Amplitude

The sensitivity specification of a counter refers to the minimum input signal amplitude that can be counted. This amplitude is usually specified as an rms value of a sine wave input. For pulse inputs, the sensitivity is:

$2 \times 1.414 \times$ specified value.

These values are based upon the input amplifier gain and the Schmitt trigger hystersis levels.

It might seem that the more sensitive the counter input, the better. The broadband amplifier, however, makes it susceptible to noise, which can result in false triggering. Optimum sensitivity, for high impedance inputs, is considered to be 25 to 100 mV, while for 50 ohm inputs, 10 mV is practical.

Equally important is the maximum voltage value that you can apply to the input without damaging the instrument. All amplifiers have a damage level and some type of protective circuit is usually provided. Those circuits can fail to protect against high speed transients or spikes, however. For this reason, many counters use a high speed fuse in the input line.

Maximum input levels are specified for high impedance inputs as the sum of a DC value plus a peak AC value. For 50 ohm inputs, the maximum signal level may be stated in dBm (decibels above one milliwatt). A + 19 to + 27 dBm (2V to 5V rms) rating is typical for 50 ohm inputs.

Since these instruments are rather expensive, you should use them carefully to avoid damage. If you are measuring a signal where DC might be present, use an external blocking capacitor. Use the smallest value that will allow reliable triggering. In cases where you are measuring excessively large AC signals, use an external attenuator or dividing probe.

Never connect a counter directly to a transmitter or other high power signal source. To measure these signals, use a short length of unshielded wire as an antenna; position it to pick up a measurable signal. You could also use an attenuator or directional coupler of sufficient power rating.

Trigger Level

The purpose of the trigger level control is to effectively shift the hysteresis level above or below the amplifier's output reference level. As you can see in Figure 35, this allows the counting of positive or negative pulse trains. It also allows the operator to position the triggering region away from portions of the signal which might cause false triggering due to noise or other factors.

Many "no frills" counters use a fixed trigger level over which the operator has no control. More sophisticated counters, however, provide a 3-position switch with the "preset" position corresponding to Figure 35A, to " + " position corresponding to Figure 35B, and a " − " position as in Figure 35C. An even more flexible arrangement is a slope switch, which moves the trigger level above or below the reference level, and a level potentiometer which allows continuous adjustment over the entire range of the input. Figure 36 shows some of the ways in which these controls can be used to minimize trigger errors.

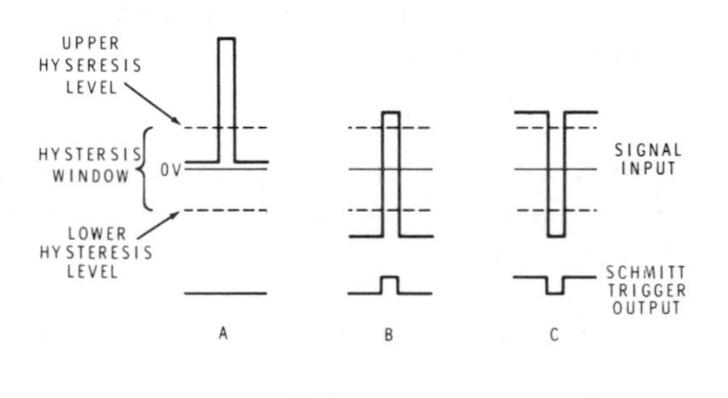

Figure 35

A — Trigger level preset to normal position. Signal will not be counted as it does not cross both hysteresis levels.

B — Trigger level switched to "+" position. Signal will now be counted.

C — Trigger level switched to "−" position for counting negative referenced signals.

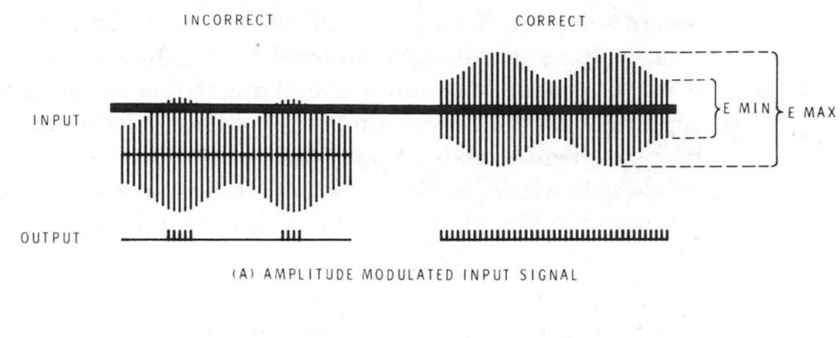

(A) AMPLITUDE MODULATED INPUT SIGNAL

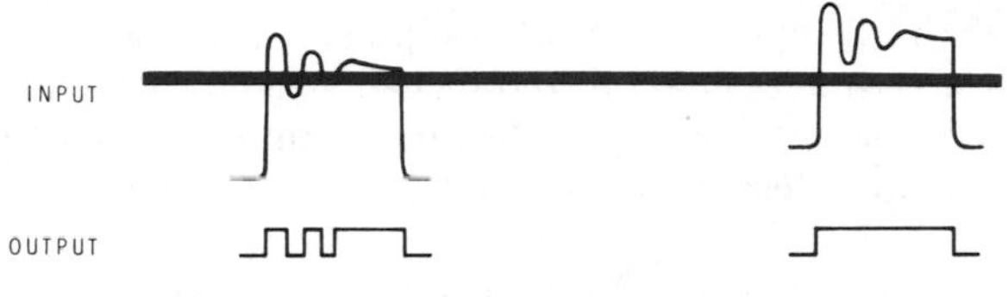

(B) PULSE INPUT WITH OVERSHOOT

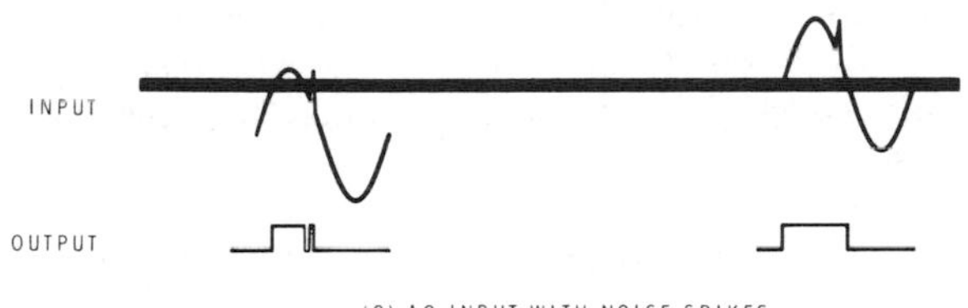

(C) AC INPUT WITH NOISE SPIKES

INPUT

OUTPUT

(D) SAWTOOTH INPUT WITH DC OFFSET

Figure 36
Adjustment of trigger level to obtain correct count.

Figure 36A shows how you could obtain an incorrect reading if you were measuring an amplitude modulated signal. By positioning the signal so that the hysteresis window is looking at the center of the waveform, amplitude modulation would not be a factor. This would hold true unless the percentage of modulation became so great that the amplitude of E min was less than the hysteresis window. In this case, only a special purpose counter with "burst" measuring capability would be able to measure it accurately.

The waveform in Figure 36B shows the result of improper termination of the coaxial input cable. Fast rise pulses will tend to ring, or oscillate, when the input cable is not terminated in its characteristic impedance. As you can see, this will not affect the count if it is positioned where the oscillations do not traverse the hysteresis window.

Figure 36C shows the effects of voltage spikes (switching transients, etc.) riding the input signal. Since these noise spikes are random in nature, it is not possible to predict what portion of the waveshape they will appear on. In the case of sinusoidal input, it is usually best to trigger on the zero-crossing point. This zero reference point has the steepest slope; thus, will traverse the hysteresis window the fastest. Should this type of noise become too great a problem, it may be necessary to filter the input signal.

If you measure the frequency of a signal that has a DC offset, it may be advisable to use AC coupling to remove this DC component. Another alternative is to counteract the DC level using the trigger level control.

The important thing to remember is that the Schmitt trigger output will go high any time its input increases through the upper hysteresis level and will go low when it decreases through the lower hysteresis level.

By knowing as much as possible about the signal you are attempting to measure, and by understanding the limitations of your counter, fairly accurate measurements are possible under very adverse signal conditions.

MAIN GATE, COUNTER, AND DISPLAY SECTION, TYPICAL CIRCUITS

Main Gate

The input signal, after being converted to a fast transition rectangular waveshape, is applied to one input of the main gate. While this circuit is represented by a simple "AND" gate, it is actually a slightly more complex logic circuit. The actual design will be somewhat different for nearly every counter and will vary from a simple, straightforward circuit on a pure frequency counter to a relatively complex circuit for time interval counters. In any case, two signal inputs will enter the gate circuit. One of these is the conditioned input signal and the other is the gate control signal. As you will remember from your digital studies, if one input of an AND gate is held high and pulses of the proper logic level are applied to the other input, those pulses will be felt at the gate output. In this condition, the gate is open. When the high input is changed to a logic low, the pulses on the other input will not pass through the circuit; thus, the gate is closed.

One source of counter error is attributable to this circuit. This is the plus-or-minus (±) one-count ambiguity. If you read the accuracy specifications for any general purpose counter it will state the time base accuracy, ± one count. This ± count is caused by the gate control signal not being synchronized with the input frequency. This is illustrated by Figure 37. In both cases, the gate is held open for exactly the same amount of time. However, when the gate is opened at time "A" two pulses are allowed to pass through to the counters, while the gate at time "B" will allow only one pulse to pass through. Thus, as we do not know when the counter is gating, relative to the input waveshape, there is a one-count uncertainty in the reading.

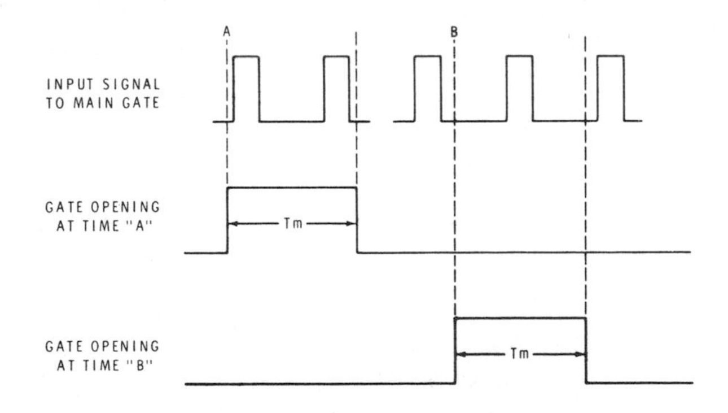

Figure 37

To summarize overall counter accuracy, a true statement of uncertainty would be; the indicated reading, ± time base error, ± trigger error, ± one count.

Counter and Display Section — Overview

Now that we have discussed the internal timing of the instrument and the conditioning of the input signal, the only remaining area to cover is the actual counting and display of the number of pulses received per unit of time.

Figure 38 shows a functional block diagram of one way this is accomplished.

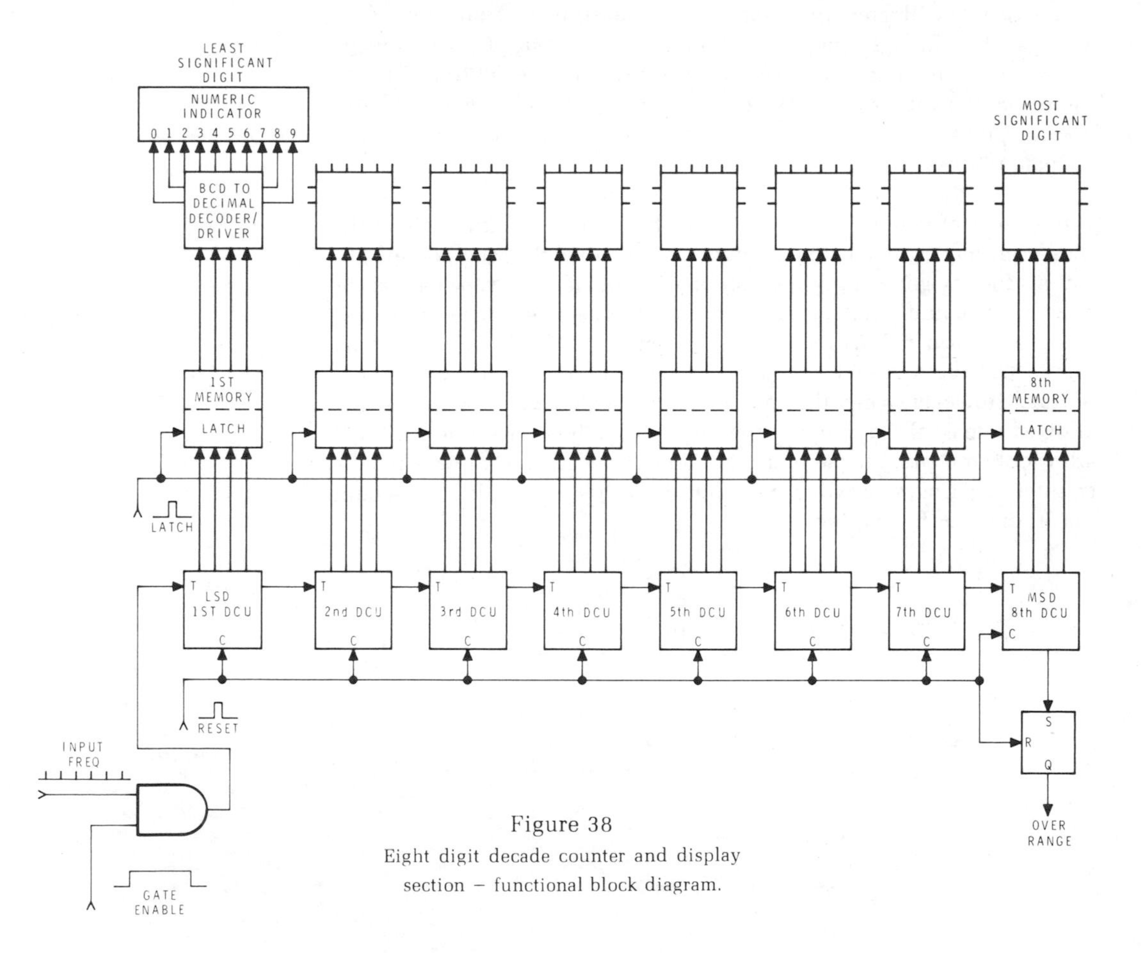

Figure 38
Eight digit decade counter and display section – functional block diagram.

As you will recall from previous discussions, when the gate enable line is high, the gate is open. This allows the input frequency pulses to reach the first decade counter unit (DCU), which acts as a decade divider. Each DCU has two outputs; a four-bit binary coded decimal number (BCD) which goes to its associated memory latch and a "carry" pulse to the next DCU. One carry pulse output is generated for every ten input pulses. The final DCU, which is the most significant digit, will reach a count of nine only when the instrument is nearing its operational limits. When it receives a tenth pulse its carry pulse will cause a set-reset flip-flop to set, lighting an overrange lamp and alerting the operator to an invalid reading.

During the counting period, the outputs of the DCU's have been applied to the BCD inputs of their associated memory latches. This will have no effect, however, as long as the "latch" input remains low. At the end of the counting period, the main gate will close and the DCU output states will represent the total number of pulses received during that period. The "latch" line will now go high momentarily, causing the BCD number contained in each DCU to be locked into memory. Once this has been accomplished, the DCU reset line receives a high pulse which resets all of the DCU's to zero in preparation for the next counting period.

The outputs of the memory latch circuits are applied to decoder/drivers which convert the BCD value into a form compatable with the type display used. In this example, an illuminated decimal number readout is shown that requires a BCD-to-decimal decoder. If a seven-segment display is used, a BCD-to-seven segment decoder is required.

Decade Counter Units

The only difference between the DCU's is the speed of counting which is required. The first DCU must be able to accurately count at the maximum input frequency rate. Subsequent stages, however, have maximum speeds of decreasing factors of ten, each stage requires ten input pulses for one output pulse. For this reason, special consideration must be taken in the design of the first counter stage in high speed instruments. This usually is in the form of more expensive, high-speed IC chips using ECL logic.

For the purpose of detailed explanation, we will discuss one decade counter stage and its associated memory, decoder, and display unit. Keep in mind, however, that one of these complete sets will be required for each digit of readout.

A single decade counter unit is made up of four flip-flops configured as a four-bit binary counter. However, instead of being allowed to count from zero through fifteen, a gate circuit is added which resets all four flip-flops to zero on the tenth input pulse, as shown in Figure 39. Notice that the AND gate inputs are connected to the Q outputs of the B and D flip-flops. As the timing diagram shows, the D output goes high at the end of the eighth input pulse. The B output has just gone low. At the end of the tenth input pulse, however, the B output will attempt to go high. As the D output is already high, this low-to-high transition will be felt, through the AND-NOR gate combination, as a high-to-low transition to the four clear inputs, resetting all the counter stages to zero. This resetting causes the output of the D flip-flop to go low, allowing the next DCU to record one count.

The reset bus is connected to each DCU through the NOR gate. The leading edge of this signal will also reset all counter stages to zero.

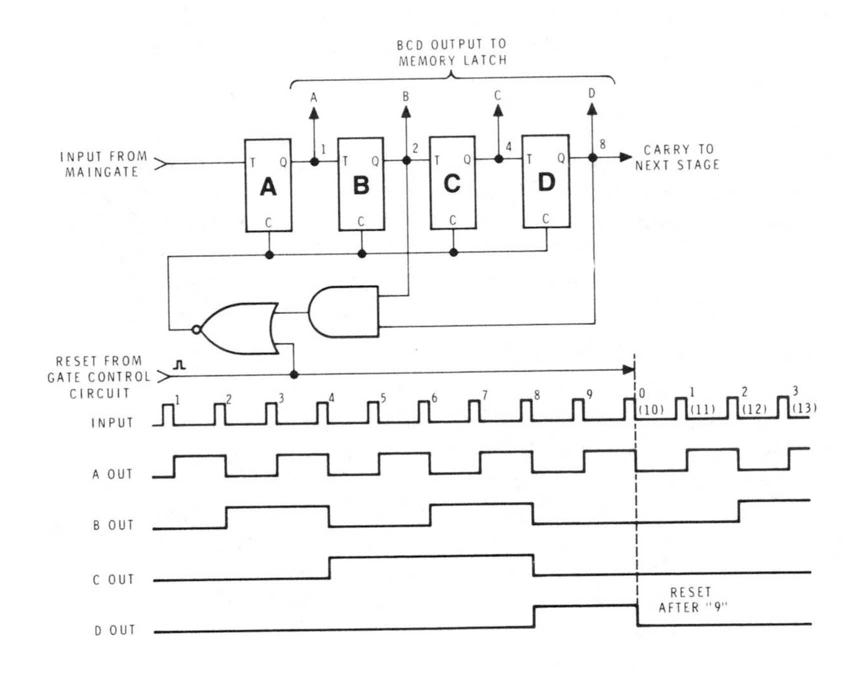

Figure 39
Decade counter stage with timing waveforms.

A data output is taken from the Q output of each flip-flop, the combination forming a binary-coded decimal word indicating the count content of the stage. These combinations are:

INPUT COUNT	BCD OUTPUT D	C	B	A	
0	0	0	0	0	
1	0	0	0	1	
2	0	0	1	0	
3	0	0	1	1	
4	0	1	0	0	
5	0	1	0	1	
6	0	1	1	0	
7	0	1	1	1	
8	1	0	0	0	
9	1	0	0	1	
0 Reset	0	0	0	0	— Carry 1 to next DCU

The BCD output is applied to the DCU's associated memory stage as shown in Figure 40.

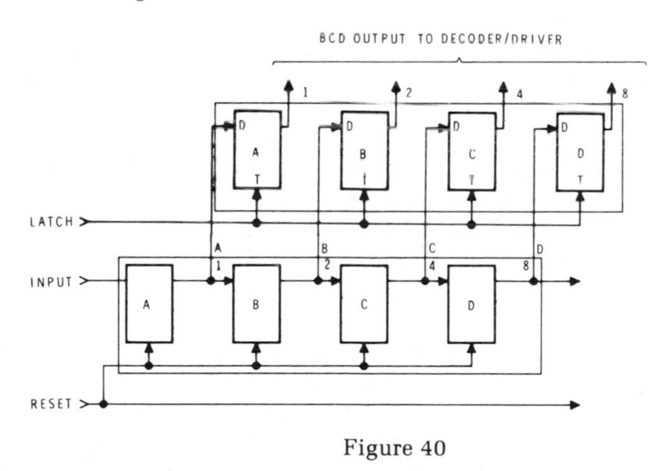

Figure 40

Memory Latch

At the end of each counting cycle, the DCU will contain some count information which will be present at the memory latch circuits D (data) inputs, but will not have any effect on the circuit until a latch command pulse is felt on the T (transfer) input. You can see the reason for this by looking in Figure 41.

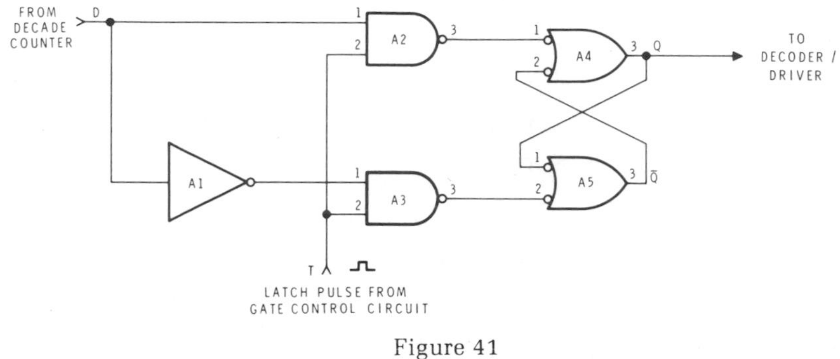

Figure 41
"D" flip-flop memory latch circuit – logic diagram.

In this circuit, A4 and A5 form an RS latch circuit, while A1, A2, and A3 make up a gating input circuit.

First, let's assume that the output of A4 is low (a binary 0 stored) and that the D input is also low. This means that a low is felt at A2 Pin 1, while A3 Pin 1 is high due to inverter A1. When the latch pulse comes in at the T input, Pins 2 of A2 and A3 both go high. Since A2 has one high input (T) and one low input (D), its output will remain high. A3 has two high inputs; thus, its output will go low. A low applied to A5 Pin 2 will have no effect on the latch circuit because A5 Pin 3 is already high (the complement of A4 Pin 3) due to the low on A5 Pin 1. Thus, if a 0 is stored from a previous count and a 0 is presented for storage from the latest count, no switching action will take place. This is also true if a 1 is stored and a second 1 presented.

Now let's look at circuit operation when the presented data is different than the stored data. With the same binary 0 stored in the latch circuit, we will now bring in a binary 1. With the arrival of a latch command pulse, both inputs of A2 will be high, causing its output to go low. This low, felt at A4 Pin 1, will cause A4 Pin 3 to go high. Both of A5's inputs will now be high. Therefore, its output will go low, locking A4 to the 1 state.

Conversely, had a 1 been stored and a 0 inputted, A5 would have flipped causing A4 to lock to the 0 state. Keep in mind that four of these circuits are required for each digit of readout on the instrument.

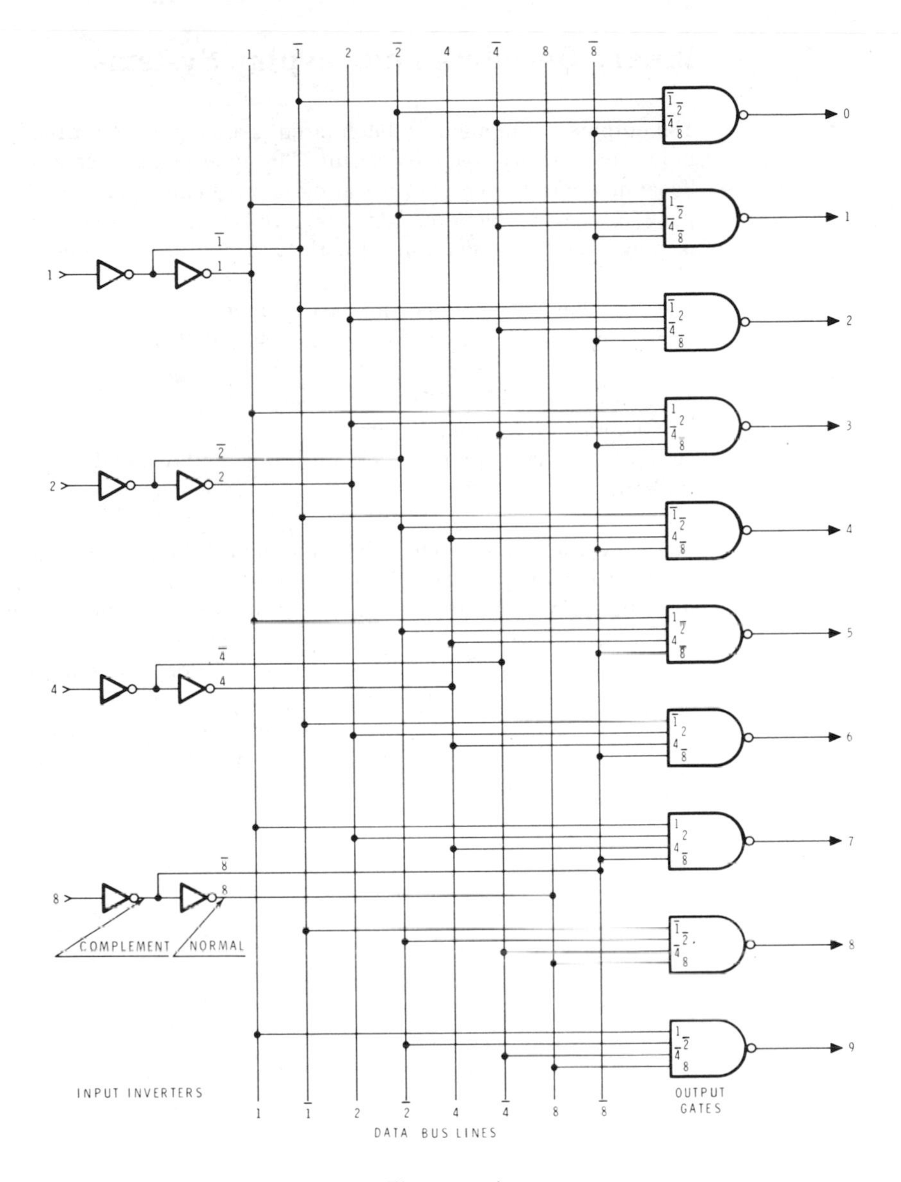

Figure 42A
BCD — to decimal decoder — logic diagram.

Binary Decoders and Display Systems

The outputs of the memory latch circuits are connected directly to the inputs of a binary decoder circuit. The type of decoder circuit used depends upon the type of indicating display employed. The two most popular types are the lighted decimal number and the 7-segment number displays. Decoder operation for both types will be discussed briefly.

The function of each decoder is to convert a binary-coded-decimal number into a decimal number. In the case of a BCD-to-decimal decoder, it converts a four bit combination of binary numbers to one specific decimal number. The BCD-to-seven-segment decoder converts the four bit combination to another combination of from two to seven bits. The end result is the same, however, an illuminated decimal number in the readout.

Figure 42A shows how ten NAND gates and eight inverter stages can be connected to form a BCD-to-decimal decoder. The two inverters on each input line provide both the normal input state and the complement of that state to data buses. In this example, a low output state (binary 0) is required to illuminate the readout numeral. In order to obtain this low output state, all four inputs of a gate must be high (binary 1). With both possible states of each input available, it is a simple matter to connect the gate inputs to bus lines which will all be high when the desirable BCD number is present.

For example, if a count of 5 is stored in a memory latch circuit, its 8421 output will be 0101 (see Figure 42B). By connecting the #5 gate to the normal 1, complement 2, normal 4 and complement 8 ($\overline{8}$ 4 $\overline{2}$ 1) bus lines, all four inputs will be high only when a BCD five is present. Any other BCD number from zero to nine will result in one or more of the #5 gate inputs being low, which will inhibit the gate.

Figure 42B is a truth table showing the input and output states of the decoder for counts zero through nine. Figure 42C shows the gate input to bus line connections which will provide the correct decoding action. If you have any doubt as to how this circuit operates, you should take a few moments, at this point, and trace through these connections. Start by selecting a number at random from the BCD input column. Then determine which four bus lines will be high with that input. Next, trace all gate connections between the high bus lines and the gate inputs. You should find that only one gate will have four high inputs.

	BCD INPUT	DECIMAL OUTPUT
	8 4 2 1	0 1 2 3 4 5 6 7 8 9
0	0 0 0 0	0 1 1 1 1 1 1 1 1 1
1	0 0 0 1	1 0 1 1 1 1 1 1 1 1
2	0 0 1 0	1 1 0 1 1 1 1 1 1 1
3	0 0 1 1	1 1 1 0 1 1 1 1 1 1
4	0 1 0 0	1 1 1 1 0 1 1 1 1 1
5	0 1 0 1	1 1 1 1 1 0 1 1 1 1
6	0 1 1 0	1 1 1 1 1 1 0 1 1 1
7	0 1 1 1	1 1 1 1 1 1 1 0 1 1
8	1 0 0 0	1 1 1 1 1 1 1 1 0 1
9	1 0 0 1	1 1 1 1 1 1 1 1 1 0

Figure 42B
Input – output truth table.

GATE	BUS LINES			
0	$\overline{8}$	$\overline{4}$	$\overline{2}$	$\overline{1}$
1	$\overline{8}$	$\overline{4}$	$\overline{2}$	1
2	$\overline{8}$	$\overline{4}$	2	$\overline{1}$
3	$\overline{8}$	$\overline{4}$	2	1
4	$\overline{8}$	4	$\overline{2}$	$\overline{1}$
5	$\overline{8}$	4	$\overline{2}$	1
6	$\overline{8}$	4	2	$\overline{1}$
7	$\overline{8}$	4	2	1
8	8	$\overline{4}$	$\overline{2}$	$\overline{1}$
9	8	$\overline{4}$	$\overline{2}$	1

Figure 42C
Gate to bus line connection chart.

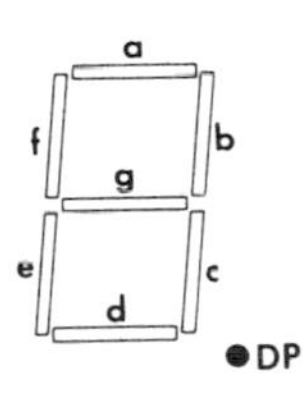

Figure 43
7-Segment display format.

The other type of decoder/display to be discussed is the 7-segment display system. This uses an indicator with seven sections within the viewing area which can be illuminated individually. By causing various sections to light, the digits 0 through 9 can be displayed. Figure 43 shows the standard 7-segment display configuration. These segments can be constructed with a variety of light emitting elements including incandescent filament, light-emitting diodes (LED's), gas discharge glow diodes, and liquid crystals. To display the digit 5, for instance, it would be necessary to illuminate segments a, c, d, f, and g. The digit 1 requires only segments b and c, while 8, requires all seven segments to be activated. The standard 7-segment numerical display format is shown in Figure 44.

Figure 44

A BCD-to-seven-segment decoder/driver circuit is used to operate these display devices. This is accomplished in much the same manner as was BCD to decimal decoding. However, as you might guess, the logic circuits employed are considerably more complex due to the fact that a combination of output levels are required for each digit. Figure 45A and B show the block diagram and truth table for a BCD-to-seven-segment decoder/driver circuit. These two items, along with the display format in Figure 43, should allow you to understand how the displayed digit is derived from the data stored in its accociated memory latch circuit.

The "segment output" section of truth table assumes a binary 0 is required to light a segment while a binary 1 will disable it.

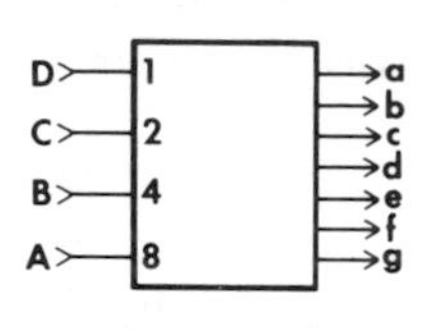

Figure 45A

INPUTS					SEGMENT OUTPUTS						
DECIMAL	A	B	C	D	a	b	c	d	e	f	g
0	0	0	0	0	0	0	0	0	0	0	1
1	0	0	0	1	1	0	0	1	1	1	1
2	0	0	1	0	0	0	1	0	0	1	0
3	0	0	1	1	0	0	0	0	1	1	0
4	0	1	0	0	1	0	0	1	1	0	0
5	0	1	0	1	0	1	0	0	1	0	0
6	0	1	1	0	1	1	0	0	0	0	0
7	0	1	1	1	0	0	0	1	1	1	1
8	1	0	0	0	0	0	0	0	0	0	0
9	1	0	0	1	0	0	0	1	1	0	0

Figure 45B

COUNTER MEASUREMENT MODES

Now that you are somewhat familiar with the circuit functions, as they apply to direct frequency measurement, let's look at how these circuits can be rearranged in order to perform other types of measurements. The most common of these measurements are:

- Totalizing or events counting
- Period
- Period average
- Time interval
- Time interval average

Totalizing

The simplest function an electronic counter can perform is that of accumulating a count of input events. A block diagram of this configuration is shown in Figure 48. Notice that the time base circuit is not used for controlling the main gate in this function. The gate is, instead, controlled by a manual switch. When the switch is placed in the "open" position, the gate is opened and will remain open until the switch is turned to the "closed" position. The latching and automatic reset features are disabled.

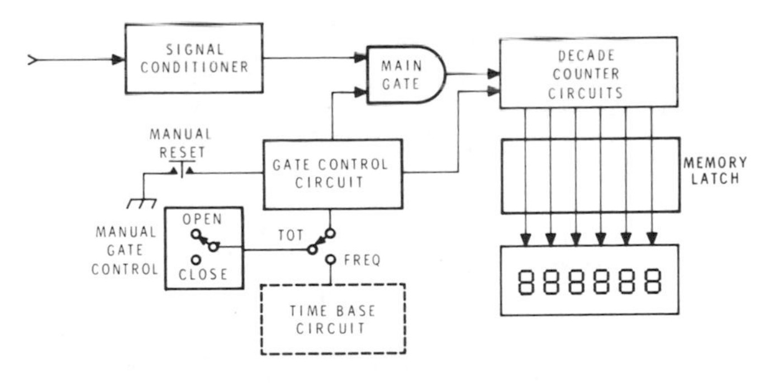

Figure 48
Electronic counter – totalizing function block diagram.

The display section continuously monitors the count-in-progress. When the switch is turned off, the total accumulated count remains in the decade counters. If the gate is opened once more, any incoming pulses will be added to the number from the previous count. When he wants a new count, the operator must press the manual reset switch, which will reset all of the decade counters to zero. The decimal point is disabled in this function since the numbers are a "raw count" and are not referenced to time or frequency.

One application of this function is the measurement of contact bounce. When a switch is operated, or a relay energized, the contacts will usually bounce one or more times. This is due to elasticity of the switch materials and can be a critical factor in digital logic circuits. Figure 49 shows a simple circuit which allows you to measure the bounce in either manual switches or relay contacts. In theory, the voltage across the 1 kΩ resistor would be zero with the contacts open and would step to 1.5 V when the contacts close. In this example, however, three bounces occurred before the voltage settled at 1.5 V. This is not at all uncommon.

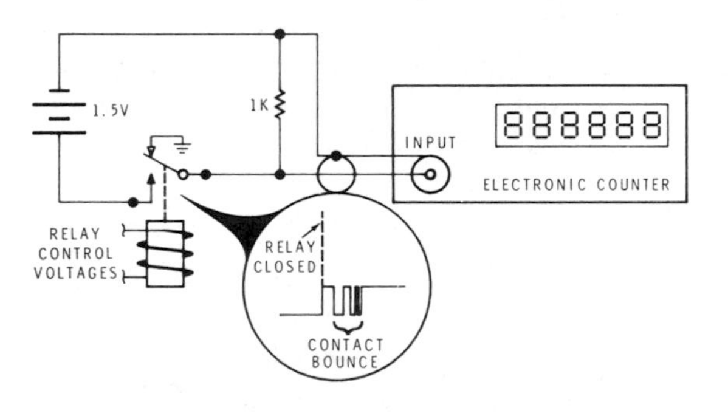

Figure 49

Period Measurement

One of the ways you can make accurate low frequency measurements is by measuring the period of the signal rather than its frequency. For example, at 40 Hz the ±1 count factor would result in an uncertainty of 0.025%. The reciprocal of 40 Hz (1/F), read out in microseconds, would be 25,000 μsec. The ± 1 count uncertainty is now reduced to 0.00004%, thus greatly increasing the accuracy by reducing the effects of ambiguity.

Figure 50 shows how the circuits are reconfigured to perform period measurements.

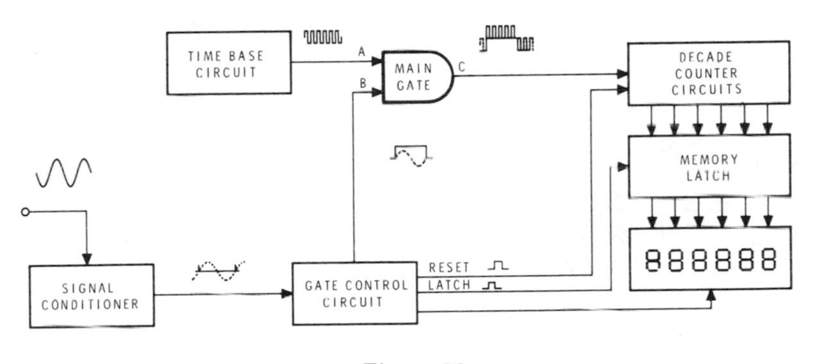

Figure 50
Electronic counter – period function block diagram.

In this mode, the roles of the time base and the signal conditioner are reversed. The gate is now controlled by the duration of the input signal, while the decade counters count the very stable clock pulses from the time base section. Therefore, if a 1 ms time base is selected, we know the pulses being counted are precisely 1 ms apart. Thus, if we read 105 on the readout display, we know the gate was held open for 105 ms. Since the gate is controlled by the input signal, its duration must also be 105 ms, or 9.524 Hz (F = 1/T).

Period Average Measurement

In order to achieve more accuracy, it is often desirable to take the average of a number of period readings. A simple way of doing this is to divide the input frequency, as shown in Figure 51. This effectively lengthens the gate time by a factor of 1000. In the example for period measurement, we received a count of 105 ms, which we said equaled 9.524 Hz. Now, by extending the gate time, we might read 104610 ms. This number must be divided by 1000 to find the average period duration. This gives us 104.610 ms or 9.559 Hz.

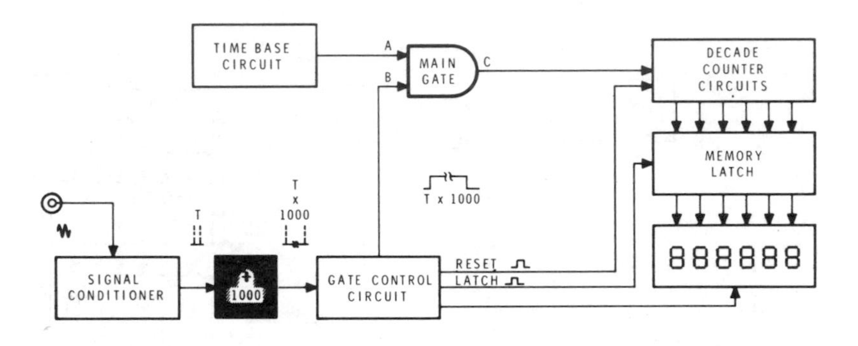

Figure 51

This sounds much more complicated than regular period measurement; but as long as the gate is extended in decade steps (i.e. 10, 100, 1000), it is only necessary to move the decimal point the correct number of places in the display to provide direct readout. This can easily be accomplished by means of the function switch. It is even simpler if a 1000 period average is used. All you need to do in this case is to read the display in ms for period

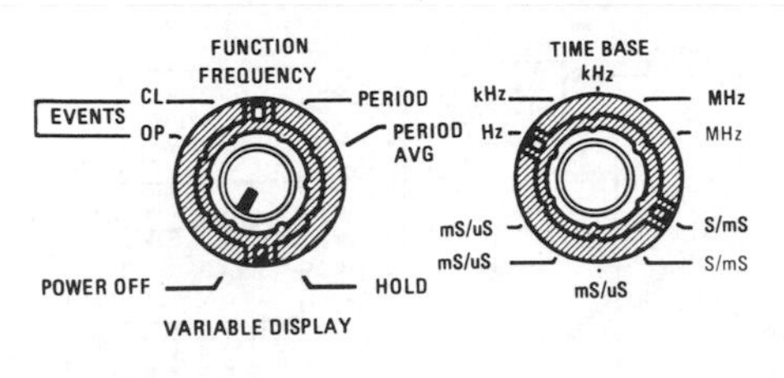

Figure 52
Function and time base controls.

and μs for period average. Notice in Figure 52 that the lower half of the time base switch is marked with two values for each position. The first value indicates the display scale for the period function while the second value is the display scale for the 1000 period average function. For example, with the time base switch in the position shown and the function switch in the "period" position, applying a 400 Hz signal to the input would result in a reading of 2.5 ms. By changing the function switch to "period average" the reading would change to 2500.0 μs. Thus, we have added three digits of resolution and minimized the effects of random errors simply by adding a 1000:1 divider stage; at the same time avoiding the necessity for additional logic circuits to move the decimal point.

Time Interval Measurement

As you can see in Figure 53, the time interval function is similar to the period function, inasmuch as the input signal controls the gate time while the time base provides the signal to be counted. However, while the period function was concerned only with the single dimension of time, time interval measurements consider both time and amplitude. Due to the fact that this mode requires duplicate signal conditioner sections, special precision control circuitry, and a more complex gate control circuit, it is found only on more expensive instruments.

The gate control circuit in Figure 53 has two inputs, one from each of the signal conditioners. The gate is opened by a signal from the "start" input, allowing time base pulses to pass through to the decade counters until a "stop" input closes the gate. Thus, the counter will display the time

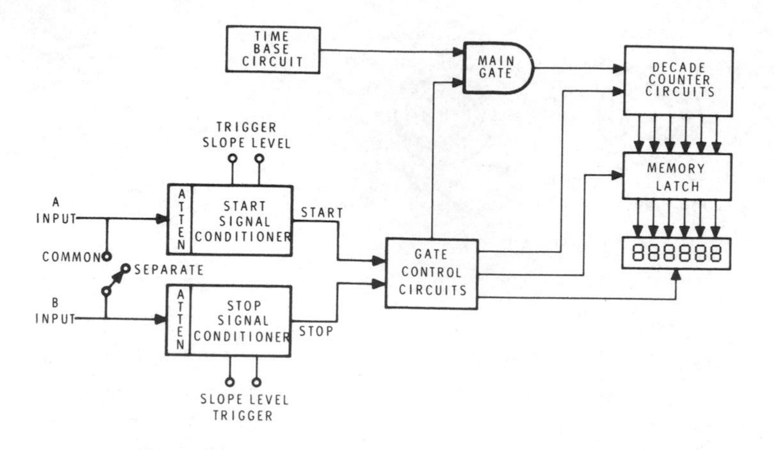

Figure 53

interval between the two input signals. The two triggering points are determined by the front panel control settings, examples of which are shown in Figure 54.

Notice that the panel is divided into two sets of identical controls and inputs.

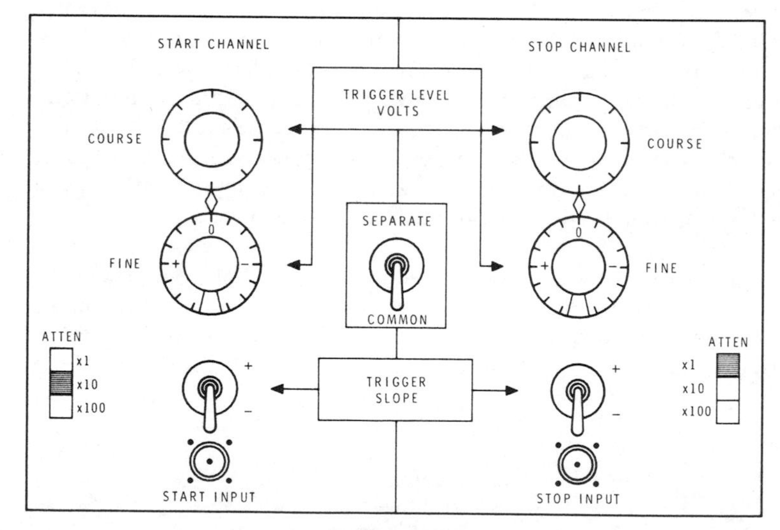

Figure 54

The only exception to this is the **separate – common** switch. In the **separate** position we can measure the time difference between selected points on two separate waveforms. In the **common** position, the time difference between two selected points on the same waveform is measured.

The **attenuator** switches are used, in conjunction with the **trigger level volts** controls, and the **trigger slope** switches, to select the trigger points for the start and stop pulses. Settings for the attenuators are self explanatory with attenuation ratios of 1:1, 10:1, or 100:1 available.

The **trigger slope** switches select the slope of the input waveform which the circuits will trigger on, either positive-going or negative-going.

The **trigger level volts** controls select the exact amplitude at which triggering occurs.

Thus, with the **attenuator** set at ×10, **trigger level** adjusted to 2.45, and **trigger slope** in the positive position, a trigger pulse will be generated whenever a positive-going waveshape passes through the 24.5 volt level. This pulse will be routed to the gate control section where it will either open or close the gate.

Examples of various trigger slope and level points are shown in Figure 55. As both channels are identical, these trigger points could generate either a start or a stop pulse, depending upon the input connection.

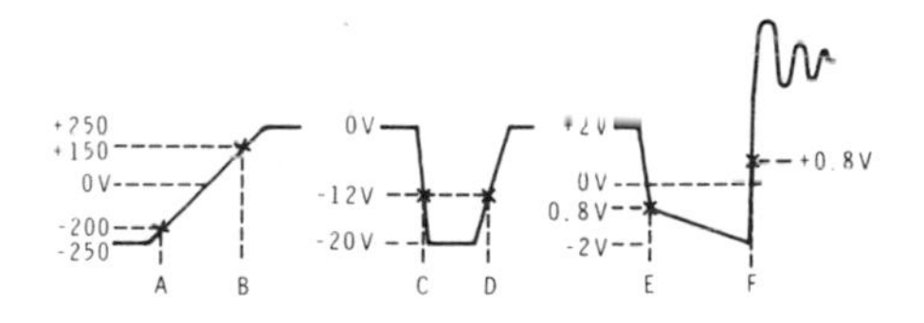

Figure 55
Time interval trigger slope and level selection.

To trigger on these points, the controls should be set as follows:

	Point A	Point B	Point C
Attenuator	×100	×100	×10
Trigger Slope	Positive	Positive	Negative
Trigger Level	−2.00	+1.50	−1.20
	Point D	Point E	Point F
Attenuator	×10	×1	×1
Trigger Slope	Positive	Negative	Positive
Trigger Level	−1.20	−0.80	+0.80

As in many counter applications, you should view the inputs on an oscilloscope to determine approximate values and to establish the desired trigger slopes and amplitudes.

Now let's discuss a few sample applications of time interval measurement. Due to its versatility, you are sure to come up with several additional ways in which you can apply this technique.

Time Measurements Between Two Signals

As an example of this, let's assume we have an oscilloscope with a delayed sweep function that we wish to calibrate for sweep delay accuracy.

First, we would connect the main sweep trigger to the start input of the counter and the delayed sweep trigger to the stop input. Assuming the triggers to be as shown in Figure 56, we would next set the time interval unit controls as follows:

Separate/common	**Separate**
Attenuator	**×10** (both channels)
Trigger Slope	**Positive** (both channels)
Trigger Level	**+0.80** (both channels)

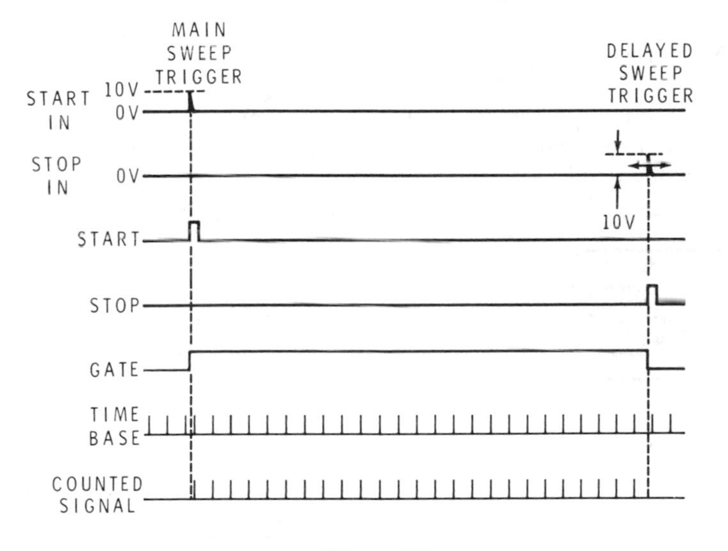

Figure 56
Calibration check of oscilloscope delayed sweep control using time interval measurement — timing diagram.

With both the counter and the scope operating, the counter should be indicating some time interval reading at this point. By selecting a combination of oscilloscope sweep rate and counter time base frequency which will allow a relatively high count to be obtained, it is now a simple matter to check the calibration of the sweep delay circuit.

The oscilloscope controls are usually designed for operator convenience. That is, a scope with 10 cm of horizontal deflection may use a ten-turn dial potentiometer as a sweep delay control. This will allow one dial revolution (major dial division) per cm of sweep delay. These dials usually provide resolution to the nearest hundredth with readings from 0.00 to 10.00. Thus, a sweep delay control setting of 1.00 and a main sweep rate of 1 ms/cm should result in a sweep delay time of 1 ms.

With these sweep triggers connected to the time interval unit, the main sweep trigger will open the gate and, 1 ms later, the delayed sweep trigger will close it. If you were to select a counter time base of 1 MHz (1 μs pulse rate), the counter should display a reading of 1,000 μs. If, however, a reading of 1,026 μs were obtained, you would know the actual delay was 26 μs longer than indicated on the dial. This would be an error of 2.6% of full scale. The linearity of the delay circuit can be determined by checking the accuracy at each major division (1.00, 2.00, 3.00, etc.).

Phase Measurements

Another application, using separate inputs, is the measurement of phase shift. The equipment setup for this measurement is shown in Figure 57A. Figure 57B shows the wave shape timing.

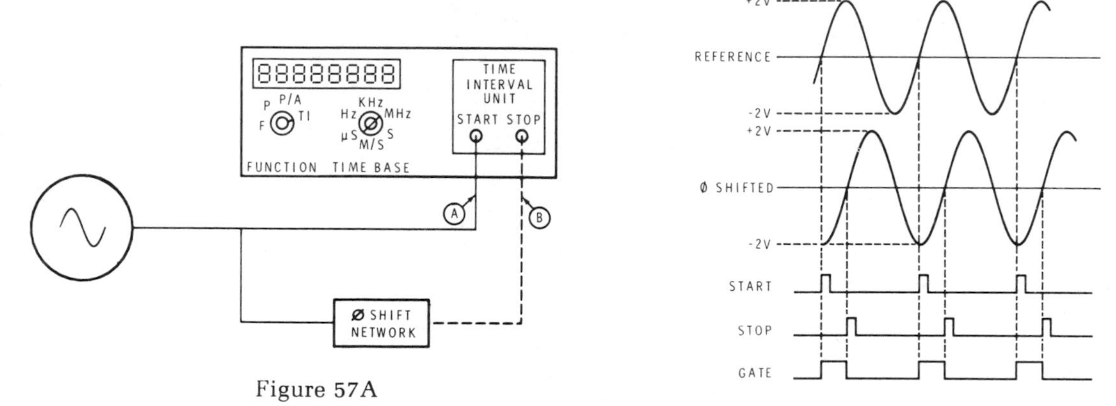

Figure 57A

Figure 57B

In order to calculate phase shift, it is necessary to obtain two separate readings. First, you will need to know the period of the measured signal. Second, you must know the amount of time by which the shifted signal lags the reference signal. With these two facts, you can calculate the phase angle with the formula: $\not{\text{O}} = \frac{360 \times TI}{T}$

Where Ø = phase angle

TI = Time interval between the same point on both sine waves (for example, at the zero crossing on the positive slope)

T = period of the sine wave

To accomplish this, you would first connect only the reference signal to the **start** channel (input A, Figure 57A) and set the controls as follows:

Separate/common	**Common**
Attenuator	**×1** (both channels)
Trigger Slope	**Positive** (both channels)
Trigger Level	**0.00** (both channels)

These settings will result in a counter reading equal to the period of the sine wave. Now, by connecting the phase shifted signal to the **stop** input (input B, Figure 57A) and changing the **separate/common** switch to the **separate** position, the counter will start counting when the reference signal passes through 0 V on the positive slope. It will continue to count until the phase shifted signal passes through the same point, generating a stop pulse. The resultant reading is a time interval proportional to the amount of phase shift.

For instance, if we are working with a 5 kHz signal, the period reading will be 200 μs. If we receive a time interval reading of 15 μs between the reference and the shifted signal, the angular phase shift is equal to $\frac{360 \times (15 \times 10^{-6})}{200 \times 10^{-6}}$ or, 27 degrees.

Time Relationships Between Two Points on the Same Waveform

When you are making measurements on this type, place the separate/common switch in the common position and use only one input connection. You have already seen one application of this method when we determined the waveshape period as part of a phase measurement. Other possible applications are pulse width and pulse spacing measurements.

Pulse Width

To measure the pulse width of a signal, like the one shown in Figure 58 for example, you would connect the signal to either the start or stop input connector. It makes no difference which, as these will be connected internally for this measurement. Set the instrument controls as follows:

Separate/common	**Common**
Start Channel	
Attenuator	**×10**
Trigger Slope	**Positive**
Trigger Level	**+0.75**
Stop Channel	
Attenuator	**×10**
Trigger Slope	**Negative**
Trigger Level	**+0.75**

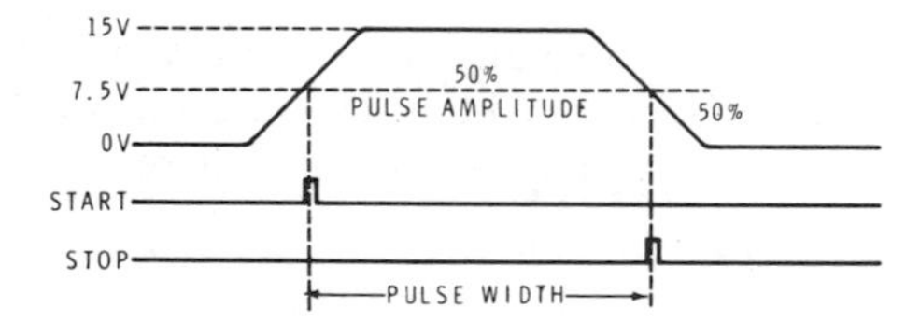

Figure 58

Now if you select a time base compatible with the period being measured, the pulse width will be displayed on the readout.

As the pulse amplitude passes through the preselected trigger level of + 7.5 V on the positive-going slope, the start channel will generate a pulse. This pulse will cause the gate control circuit to open the main gate and allow timing pulses, from the time base section, to reach the decade counters. At the end of the input signal pulse, the stop channel will sense its preselected trigger point (+ 7.5 V) on the negative-going slope, and generate a stop pulse which will close the main gate, ending the count. The resultant readout will be equal to the duration of the input pulse width.

Pulse Spacing Measurement

To measure the spacing between pulses, set the time interval controls to trigger the start channel at the 50% point of the input signal's trailing edge, as shown in Figure 59. The counter will then count time base pulses until the stop channel is triggered at the 50% point on the next leading edge. The resultant reading will indicate the elapsed time between pulses.

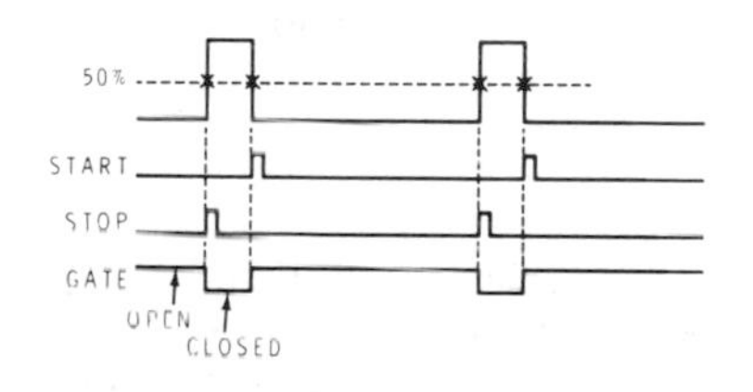

Figure 59
Pulse spacing measurement – timing diagram.

Time Interval Averaging

As you remember, in period measurement, it was possible to increase accuracy by reading the average value of a number of periods. This is also true of time interval measurements; however, you will have to use a slightly different method.

In the period function it was only necessary to extend the gate length by a multiple of the input frequency because we were measuring the entire period of the input signal. In the time interval function, however, we are only concerned with a portion of the input period. Therefore, we cannot simply hold the gate open. It must be opened and closed at the proper times if we are to measure only that period of the input signal in which we are interested. Figure 60 shows one method of accomplishing this task.

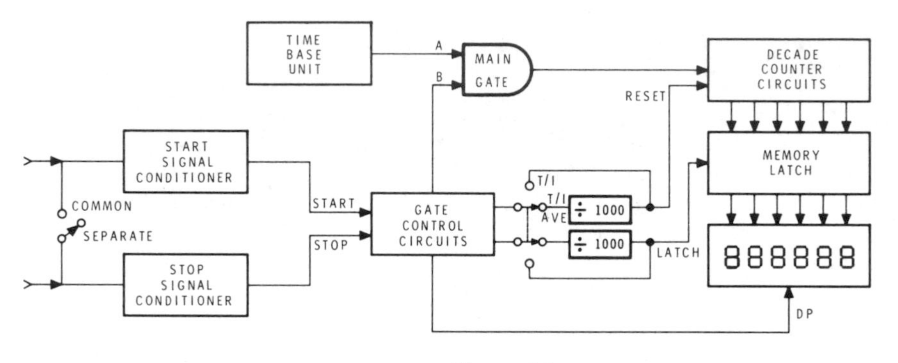

Figure 60

As you can see, no change has been made to the signal conditioner or gate control circuits. These will continue to operate as in the regular time interval mode. Notice, however, that the latch and reset pulses are now routed through a 1000:1 divider. Thus, 1,000 time interval readings will be accumulated in the decade counters before the count is latched into memory.

For example, if the time interval we were measuring was actually 29.25713 μs, a straight time interval measurement would read 29.3 μs, due to the maximum 0.1 μs resolution of the 10 MHz time base. By accumulating this count for one thousand intervals, the decade counters would have counted 292,571 pulses. To put this value in the proper units for readout, it is only necessary to move the decimal point three places to the left (29.257.1). This will provide direct readout of the mathematical average of one thousand readings while providing an additional three digits of resolution.

EXTENDING THE COUNTER'S LIMITS

Counters, being digital instruments, are limited in their frequency response by the switching speed of their logic components. Current state-of-the-art allows direct frequency counting to about 500 MHz, although anything over 250 MHz becomes quite expensive. Therefore, most counters rated over 250 MHz, and all over 500 MHz, use some type of down-conversion to bring the input frequency within range of the basic counting system. Three methods of down-conversion are currently popular. These are prescaling, heterodyne converters, and transfer oscillators.

Prescaling

Prescaling simply means dividing, or scaling down, the input frequency before it is applied to the counting circuits. For instance, a 50 MHz counter using a 4:1 prescaler is capable of measuring frequencies up to 200 MHz. In principle, any division ratio may be used but, because high speed counter circuits are binary devices, they are usually designed to divide the input signal by an integral power of 2 (2, 4, 8, etc.).

As you can see, without other circuit modifications this would lead to complications. In order to determine the input frequency, it would be necessary to multiply the counter reading by the scale factor. This would be a major inconvenience; therefore, simple external prescaling units usually use decade counters which provide scale factors of 10 or 100. This only requires the mental movement of the decimal point in the readout. These external prescalers are an economical way to increase the range of existing low frequency counters. They do have one disadvantage however. One digit of resolution is lost for each decade of prescaling used. For example, a 145,600.0 kHz signal measured with a 10:1 prescaler will read 14560.0 kHz. Mental movement of the decimal point yields 145,600 kHz. The 0.1 kHz of resolution is lost unless the gate time is increased by a factor of ten, if the counter has that capability.

Counters which contain built-in prescaling units eliminate both the problem of decimal point movement and/or resolution loss by automatically extending the gate length by the scale factor. If we count a signal at half speed for twice as long, the scale factor is self-cancelling and the counter displays the correct input frequency. The disadvantage of this method is the speed of obtaining a reading. A one-second gate time, using a 10:1 prescaler, requires 10 seconds to obtain a reading. Therefore, prescaling has a built-in tradeoff of speed vs resolution.

By prescaling, a counter's input frequency range may be extended to somewhat over 1 GHz. Above this frequency it is necessary to use either the heterodyne frequency conversion or transfer oscillator method. These may be an integral part of the counter's signal conditioner section, as with special purpose microwave counters, or plug-in adapter units for general purpose universal counters.

Heterodyne Frequency Converter

Heterodyne frequency converters operate on much the same principle as the heterodyne frequency meters you studied earlier. There are a few important differences, however. Figure 61 shows a simplified block diagram of a typical manually tuned heterodyne converter. For explanation purposes, let's assume we are using a counter with a frequency range of at least 250 MHz.

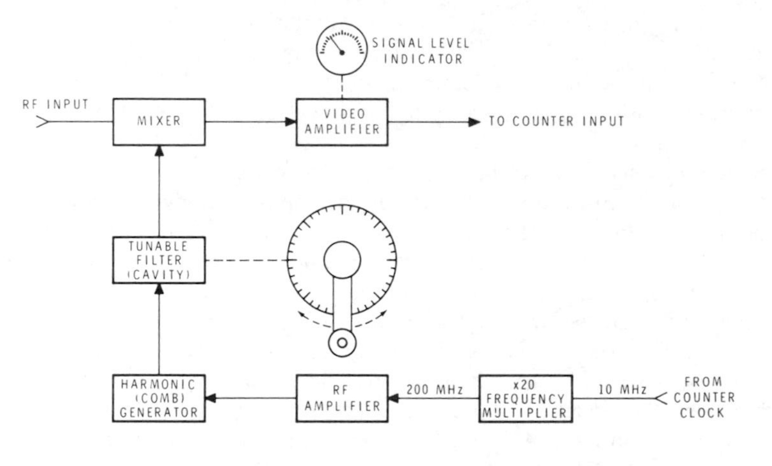

Figure 61
Heterodyne frequency converter – simplified block diagram.

As you can see, this unit uses the counter time base oscillator in place of a variable frequency oscillator circuit. The highly stable 10 MHz time base frequency is applied to a × 20 frequency multiplier which produces an output of 200 MHz with the same frequency stability as the time base oscillator.

This signal is then amplified to a sufficient power level to drive a harmonic generator. This circuit is sometimes referred to as a "comb" generator, as a spectrum plot of its output resembles a comb whose teeth are evenly spaced at intervals equal to the input frequency; in this case, every 200 MHz. Figure 62 depicts this output spectrum. If our heterodyne converter is designed to operate over a range of 200 MHz (0.2 GHz) to 12.4 GHz, the harmonic generator will put out some sixty discrete frequencies. All of these frequencies are applied to a tunable filter.

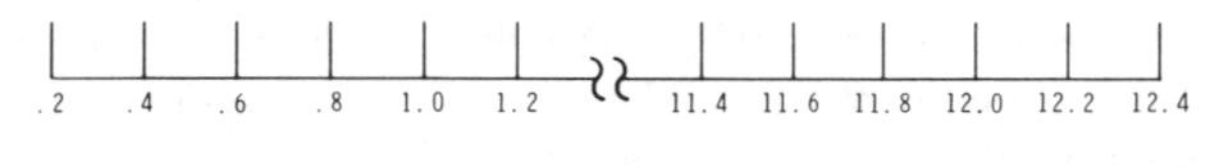

Figure 62

The tunable filter is nothing more than a variable resonant cavity, which you will remember from your study of microwave wavemeters. The front panel tuning control varies the physical size of the cavity, which causes its resonant frequency to vary from the low end of its specified frequency range to its high end. In this case, from 0.2 GHz to 12.4 GHz. If we varied the tuning dial through its entire range, we would apply each of the harmonics, one at a time, to the mixer circuit.

Keep in mind that, by tuning the cavity, we are not changing the *frequency* of the signal applied to the mixer, only selecting *which frequency* is applied. The actual frequencies are multiples of the counter's time base oscillator.

At the oscillator input to the mixer, the signal will appear to be in discrete frequency steps of 0.2 GHz, separated by periods of no signal, as shown in Figure 63.

SIGNAL AMPLITUDE

1.2 GHz

1.3 GHz

TUNING CONTROL

Figure 63

Now, that a multiple of 200 MHz is applied to one mixer input, let's apply an external RF signal to its other input. The two input frequencies will beat together and produce two additional frequencies which are equal to the sum and the difference of the two inputs. These four frequencies will be felt at the video amplifier input.

The video amplifier stage serves a dual purpose. In addition to amplifying the signal, it acts as a low-pass filter circuit. Due to its bandwidth, which is approximately the same as the counter's input frequency limit, only frequencies below this limit will be amplified and passed on to the counter's signal conditioner. A built-in signal level indicator shows when a measurable RF signal is present at the amplifier output. Any signal amplified by the video amplifier will be passed on to the counter input and counted in the normal manner.

Operation of a heterodyne converter can best be shown if we assume an RF input value and discuss the resultant action. By injecting an RF signal of 3.255 GHz and tuning the converter upward from its low frequency end, we will reach a point where the difference frequency will fall within the bandwidth of the video amplifier (that is, 250 MHz). Figure 64 shows the difference frequencies obtained when the cavity is tuned upward through the points where indications are received on the signal level meter.

INPUT FREQUENCY (GHz)	3.245	3.245	3.245	3.245	3.245	3.245
HARMONIC FREQUENCY (GHz)	2.600	2.800	3.000	3.200	3.400	3.600
DIFFERENCE FREQUENCY (GHz)	.645	.445	.245	.045	.155	.355

Figure 64

Difference frequency versus harmonic frequency chart.

As you can see, more than one harmonic will produce a different frequency which falls within the 250 MHz bandwidth limit of the video amplifier and, in each case, the counter will display the different value. This points up why you should always start at the low end of the converter's frequency range and search upward. By doing this, the first indication will always occur when the harmonic frequency is below the input frequency. Therefore, the counter reading can be directly added to the converter frequency dial. If you tuned downward from the high end you would have to subtract the first counter reading from the frequency dial, which is much less convenient. The center indication is of little use by itself, as it could be slightly above or below the RF signal, and we would be unsure if we should add or subtract it from the converter's tuning dial reading.

Heterodyne converter units are currently available in many frequency ranges, from 50 MHz to over 20 GHz. Due to physical design considerations, however, no one converter can cover this entire frequency range. Therefore, in order to measure all of these frequencies, as many as four separate converters might be necessary. This could prove to be quite expensive, as these units cost between $1,000 and $3,000 each.

If you have a requirement for this broadband measurement capability and slightly lower accuracy is acceptable, you might consider using a transfer oscillator unit.

Transfer Oscillator Converter – Basic Principles

Like the heterodyne converter, the transfer oscillator mixes the incoming unknown frequency with an internally generated signal. The basic difference in the two methods is the internal standard frequency used and the frequency which is counted. While the heterodyne converter uses the counter time base as its standard, the transfer oscillator employs a variable frequency oscillator. This is shown in Figure 65. Notice also that the reference frequency is the signal which is counted rather than the beat frequency. This is the same basic principle as the transfer oscillator instrument you studied earlier in this section.

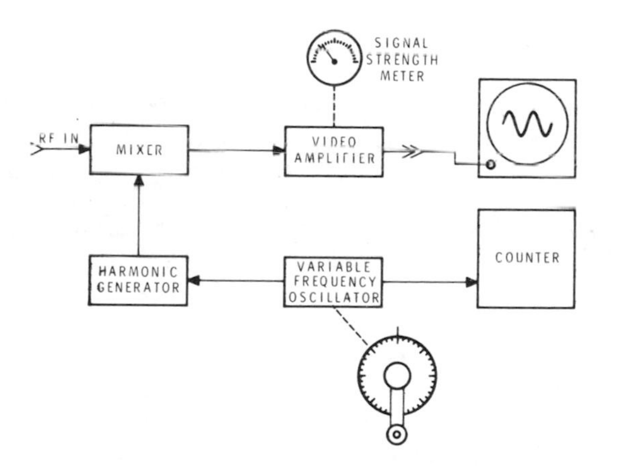

Figure 65

VFO

The variable frequency oscillator (VFO) is tunable over a typical frequency range of 80 to 220 MHz. One VFO output is connected to the counter's input stage while the other drives a harmonic generator.

HARMONIC GENERATOR

The harmonic generator produces a comb of harmonic frequencies. However, while the heterodyne converter's harmonics were constantly spaced, the transfer oscillator's harmonic spacing varies as the VFO frequency is changed. The generator must be capable of producing strong harmonics up to at least the 100th harmonic of the VFO frequency. If the VFO is tunable to 200 MHz, the 100th harmonic would be 20 GHz, which would be the upper measurement limit for this unit.

MIXER AND VIDEO AMPLIFIER

The entire harmonic frequency comb is injected into the mixer stage and mixed with the unknown RF signal to be measured. As the VFO is slowly tuned, one of the harmonic frequencies will approach the unknown frequency close enough to bring the beat frequency within bandwidth range of the video amplifier. This will provide an indication on the signal strength meter and also on the oscilloscope. The unit is now carefully tuned for a zero-beat indication on the oscilloscope.

A precise zero-beat indicates that some harmonic of the VFO frequency, which is now displayed on the counter, is exactly equal to the input frequency. If you know the approximate value of the RF input frequency, you can determine the measured value. First, identify the harmonic number, and then multiply the VFO frequency by the harmonic number:

1. $\text{Harmonic number} = \dfrac{\text{approximate RF input}}{\text{VFO frequency}}$ (round off answers)

2. $\text{Input frequency} = \text{harmonic number} \times \text{VFO frequency}$

If the input frequency is completely unknown, you must locate two adjacent zero-beats. As in the earlier discussion of transfer oscillators, you can calculate the input frequency as follows:

1. $H1 = \frac{F2}{F1 - F2}$ (round off answers)

2. Input frequency $= H1 \times F1$,

OR

1. $H2 = \frac{F1}{F1 - F2}$

2. Input frequency $= H2 \times F2$

Where F1 = VFO frequency at first zero-beat

F2 = VFO frequency at next adjacent zero-beat

H1 = harmonic number associated with F1

H2 = harmonic number associated with F2

The greatest sources of error, when you are using this method, will lie in your ability to obtain a precise zero-beat, coupled with the instability of the VFO. Even a very good VFO has a stability of only about ±10 ppm, which is well below the stability of the average counter.

Typical Transfer Oscillator Unit

A large number of transfer oscillator plug-ins are currently available from several manufacturers. Each use slightly different techniques to provide improvements in speed, accuracy, and ease of operation. Because of the wide array of equipment on the market, it would be impractical to include a detailed description of every unit you might encounter. Therefore, we will briefly describe some of the general methods used to enhance the versatility of these units. Keep in mind that the circuits discussed may be packaged as a plug-in module designed for specific counters or may be incorporated into the counter's signal processing section.

AUTOMATIC PHASE CONTROL (APC)

In order to reduce the effects of operator error, when he is tuning for a zero-beat, a "phase-locked loop" circuit is usually employed. Figure 66 shows the basic transfer oscillator circuit with a phase-locked loop added.

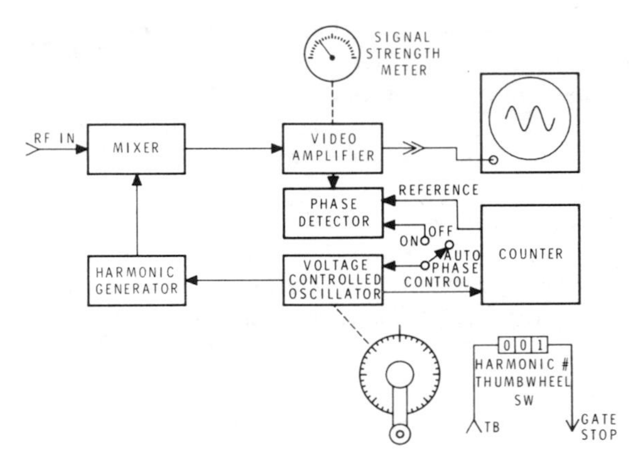

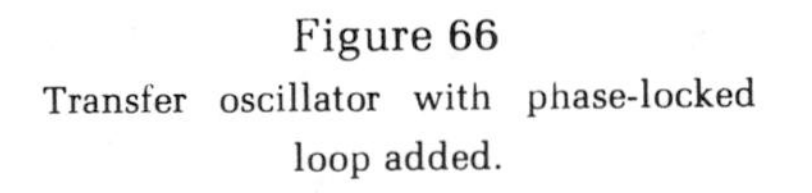

Figure 66
Transfer oscillator with phase-locked loop added.

This may appear quite complicated but its principle of operation is simple. Notice that a voltage-controlled oscillator (VCO) is used as a local oscillator and that a phase detector circuit monitors the output of the video amplifier.

The phase detector compares the video amplifier output with a standard reference signal. This is usually a submultiple of the counter time base oscillator. Any difference in frequency between the two signals will produce a corresponding change in phase detector output voltage. The output of this stage is used to control the frequency output of VCO local oscillator over a narrow range.

To illustrate the operation of this circuit, let's assume that we are measuring an RF signal of approximately 2 GHz. Starting at the low end of the frequency dial, with the APC switch off and tuning slowly upward, we would receive a signal indication somewhat below 100 MHz. This would be caused by the 20th harmonic of the local oscillator frequency beating with the 2 GHz RF input (100 MHz × 20 = 2000 MHz or 2 GHz). With the oscillator still tuned below the zero-beat point and an indication on the signal strength meter, you would turn the APC switch on. This would allow the phase detector to take over fine tuning control of the VCO.

Returning to the phase detector circuit for a moment, let's assume we are using a 1 MHz reference signal from the counter's time base section. With the mixer receiving a 2,000 MHz RF input signal and a 1,998 MHz harmonic signal (20th harmonic of the VCO, if tuned to 99.9 MHz), a difference frequency of 2 MHz will be passed through the video amplifier to the phase detector.

The phase detector will compare the 1 MHz reference to the 2 MHz difference signal and put out a voltage level which will cause the VCO to increase in frequency. When the difference frequency exactly matches the reference, the phase detector output will stabilize, causing the oscillator to remain at that frequency. In this case, the harmonic frequency would have to be 1999 MHz, or the 20th harmonic of 99.950 MHz. The counter will indicate the oscillator frequency; 19.950 MHz. A phase-locked loop now exists. If the RF input frequency should drift slightly, the phase detector will sense the change in difference frequency and force the VCO to track the input. Conversely, should the VCO frequency attempt to drift, due to temperature or voltage changes, the phase detector will again sense the change and counteract it.

By effectively locking the local oscillator to the input frequency, we have eliminated two sources of error but have produced a third. The problem of achieving a precise zero-beat no longer exists due to automatic lock-on. Also, the inherent instability of the local oscillator has been overcome by referencing it to the counter's time base oscillator stability. However, since phase lock cannot be achieved with zero difference frequency, the oscillator frequency will always be offset by the value of the reference frequency, divided by the harmonic number being used. This can be calculated and added to the counter reading, but it is time consuming and provides another possible source of human error. Therefore, a method is usually provided to allow the operator to dial-in the harmonic number, commonly in the form of a thumbwheel switch.

This switch is part of a gate extender circuit. The counter gate will be extended by a factor of the number set into the switch. In our previous example, where the VCO was locked on at 99.95 MHz and with the gate extender switch set at 001, the counter would read 99.95 MHz. However, it was the 20th harmonic of this frequency which was actually beating with the incoming RF signal. Therefore, if 020 (20th harmonic) were set into the gate extender switch, the counter gate would be held open twenty times longer than normal, resulting in a reading of 99.95 MHz × 20 or 1999 MHz. Thus, the counter now directly indicates the incoming RF signal minus the reference frequency. This makes it much more convenient to monitor or adjust a frequency source once lock-on is achieved and the harmonic number established.

In counters where the transfer oscillator is an integral part of the counters signal conditioner section, the reference offset error can be eliminated by causing the decade counter stages to reset to the value of the reference frequency instead of to zero. Thus, the offset error is automatically added to the counter reading, causing the counter to read out the input frequency directly.

CALIBRATING THE COUNTER

All counters should have their time bases recalibrated periodically in order to maintain their accuracy. An expensive, double oven oscillator will require less frequent calibration than will a room temperature oscillator, but even the best counters should be recalibrated at periods ranging from ninety days to one year, depending upon operating environment and accuracy requirements. You should also do this before you make any critical precision measurement, the required accuracy of which approaches the accuracy of the counter.

Several methods are used to calibrate counter time base oscillators. These are based upon the degree of accuracy and stability of the counter oscillator, but all involve the comparison of the counter's oscillator frequency with some type of standard frequency source. This standard frequency source should be several times more accurate than the oscillator to be calibrated. For this reason, calibration of ultra-stable oscillator circuits involve complicated techniques using special purpose receiver-comparitors. The more common counters, with time base specifications on the order of $\pm$ 1 to 10 ppm, are quite easily calibrated using relatively common test equipment.

First, you will need either a standard oscillator of known accuracy or a communications receiver capable of receiving WWV. Before attempting any calibration adjustment, be sure both the counter and any test equipment you use are fully warmed up.

Direct Comparison

When the standard is greater than five times more accurate than the best resolution of the counter at its time base frequency, its output can be directly applied to the counter input. Then adjust the time base oscillator frequency for the correct frequency readout on the counter. The adjustment of this control will sometimes cause the oscillator circuit to become unstable for a short period of time. Therefore, you should allow several moments, after you make the adjustment, for the oscillator to settle down to its new operating frequency.

Calibration With An Oscilloscope

A more accurate adjustment can be made if your counter has a time base output connector. This will usually provide a 1 MHz signal referenced to the time base oscillator. Connect this output to the vertical input of an oscilloscope and connect the output of the standard oscillator to the horizontal input. (This assumes that both the vertical and horizontal amplifiers of the oscilloscope will pass 1 MHz.) The oscilloscope will now display a Lissajous pattern. You can now adjust the time base frequency for as stable a pattern as possible.

Calibration Using WWV

To calibrate the time base oscillator using WWV as a frequency standard, connect the equipment as shown in Figure 67. Refer to Appendix A for frequencies and broadcast format. If the counter has no output connector, attach a piece of insulated wire to the receiver antenna, and place the other end of the wire near the counter's time base oscillator. In either case, an oscillator harmonic should beat with the WWV signal. This beat will not be audible, unless the oscillator is far off frequency, but will appear as a rhythmic variation in the S-meter reading.

At this point, you should adjust the potentiometer in Figure 67 (or reposition the insulated wire) for the most pronounced beat indication on the S meter. Then you can adjust the oscillator for the slowest possible swing in the S-meter reading. It will be impossible to stop its movement completely, but a period of a few seconds per cycle will result in excellent short term accuracy. For instance, if you see six cycles of S-meter oscillation in a sixty second period, the beat frequency is equal to 6/60 or 0.1 Hz.

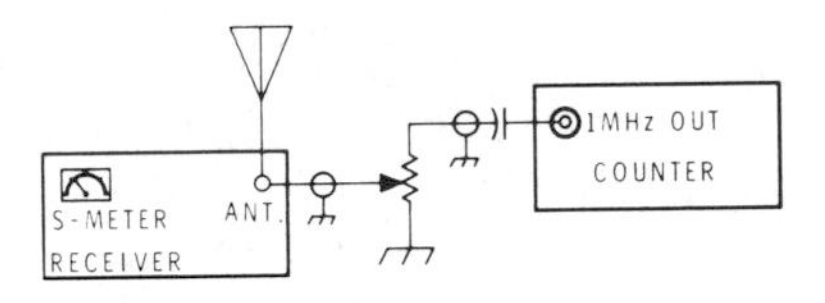

Figure 67
Counter calibration setup using WWV.

Assuming the 15 MHz WWV signal is being used, the oscillator error is $\frac{0.1}{15 \times 10^6}$ or 6.67×10^{-9}. As you can see from this example, the highest usable WWV frequency will provide the highest degree of accuracy.

Another factor to consider, when you use this technique, is transmission quality. Choose time of day when you can receive WWV with a minimum of fading, as varying signal strength readings will tend to hide beat indications on the S meter.

Unit 5

FREQUENCY GENERATION

INTRODUCTION

Signal generators play a vital role in the design, construction, and maintenance of electronic equipment. These generators vary from a simple fixed-tone oscillator for checking continuity to ultra-sophisticated generators having multiple waveform outputs, tremendous frequency range and exotic specifications. The vast majority of signal generators, however, lie somewhere between these extremes and are the instruments which will be covered in this unit.

A source of known frequency and amplitude, used with an accurately calibrated meter or scope, can provide precise information on circuit operation. By knowing exactly what is going into a circuit, and comparing this to the output, it is possible to measure the circuit's gain, frequency response, sensitivity, maximum output, distortion, and many other parameters.

In this unit, you will acquire a basic knowledge of the types of signal generators, how they operate, and how to apply them.

BASIC SIGNAL GENERATOR CONCEPTS

This section will briefly discuss some common characteristics of generators. It will examine generator frequency ranges, modulation, and waveforms. This will give you an overview before we discuss the internal operation of these instruments.

Frequency Ranges

One method of grouping, or classifying generators is by their output frequency range. Under this system, three classes are normally used. These are audio, RF, and microwave. The actual frequency ranges of each type are very general and refer more to the type of equipment they are used with than to their specific output frequency.

Audio generators are available in the 0.1 Hz to 1 MHz range, and the majority are tunable from about 5 Hz to 100 kHz. RF generators may start as low as 50 KHz and range upward to 1 GHz. Above 1 GHz is generally considered to be the microwave region. Notice that there is considerable overlap in frequency coverage. The high frequency end of audio generator coverage is well into the RF region, while the RF generator may be used for signals below the upper audio generator limits. This occurs for a number of reasons. Generator design is based upon cost, efficiency, convenience, and intended use. These factors will become more apparent after studying the types of generators and their applications. For the moment, however, remember that frequency classifications are very general and not strictly defined.

Modulation Characteristics

Many, if not most, sine wave generators provide an amplitude modulated output. As you remember from previous studies, amplitude modulation consists of varying the amplitude of a high frequency carrier wave in accordance with a relatively low frequency signal. This is shown in Figure 1. If the high frequency carrier and the low frequency tone are mixed across a nonlinear device, the resultant waveform would appear much as waveform C in Figure 1. That is, the original carrier frequency remains unchanged but its amplitude varies at the frequency of the modulating tone.

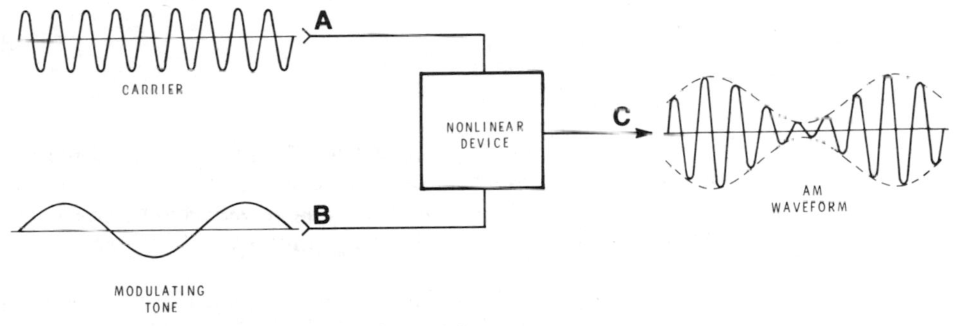

Figure 1

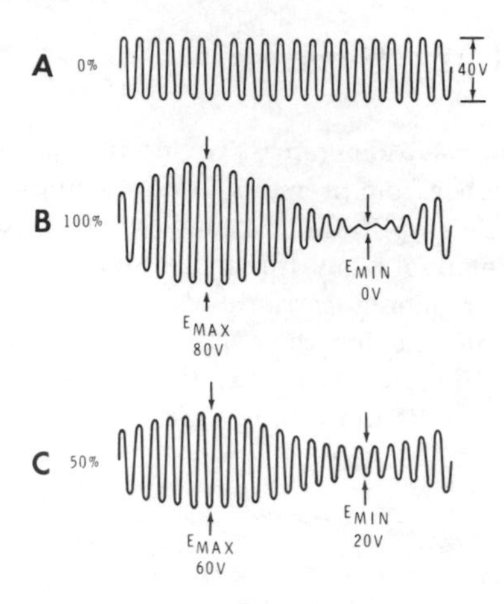

Figure 2
Percent of modulation.

The amount of amplitude change determines the percentage of modulation. This is an important characteristic of an AM waveform. The degree of modulation is expressed as a percentage between 0% and 100%. To review how this is computed, refer to Figure 2.

Waveform A shows the unmodulated carrier signal at a constant amplitude of 40 volts peak-to-peak. This is an example of 0% modulation.

Waveform B shows the same carrier modulated to 100%. Here, the amplitude of the modulated waveform falls to zero volts for an instant during each cycle of the modulating wave. Also, the amplitude increases to 80 volts peak-to-peak once during each modulation cycle. The average peak-to-peak amplitude is still 40 volts.

Waveform C shows the carrier modulated at 50%. The peak-to-peak amplitude varies from a maximum of 60 volts to a minimum of 20 volts peak-to-peak.

The equation for computing the percent of modulation is

$$\% \text{ mod} = \frac{E_{max} - E_{min}}{E_{max} + E_{min}} \times 100$$

For example, in Figure 2C, E_{max} equals 60 volts and E_{min} equals 20 volts.

Thus:

$$\% \text{ mod} = \frac{60 \text{ V} - 20 \text{ V}}{60 \text{ V} + 20 \text{ V}} \times 100$$

$$\% \text{ mod} = \frac{40 \text{ V}}{80 \text{ V}} \times 100$$

$$\% \text{ mod} = 0.5 \times 100$$

$$\% \text{ mod} = 50\%$$

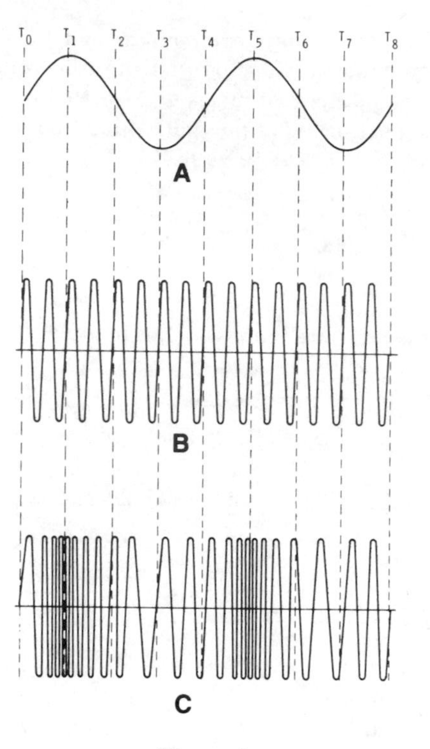

Figure 3
Frequency modulation.

Another form of modulation, incorporated in many generators, is frequency modulation (FM). With frequency modulation, the intelligence is impressed on the carrier as frequency variations rather than amplitude variation. Figure 3 illustrates the FM waveform. The intelligence, or modulating waveform, is shown in Figure 3A while the unmodulated carrier is shown in Figure 3B. The resulting frequency-modulated waveform is shown in Figure 3C.

At time T_0, the modulated waveform (C) is at its center frequency. As the modulating waveform (A) swings positive, the frequency of the carrier is increased. At T_1, the modulating signal reaches its maximum amplitude, resulting in the carrier reaching its maximum frequency. From T_1 to T_2, the modulating signal amplitude swings back to zero, which causes the carrier to decrease in frequency until, at T_2, it is back to its center frequency.

As the modulating signal swings negative, the carrier decreases below its center frequency until, at T_3, the modulating signal is at its maximum negative value and the carrier is at its minimum frequency. After T_3, the carrier again returns to center frequency as the modulating signal returns to zero volts. From time T_4 to T_8, the modulating signal repeats its cycle. As a result, the carrier is again shifted in frequency; first above, then below its center frequency. Notice that it returns to its center frequency each time the modulating signal passes through zero volts.

The carrier frequency changes equally above and below its center frequency. The magnitude, or amount, of change from the center frequency is known as **frequency deviation**. For example, if we assume a signal with a carrier center frequency of 100 MHz, a minimum frequency of 99.9 MHz, and a maximum frequency of 100.1 MHz, the frequency deviation would be 0.1 MHz or 100 kHz.

As you can see, the amount by which the carrier deviates from the center frequency is determined by the amplitude of the modulating signal. The rate at which it deviates is a function of the modulating signal frequency.

Frequency Domain Analysis

Up to this point, signals have been shown and discussed as they would appear on an oscilloscope, i.e. amplitude versus time. This is "time domain analysis". This method considers only voltage or current variations per unit of time. Although time domain analysis is convenient, inexpensive and very informative, it does not present the entire picture. In order to gain greater insight into signal composition, it is necessary to add another dimension; frequency.

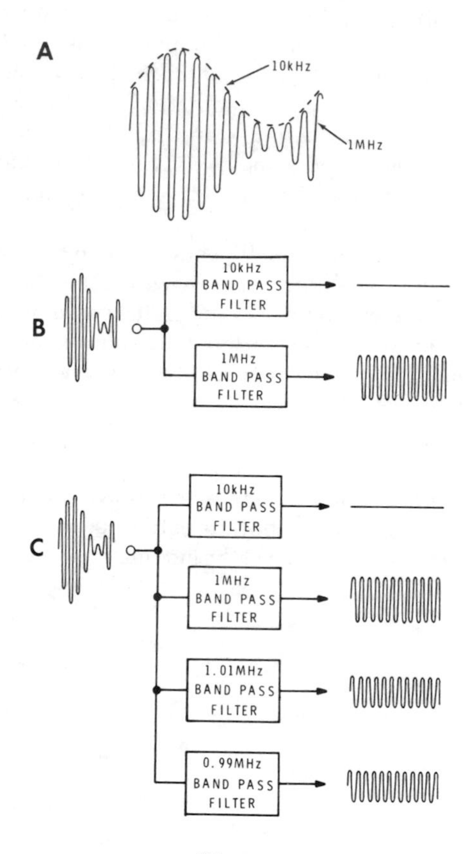

Figure 4
Frequency components of an AM signal.

As you recall from the discussion of amplitude modulation, the scope presentation of an AM waveform revealed an envelope or carrier signal which varied in amplitude at the rate of the modulating signal. It would appear from this waveform that these are the only two signals present. For example, the waveform shown in Figure 4A appears to have only a 10 kHz component and a 1 MHz component. Therefore, it might seem reasonable to assume that if you were to pass this waveform through a sharply-tuned 10 kHz bandpass filter, it would be possible to recover the modulating signal. In practice, however, you would find that no 10 kHz signal exists at the output of the filter. Equally interesting, is the result of passing the signal through a 1 MHz bandpass filter. The output of this filter would yield the original carrier signal with no trace of modulation. This is illustrated in Figure 4B. Where then, is the information which was contained in the modulating signal?

Remember from your studies of heterodyne, or mixing action, that mixing two frequencies create additional frequencies; the sum and the difference of the two original frequencies. Figure 4C shows the result of adding two additional filters; one which will pass only 1.01 MHz (the sum) and the other which will pass only .99 MHz (the difference). The result of amplitude modulation, then, is much like mixing action with one notable exception. The modulator output contains no power at a frequency equal to the modulating signal. Only the carrier, the sum, and the difference are present. These sum and difference frequencies are known as sidebands. Notice, in Figure 4C, that the sidebands have less amplitude than the carrier and that all three signals have constant amplitudes.

Carrying this line of reasoning one step further, assume you have a tunable bandpass filter similar to the one you studied in the section on Counter Heterodyne Frequency Converters. With this device, you could search the spectrum and determine the location of all frequencies contained within the modulated signal. This is the principle of frequency domain analysis.

Using a highly selective tunable filter, in conjunction with a frequency counter and RF voltmeter, you could plot a graph of all signals present. Instrument technology has made this task much simpler, however. By using a spectrum analyzer, it is possible for you to view a section of the frequency spectrum directly on the face of a cathode ray tube (CRT). This display is much like that of an oscilloscope except that the horizontal axis represents increments of frequency rather than time.

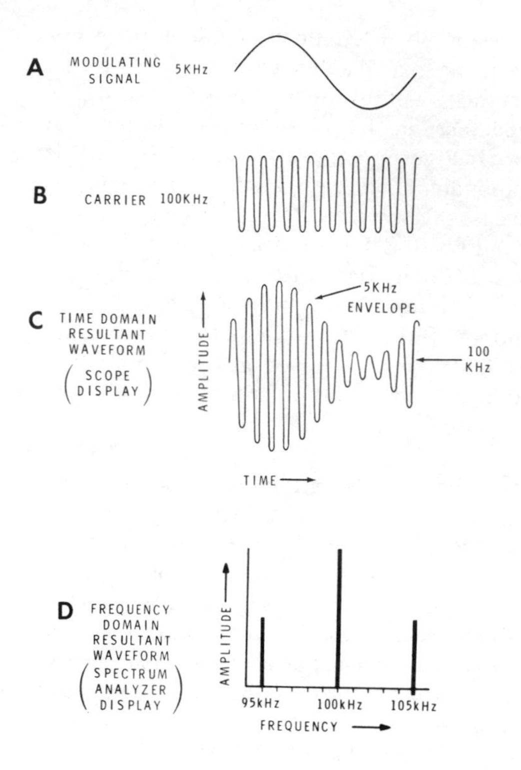

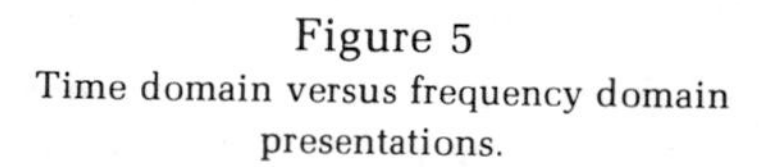

Figure 5
Time domain versus frequency domain presentations.

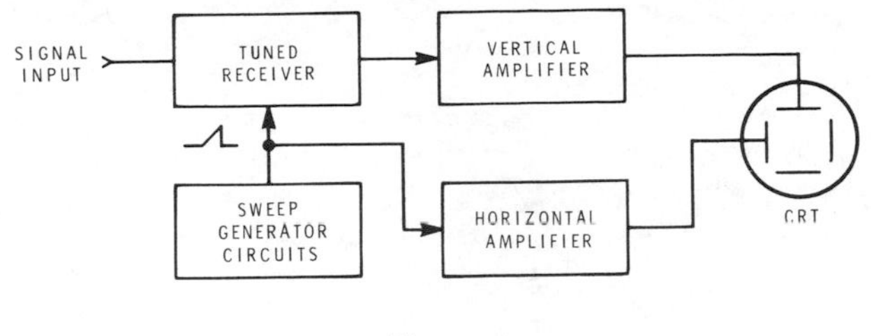

Figure 6

Figure 5 compares time domain and frequency domain presentations of the same waveform. Figure 5C shows the familiar oscilloscope display of an amplitude modulated carrier wave; in this case, 100 kHz modulated by 5 kHz. Figure 5D shows the same signal displayed on a spectrum analyzer. Notice that the horizontal scale is calibrated in frequency units, while the vertical axis represents relative signal strength, or amplitude.

Spectrum analyzers are discussed in detail in another unit of this course. However, a general understanding of their operating principle and CRT display will greatly increase your understanding of frequency domain analysis.

Figure 6 shows the block diagram of a very simplified spectrum analyzer. The sweep generator and horizontal amplifier circuits operate basically as they did in the conventional oscilloscope. Notice, however, the same voltage ramp which drives the horizontal amplifier is also routed to the tuned receiver. This ramp causes the receiver to scan a range of frequencies at the same speed that the beam sweeps across the CRT. Thus, the horizontal axis of the CRT is referenced to the receiver frequency.

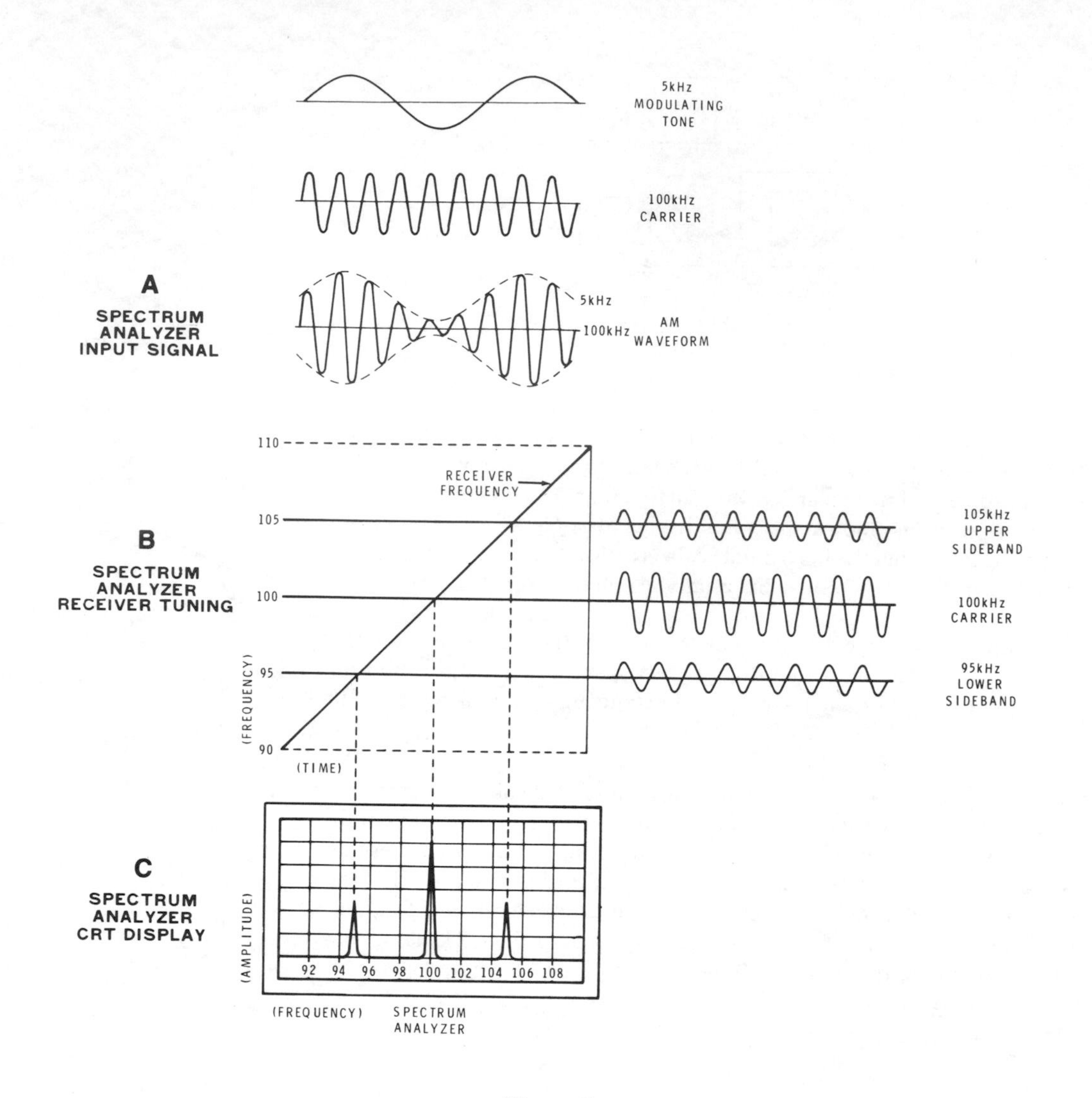

Figure 7
Producing the spectrum analyzer display.

Assume that the modulated waveform shown in Figure 7A is applied to the input of the basic spectrum analyzer in Figure 6. Further assume that the spectrum analyzer is adjusted so that its receiver is tuned to 90 kHz and its scanning range is from 90 kHz to 110 kHz.

The AM waveform in Figure 7A is shown as it would appear on an oscilloscope, i.e. in time domain. As you will recall, no signal exists, as such, at 5 kHz. However, signals equal to the sum and the difference of the modulating and carrier frequencies are present. Therefore, a strong signal should be detected at 100 kHz, and weaker signals of equal amplitude at 95 kHz (100 kHz − 5 kHz) and 105 kHz (100 kHz + 5 kHz). With this in mind, look at Figures 7B and C.

Figure 7B shows a receiver tuning chart, with frequency plotted against time. With no signal from the sweep generator, the receiver is tuned to 90 kHz. As the sweep generator starts its positive ramp, the receiver will begin tuning upward. The ramp also begins to drive the CRT electron beam across the screen at the same linear rate. As the receiver is tuned through 95 kHz, the lower sideband frequency will be detected and a signal sent to the vertical amplifier circuit. At this time, the CRT sweep will have traveled 25% of the total distance across the screen. Thus, a spike will be displayed on the CRT at this point, as shown in Figure 7C.

As the receiver continues to tune upward, it will next detect a signal at the 100 kHz carrier frequency. As this is a stronger signal than was the sideband signal, a larger signal will be felt by the vertical amplifier, causing greater CRT beam deflection. Notice that in this case, 100 kHz occurs in the center of the receiver tuning range. Thus, the 100 kHz detected signal appears in the center of the CRT display.

At a point equal to 3/4 of the sweep generator ramp amplitude, the receiver will tune through 105 kHz. The upper sideband will be detected and displayed on the CRT. Notice that the displayed sidebands are of equal amplitude and that they are lower in amplitude than the carrier. The relative amplitude between carrier and sideband is a function of the percentage of modulation.

The preceding discussion should give you a basic understanding of spectrum analyzer operation. Of course, an actual spectrum analyzer is much more complex than that depicted by Figure 6. The important thing to remember, however, is the result of the spectrum analysis; that the sideband frequencies are very real and measurable. We will build and expand on this point during the discussion of non-sinusoidal waveforms.

Before we discuss that however, we will consider one more aspect of the amplitude modulated waveform. As you have learned, the process of amplitude modulating a carrier wave produces two sidebands. By frequency domain analysis, using a spectrum analyzer, you saw that the resultant AM waveform consisted of a constant amplitude carrier, a constant amplitude lower sideband, and a constant amplitude upper sideband.

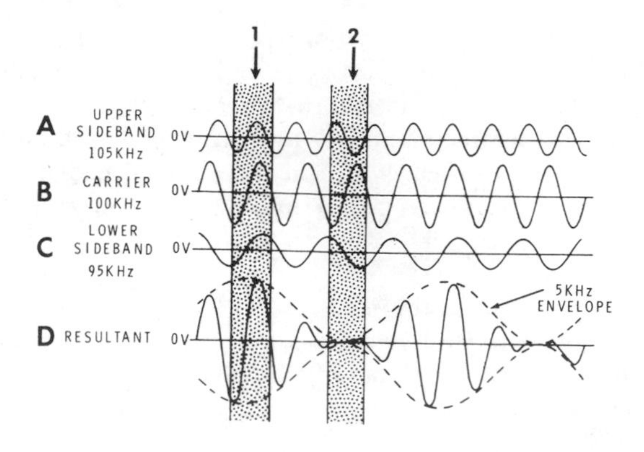

Figure 8
Developing the AM waveform envelope.

This poses an interesting question. If the signals that combine to make up the waveform are all constant in amplitude, why does an oscilloscope presentation of the resultant waveform vary in amplitude? The reason for this is easy to understand if you look at several cycles of the carrier and its sidebands. Figures 8A, B and C shows the upper sideband, the carrier, and the lower sideband respectively. Note that the carrier is twice the amplitude of either sideband.

Now, look at the phase relationship of the three signals during the shaded area labeled 1. For a brief period, they are all more or less in phase. Thus, the three waveforms add to produce a resultant wave which is about twice the amplitude of the carrier. This resultant waveform is shown in Figure 8D.

In shaded area 2, the two sidebands are in phase with each other but 180 degrees out of phase with the carrier and the three signals tend to cancel. The amplitude of the resultant waveform drops off to practically nothing.

For this reason, the resultant waveform varies in amplitude at a regular rate, forming an envelope. The frequency of this change is equal to the frequency difference between either sideband and the carrier. It is also the frequency of the intelligence that produced the modulation in the first place.

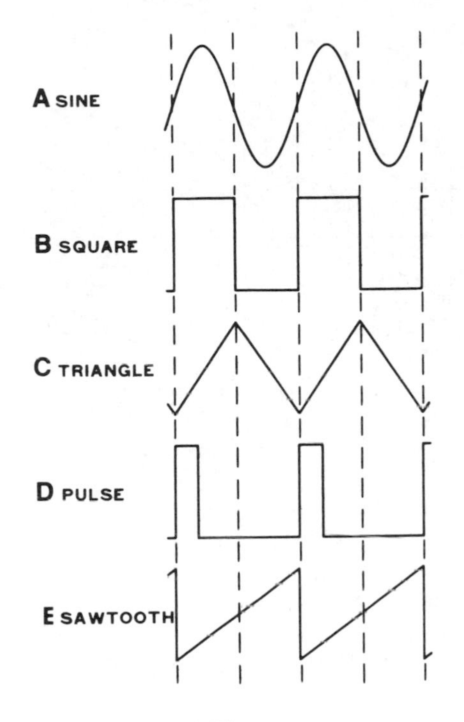

Figure 9

Waveforms

One of the most obvious characteristics of a generator is its output waveform. Some generators produce a single type signal such as a sine wave, square wave, or pulse. Many audio frequency generators allow you to select either a sine wave or a square wave. Others, such as the function generator, provide you with a wide choice of output waveforms. These may include sine waves, triangular waves, sawtooth waves, square waves, pulses, and others, commonly depending upon the instrument. Figure 9 illustrates these waveforms.

Notice that there is no difference in frequency or amplitude between these signals. The only difference is in their shape.

These waveforms and their characteristics will be covered in detail in the following discussion.

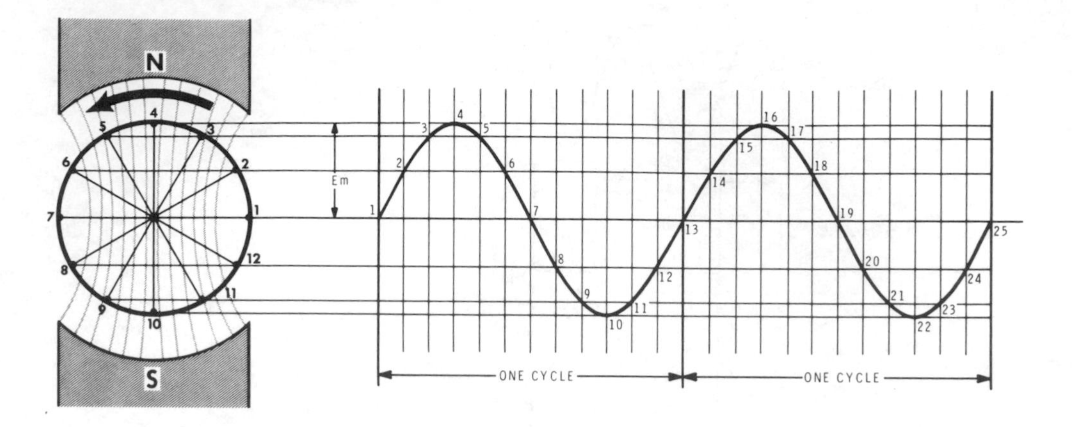

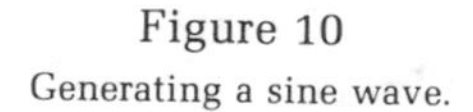
Figure 10
Generating a sine wave.

Sine Waves: The sine wave is one of the most common test signals. It is the natural output waveform of a rotary mechanical generating device or an LC tank circuit. By definition, a sine wave is a "wave shape, the amplitude of which, varies as the sine of a linear function of time". This may sound very confusing. However, a quick review of how sine waves are mechanically produced should make the meaning clear.

Figure 10 represents a conductor suspended in a magnetic field. The light lines represent lines of magnetic flux. Remember from your studies of induction and magnetism that whenever a conductor cuts a line of flux, a current is induced into the conductor. The amount of current depends upon the number and strenght of flux lines cut. The direction of current is determined by the direction of movement through the magnetic field. Thus, if the conductor is located at point 1 and being rotated counterclockwise about the center point, it is moving parallel to the lines of flux and no current is induced. Notice that this corresponds to the zero point on the voltage plot at the right.

As the conductor rotates to point 2, it is beginning to move across the magnetic field. This action continues, resulting in increasing current flow until at point 4, the conductor is moving at right angles to the

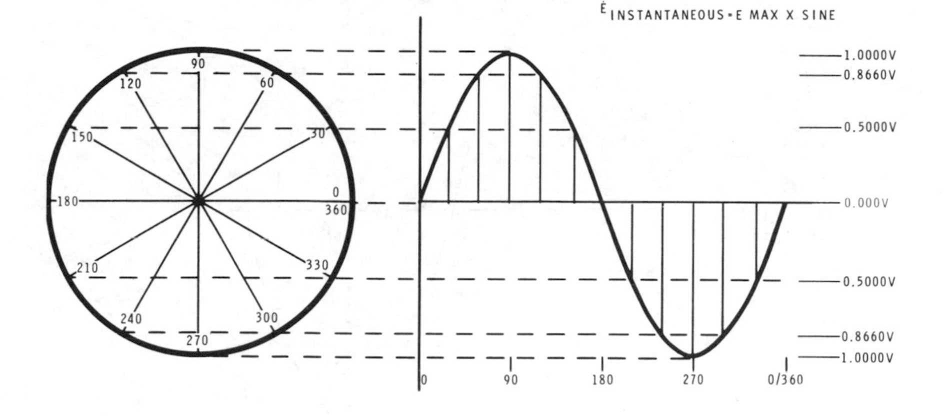

Figure 11

magnetic field. At this point, maximum current is induced. From point 4 to point 7, the sequence reverses itself until, at point 7, the conductor is again moving parallel to the lines of flux.

The movement from point 1 to point 7 generates one alternation or 1/2 cycle. As the conductor moves on from point 7, the same action occurs, except now, the conductor is moving across the force field in the opposite direction. This will result in reversing the direction of induced current and producing the alternate half cycle.

To determine the instantaneous value of voltage at any point on a pure sine wave, it is necessary to know the amount of angular rotation from zero. Figure 11 is a graphic representation of this voltage plot. Instantaneous voltage is found by multiplying the value of maximum induced voltage by the sine of the angle of rotation, ($E_i = E_{max} \times \sin \phi$). Thus, we can assume zero degrees to be at the point of zero voltage; the sine of zero degrees being 0.0000. The sine of 90 degrees is 1.0000, therefore, E max occurs at 90 degrees of rotation from zero degrees. In Figure 11, voltages plotted every 30 degrees form an approximation of a sine wave. By plotting the voltage at more points, a more accurate sine wave could be reproduced.

Some sine wave generators are also known as standard voltage generators because of their capability to provide an AC voltage of known frequency and amplitude. The output of these instruments may be calibrated in peak-to-peak, peak, or RMS values. Figure 12A and B will provide a quick review of conversion between the various values.

A perfect sine wave (one whose amplitude change exactly follows the mathematical sine function) has no distortion. Distortion results in the presence of harmonics; therefore, a pure sine wave has no harmonics. This is important when using the signal to test audio equipment because a harmonic of the test signal could feed through the circuit being tested and cause false indications at the output. Thus, if the sine wave is being used for distortion measurements, the amount of test signal distortion must be much less than the distortion caused by the circuit under test.

Sine wave generators are specified in percentage of distortion. Since "spectral purity" is the inverse of distortion, a generator with a low percentage of distortion must possess a high degree of spectral purity.

In addition to the distortion measurements, a sine wave may be used to test frequency response, gain, attenuation, sensitivity, selectivity, automatic gain control (AGC) and many other characteristics.

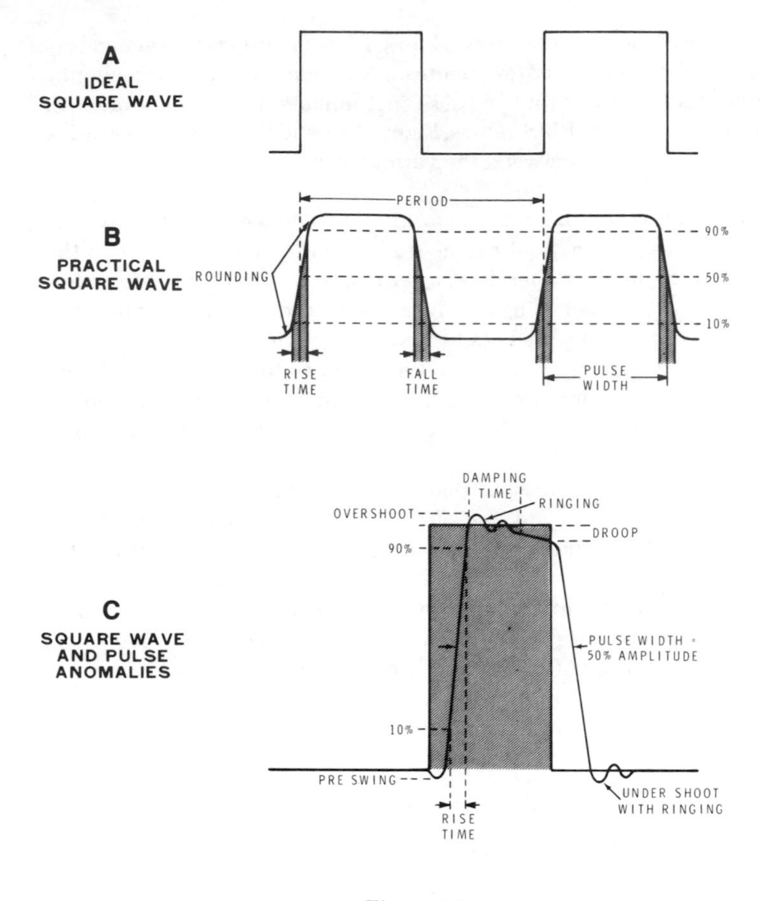

Figure 13
The square wave.

Square Waves: As the name implies, a square wave is a signal which alternately assumes one of two fixed levels, for equal lengths of time. The waveform shown in Figure 13A is an ideal square wave, switching immediately from one level to the other.

In reality, all voltage transitions require a certain amount of time to change levels. This is known as rise or fall time. The rise time is defined as the time required for the signal to rise from 10% to 90% of its maximum amplitude. The fall time is the time required for the signal to fall from 90% to 10% of its maximum amplitude.

The pulse width is measured from the 50% amplitude point on the rising edge to the 50% amplitude point on the falling edge. The period of the waveform is measured from the 50% amplitude point on one rising edge to the same point on the next rising edge. These characteristics are illustrated by Figure 13B.

The waveform frequency is defined as the number of times per second that the waveform repeats itself, i.e. the number of periods per second. Frequency as you know, is the reciprocal of time, thus,

$$\text{frequency} = \frac{1}{\text{period}}$$

In addition to rise and fall time, several other conditions may cause a square wave to deviate from the ideal shape shown in Figure 13A. Overshoot, undershoot, ringing, preswing, and droop frequently are encountered. These conditions are illustrated in Figure 13C. Notice that the leading, or rising, edge of this pulse swings in a negative direction before starting to rise. This condition is known as preswing or preshoot. The leading edge then overshoots its normal maximum value. The overshoot is often followed by damped oscillations known as ringing. Droop, or tilt, refers to the pulse top not being flat. Finally, upon returning to its normal minimum value, the trailing edge undershoots and again some ringing may occur. These conditions are usually unwanted and are caused by faulty generator circuits, mismatched output terminals or cabling problems.

Up to this point, we have considered these waveforms only in the time domain. In order to realize the full potential of a non-sinusoidal waveform as a test signal, it is necessary to analyze them in the frequency domain as well.

The concept of frequency domain analysis is based on the fact that any periodic waveform is equivalent to the sum of a number of sine waves having specific frequency, amplitude, and phase relationships.

A periodic wave is one which has the same wave shape from one cycle to the next and repeats itself at regular intervals. Any periodic wave can be formed by superimposing a number of sine waves having the proper frequency, amplitude, and phase. Mathematic proof of this statement will be left to more advanced texts, however, the principles involved can be shown by examples and viewed with a spectrum analyzer if one is available.

Figure 14A shows one cycle of a 1 kHz square wave and, according to the previous statement, is equivalent to a large number of sine waves. To be more specific, a perfect square wave is composed of a fundamental frequency and an infinite number of odd harmonics beginning in phase.

The fundamental frequency, also called the first harmonic, is the frequency of the square wave. In this example, the fundamental frequency is 1 kHz. Harmonic frequencies are exact multiples of the fundamental frequency. Thus, the harmonics of 1 kHz are:

First harmonic = 1 kHz

Second harmonic = 2 kHz

Third harmonic = 3 kHz

. . . . etc . . . .

In terms of time domain analysis, you can consider the square wave in Figure 14A as a voltage which jumps from a low level to a high level at T_0 and remains there until T_1. At T_1, it drops to its low level again and remains steady until T_2.

To analyze the same square wave, in terms of frequency domain, you must determine what sine waves are required to reproduce it. You will start with a sine wave which is equal to the square wave frequency. This is represented in Figure 14B as waveform "A". By adding sine wave "B", which is the third harmonic, in phase and at 1/3 amplitude you arrive at the resultant waveform, "C". Notice that both "A" and "B" are sine waves. However, when the instantaneous values of these two sine waves are added, the result is not a sine wave. Instead, it is a complex waveform which shows the beginnings of being a square wave.

Figure 14C shows the resultant waveform from Figure 14B added to the fifth harmonic at 1/5 amplitude. This results in much sharper corners and a somewhat flatter top.

In Figure 14D, the seventh harmonic is added at 1/7 amplitude to form the resultant waveform "G". The curve is now fairly smooth across the top and quite sharp on the corners. Adding the ninth harmonic at 1/9 amplitude and the eleventh harmonic at 1/11 amplitude would further sharpen the corners and flatten the top.

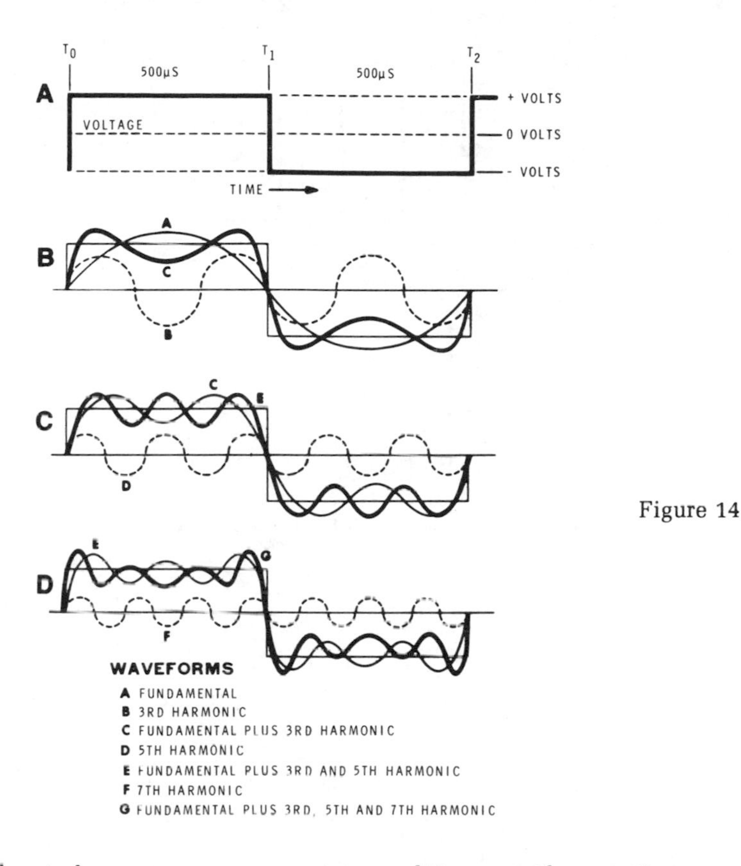

Figure 14

Remember, to form a square wave, certain conditions must be met. First, the correct harmonics must be present; in this case, the odd harmonics. If any even-numbered harmonic is present, the resultant wave shape will not be square. Second, the harmonics must be of the proper amplitude. If any required harmonic is too high or too low in amplitude, the resultant waveform will be distorted. Finally, the harmonics must have the proper phase relationship. To form a square wave they must all be in phase. On the positive-going edge of the square wave, all of the harmonic sine waves pass through zero on a positive swing. On the negative-going edge of the square wave, the harmonic sine waves all pass through zero on a negative swing. If one or more harmonics are shifted in phase, the waveform will be distorted.

Viewing the 1 kHz sine wave on two different instruments further illustrates the difference between time domain and frequency domain analysis.

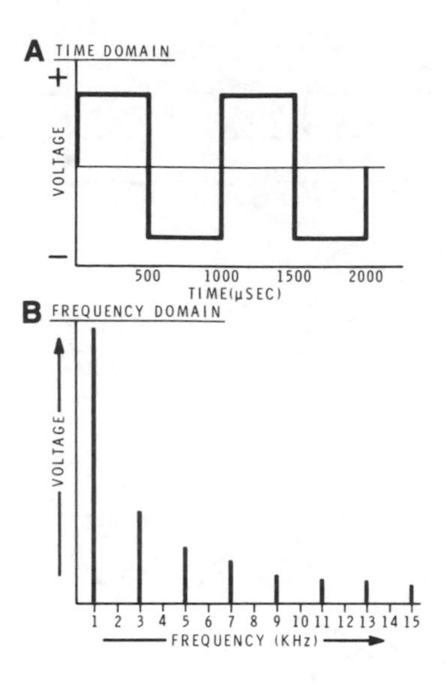

Figure 15
Time domain versus frequency domain analysis of a 1 kHz square wave.

By connecting a square wave generator to an oscilloscope, the display would be similar to that shown in Figure 15A. This is the familiar time domain display. However, by observing the same signal on a spectrum analyzer, the display would be as shown in Figure 15B. This is a frequency domain presentation of our 1 kHz square wave. Notice that a signal is present at each odd harmonic frequency point. Also note that the amplitude of each harmonic signal is the reciprocal of its harmonic number, relative to the amplitude of the first harmonic, or fundamental frequency. That is, the amplitude of the third harmonic equals 1/3 of the first, the fifth equals 1/5 of the first, etc.

As you can see, a spectrum analyzer can be a valuable tool for testing such things as amplifier distortion and frequency response. For instance, by injecting the 1 kHz square wave into an audio amplifier and observing its output with a spectrum analyzer, you could determine the flatness of response (equal amplification of all frequencies) between 1 kHz and 15 kHz.

Figure 16A shows a typical hookup for testing an audio amplifier, using a square wave generator as an input simulator and, either an oscilloscope or spectrum analyzer as an output monitor.

Whenever testing with any generator, it is important to note and remember the input waveform so that you can compare it to the waveform out of the amplifier. For this reason, it is much more effective to use a dual-trace oscilloscope for time domain analysis. By displaying the generator output on one trace and the amplifier output on the other, it is possible to simultaneously view the waveform before and after being affected by the amplifier.

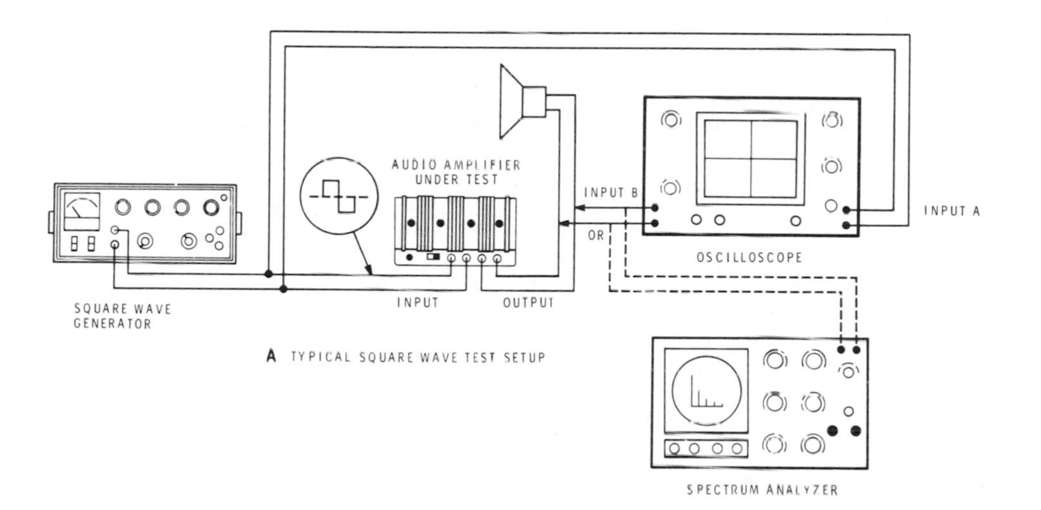

Figure 16A

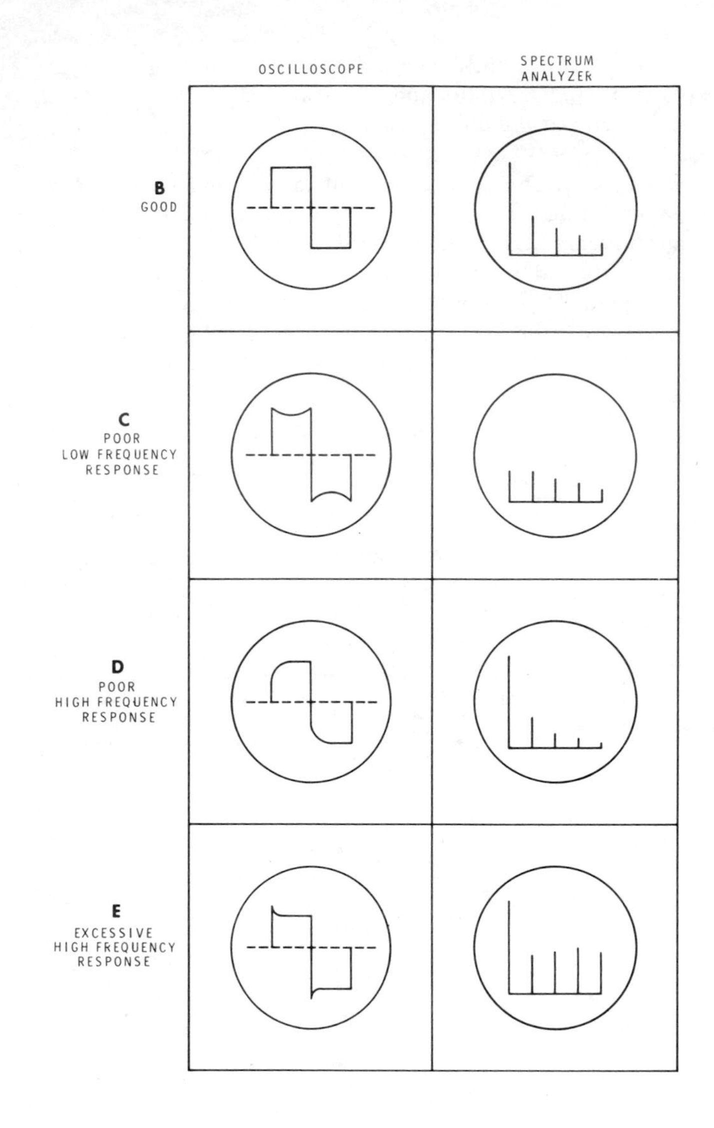

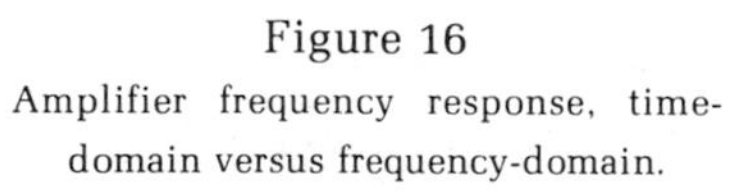

Figure 16
Amplifier frequency response, time-domain versus frequency-domain.

Figure 16B shows the results of both time domain (oscilloscope), and frequency domain (spectrum analyzer) analysis of the amplifier's frequency response. In this example, the amplifier shows a flat response from this fundamental frequency of the square wave to at least nine times this frequency. This is indicated by the undistorted waveform on the oscilloscope and the relative amplitude of the harmonics on the spectrum analyzer.

Figure 16C shows the output of an amplifier with poor low frequency response. This is indicated on the oscilloscope by the tilting or sagging wave tops. Notice, the spectrum analyzer display shows that both the first and third harmonics are greatly reduced in amplitude from example B.

In Figure 16D, the low frequency response is normal, but the high frequency response is poor. On the oscilloscope, this is shown by the rounded corners on the leading edges. The spectrum analyzer shows that the higher harmonics are greatly reduced in amplitude.

By adding high frequency boost, the amplifier will amplify high frequencies more than the lower ones and you would observe outputs similar to those shown in Figure 16E. The leading edges of the oscilloscope waveform develop overshoot, while the spectrum analyzer shows the upper harmonics to have greater amplitude than normal for a square wave.

In addition to amplifier frequency response tests, the square wave is a very useful tool in testing and adjusting a wide range of circuits and components. A number of these applications will be covered later.

Figure 17A
Sine versus triangle wave.

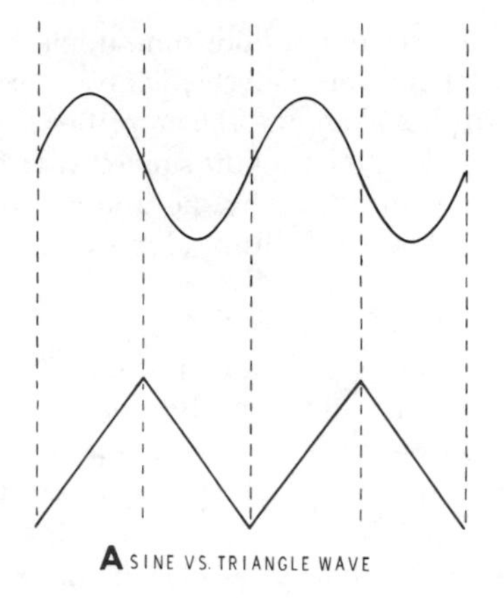

Figure 17B
Triangle wave generation.

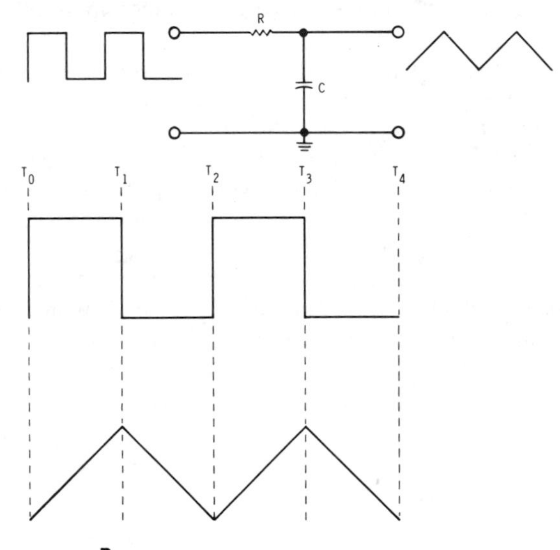

Triangle Wave: The triangle wave is a versatile waveform for testing electronic equipment. In many applications, it highlights circuit problems which would go unnoticed with a sine wave test signal. To see why this is true, we will compare the oscilloscope presentation of the two waveforms. Figure 17A shows two cycles of each waveform.

An amplifier circuit, driven by a triangle wave, can be quickly checked for gain and distortion. While the same checks can be made using a sine wave, defects are much more visible with the triangle wave.

For simple gain measurements, the sharp peaks make exact amplitude measurements easier than the rounded peaks of the sine wave.

Distortion is also much easier to detect. Sine wave distortion may reach 5 to 10 percent before it becomes apparent, while as little as 2 percent distortion on a triangle wave is easily discernable. The reason for this is the limitation of the human eye to act as a comparator. It is far easier to visually detect a deviation from a straight line than to recognize a bulge on a curved line.

The triangle wave can also be used as a source of swept DC voltage. By setting the generator frequency to 1 Hz, you have a DC level which varies, at a linear rate, from minimum to maximum and back, once each second. This voltage can be used as an external sweep signal for an X-Y plotter, pen recorder, or other relatively slow electromechanical device.

A simple method of generating a triangle wave is to integrate a square wave as shown in Figure 17B. By selecting the values on R and C to provide a long time constant at the square wave frequency, only a very small portion of the charging curve will be used. Thus, the slopes of the triangle wave will be quite linear.

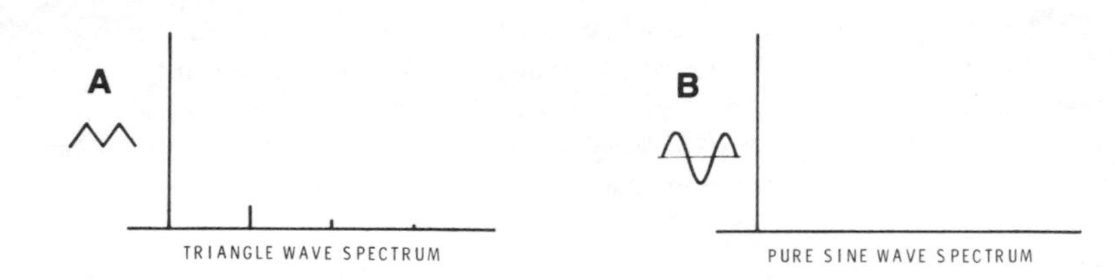

Figure 18
Time versus frequency domain comparison of triangle and sine waves.

As it is not a sine wave, the triangle wave does have a harmonic content. However, since it consists of the fundamental frequency plus the third harmonic at 1/9 amplitude, the fifth at 1/25 amplitude, etc. its value as a test wave is primarily in time domain. This spectrum is shown in Figure 18A. The spectral line of a sine wave of the same frequency is shown in Figure 18B for comparison.

Sawtooth Wave: The time domain presentation of a sawtooth wave on an oscilloscope shows some similarity to that of a triangle wave. Figure 19 compares the sawtooth and triangle waveforms. Notice the sawtooth voltage proceeds, in a linear ramp, from one level to another. However, instead of reversing and ramping back to its starting level, it "flies" back immediately and repeats the previous ramp. Notice also that this waveform may be made to slope in either the positive or negative direction.

Figure 19
Sawtooth, or ramp, waveshapes.

This waveform, like the triangle wave, may also be used as a source of sweep voltage. Unlike the triangle wave, however, it is quite rich in harmonics which makes it useful as a frequency domain test signal. Figure 20 shows the frequency composition of a negative ramp 1 kHz sawtooth wave.

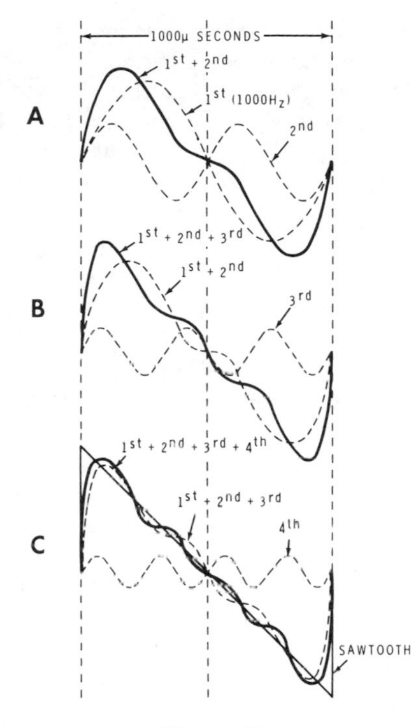

Figure 20

The sawtooth wave contains both odd and even numbered harmonics. Figure 20A shows the result of adding the 1st harmonic (fundamental) with the second harmonic at 1/2 amplitude. Notice that this waveform begins to assume its desired shape much faster than did the square wave. Figure 20B shows the result of adding the third harmonic at 1/3 amplitude to the previous resultant. This sharpens the peaks and smooths out the ramp portion of the wave shape. In Figure 20C, the fourth harmonic is added at 1/4 amplitude which brings the wave even closer to ideal shape. By continuing this process (i.e. adding the fifth harmonic at 1/5 amplitude, the sixth at 1/6, etc) a very usable sawtooth is obtained at about the ninth harmonic. Thus, if a circuit will pass this 1 kHz sawtooth signal without distortion, we know the circuit has a bandwidth of at least 9 kHz.

As you have seen, a sawtooth contains all harmonics, both even and odd numbered. In this case all of the harmonics are in phase with the fundamental frequency.

In order to reconstruct a positive-ramp sawtooth, it is only necessary to reverse the phase of the even numbered harmonics. Figure 21 shows the composition of this wave shape. Notice that the odd numbered harmonics (the first and third) are shown in the same phase as the harmonic frequency in Figure 20. However, the even numbered harmonics (second and fourth) are 180 degrees out of phase. This results in a reversal of the ramp slope. Figure 21D shows how the sawtooth would appear if it contained all harmonics up to the seventh harmonic of the fundamental frequency.

Looking at the frequency spectrum of this wave shape reveals that it is much richer in harmonics than a square wave. Figures 22A and 22B show the comparative harmonic content of the sawtooth and square wave.

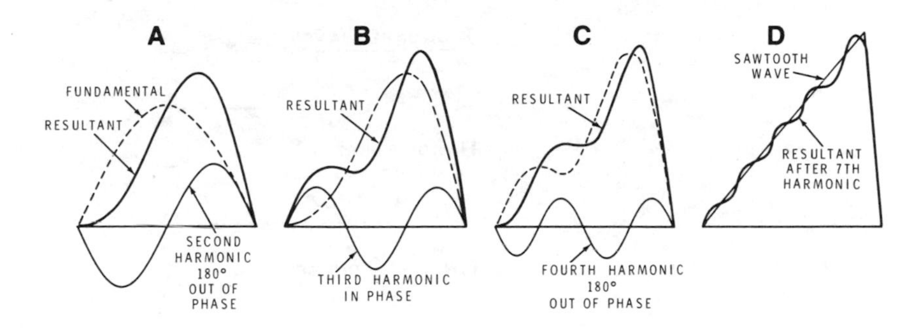

Figure 21
Composition of a positive ramp sawtooth wave.

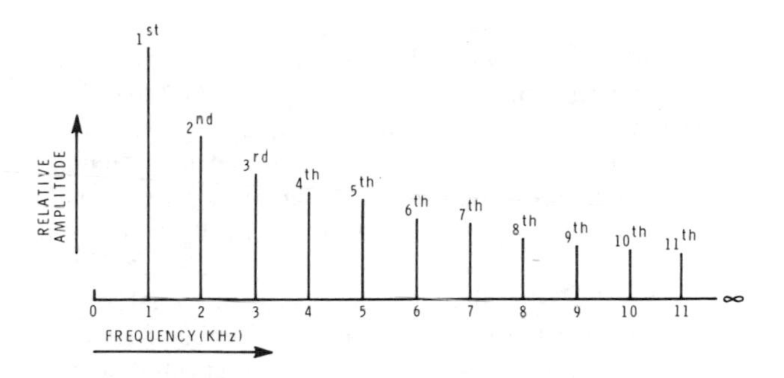

A FREQUENCY COMPOSITION OF A 1KHz SAWTOOTH WAVE

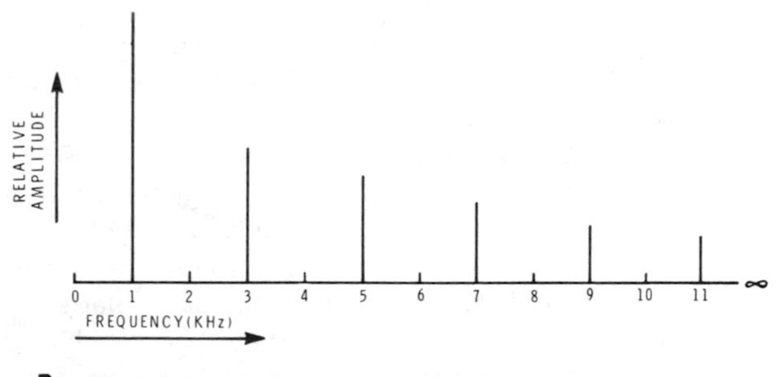

B FREQUENCY COMPOSITION OF A 1KHz SQUARE WAVE

Figure 22

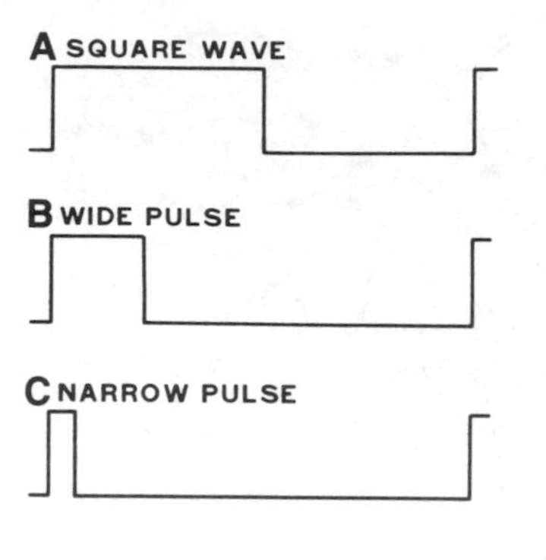

Figure 23
Time domain comparison of a square wave versus a pulse.

Pulse: The pulse wave shape is closely related to the square wave. Viewed in the time domain, as shown in Figure 23, it would appear that only the symmetry changes as the pulse width is varied. Figure 23A shows a square wave which, of course, is symmetrical. That is, the amount of time that the signal is high equals the amount of time it is low. Figure 23B and C show pulse signals of the same amplitude and repetition rate.

When a pulse is displayed on a spectrum analyzer, however, it loses its resemblance to the square wave. Remember, a square wave is made up of the fundamental frequency plus a large number of odd numbered harmonics, in phase. The amplitude of these harmonics decrease at the reciprocal of their harmonic number. (i.e. the third at 1/3, the fifth at 1/5, etc.) This frequency versus amplitude relationship is shown in Figure 24A.

As the waveform deviates from the perfect time symmetry of the square wave, however, several changes occur in its harmonic content. One is the appearance of even numbered harmonics. Also, the amplitudes of the harmonics do not diminish as quickly as do those of the square wave. This can be seen in Figure 24B. The composition of the pulse is too complex to be shown clearly by the addition of individual sine waves. The exact amplitude and phasing of these sine waves depend upon the pulse width and repetition rate of the particular signal. One important characteristic of pulse spectrum should be noted; as the pulse becomes narrower, the number of harmonics increase and their relative amplitudes become more consistent. This can be seen by comparing the two spectrum presentations shown in Figure 24B and C. These illustrations are only for comparison purposes and no attempt should be made to

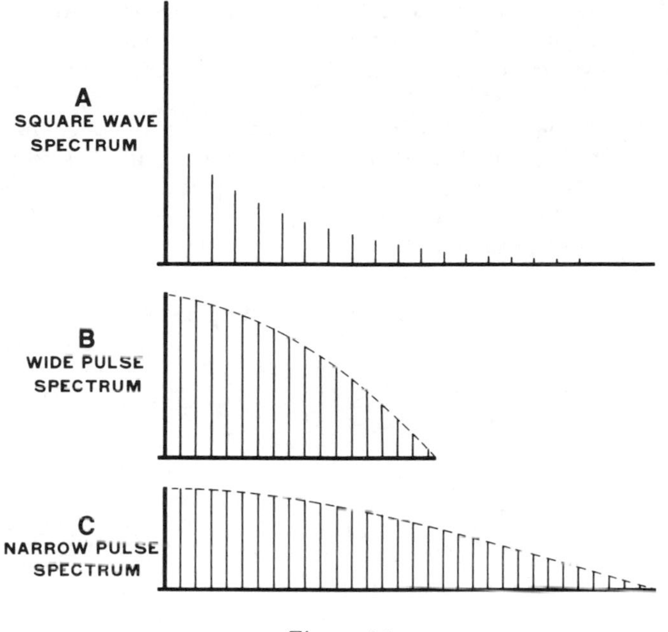

Figure 24

correlate these spectrum presentations to specific pulse widths or repetition rates. However, as you can see, a rather narrow pulse width may provide a spectral comb of harmonics similar to the output of the comb generator circuit used in the counters heterodyne frequency converter unit. The amplitudes of these harmonics are so close to that of the fundamental signal, that the difference is nearly indistinguishable over a wide frequency range. For example, a 100 nsec pulse at a repetition rate of 1 MHz produces both odd and even harmonics of 1 MHz whose amplitude is nearly equal to that of the fundamental frequency up to 15 MHz or more.

This makes the pulse an excellent waveform for testing the high frequency response of circuits, cables and connections. As you can see from the example, equipment which will pass this 100 nsec, 1 MHz, pulse train without distortion must be capable of handling a range of frequencies from 1 to 15 MHz.

Summary: Figure 25 provides you with a quick review of the waveforms covered in this section. The voltage waveforms are shown on the left with the frequency spectrum analysis of that waveform at the right.

The spectrum analyzer is still a very expensive piece of test equipment with comparatively limited applications. Therefore, it is unlikely you will be using frequency domain testing techniques directly. It is necessary, however, for you to be aware of the frequency content of these waveforms in order to intelligently evaluate the time domain test results.

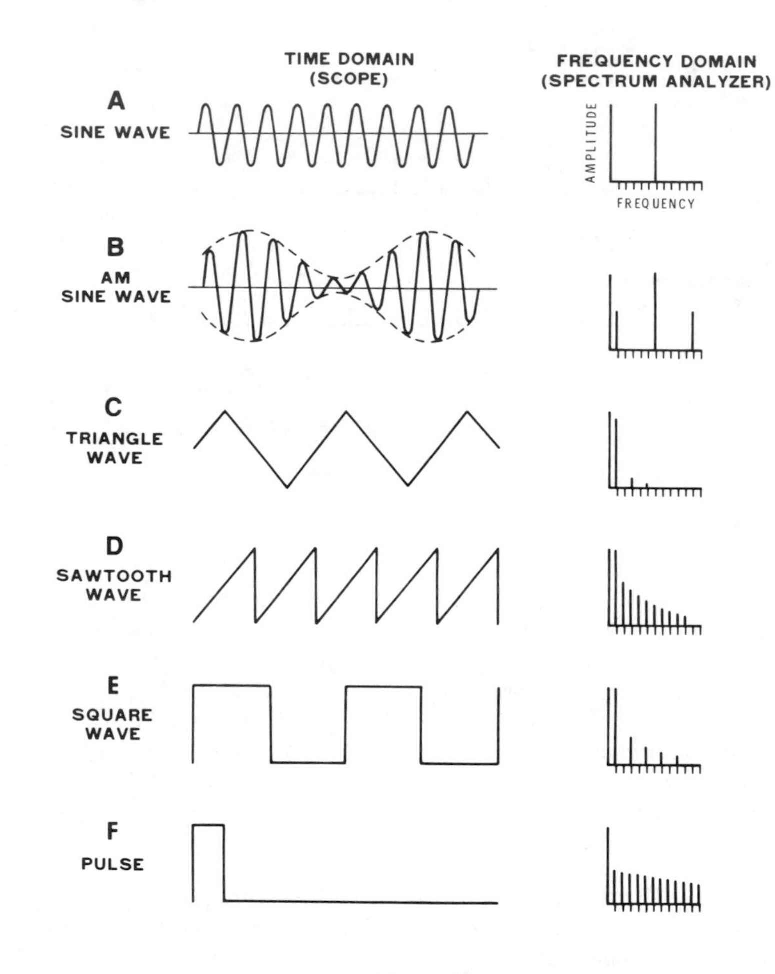

Figure 25
Comparison of wave shapes and harmonic content.

AUDIO FREQUENCY GENERATORS

The first instrument we will examine is the audio frequency generator. This versatile, general-purpose test instrument is capable of producing sine wave signals that are variable in frequency from 5 Hz or less to over 100 kHz with a typical amplitude range of 1 mV to 10 V. In addition to the sine wave output, many audio generators also provide a square wave output which is available simultaneously with the sine wave.

Figure 26 shows two examples of audio generators that are typical of the type of instrument currently available in the low to middle price range ($100 to $250).

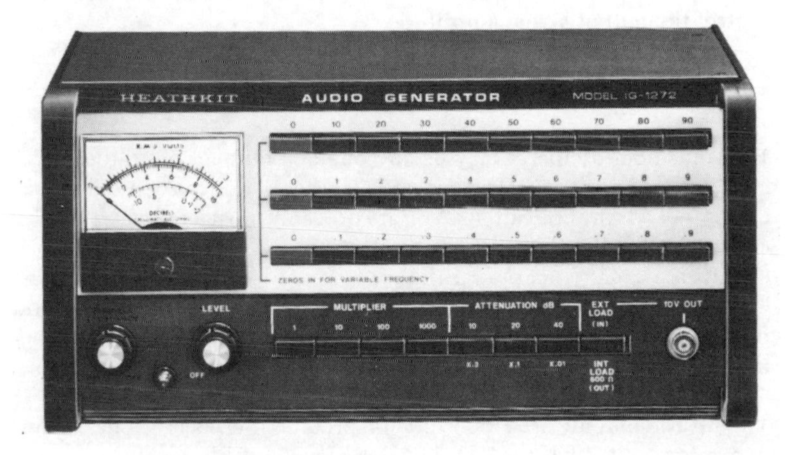

Figure 26

The prime requirements for a general-purpose audio generator are an adequate frequency range, good frequency stability, a high degree of spectral purity, and a stable controllable output amplitude. Other features that are highly desirable include the ability to maintain a constant output level throughout its tuning range, a meter to monitor output voltage, and a square wave output. With these points in mind, we will discuss some basic audio generator circuits.

Figure 27A shows a simplified block diagram of a basic audio generator. The sinusoidal oscillator produces a high quality sine wave signal. This signal is routed through the buffer amplifier stage which serves a dual purpose. It isolates the oscillator from loading effects and also reduces the output impedance to a desirable level. The attenuator allows you to control the output signal amplitude.

Oscillator Circuits

The sinusoidal oscillator may be any one of a number of oscillator circuits. Each of these circuits has characteristics which make it the logical choice for certain applications. You are probably most familiar with LC or crystal oscillators. However, in the low audio frequency range, these circuits are usually not practical. Inductors for low frequencies (around 60 Hz) would be very large and expensive, while the practical low frequency limit for crystals is approximately 50 kHz. For these reasons, most audio frequency generators use some form of RC oscillator circuit.

The most commonly used RC oscillator circuit is the Wien-bridge or some derivation of it, such as the bridged-T or notch-filter oscillator. These oscillators employ both positive and negative feedback, which provides good frequency stability with low distortion. Their design usually incorporates automatic gain control (AGC) and temperature compensation to further improve their stability.

Two methods of frequency tuning are used in modern current audio generators. The most common method uses a switch to select a frequency range and a tuning dial to vary the frequency within this selected range. These are known as continuously tunable generators, and use a ganged variable capacitor as the tuning element. Range changing is accomplished by switching in sets of fixed resistors. Figure 27B shows a simplified Wien-bridge oscillator circuit using this type tuning.

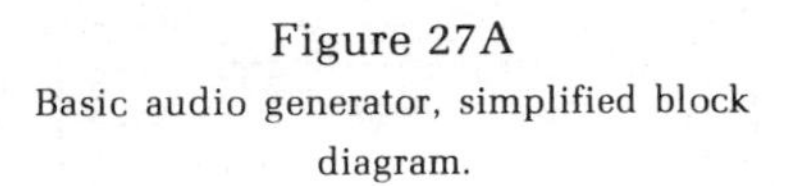

Figure 27A
Basic audio generator, simplified block diagram.

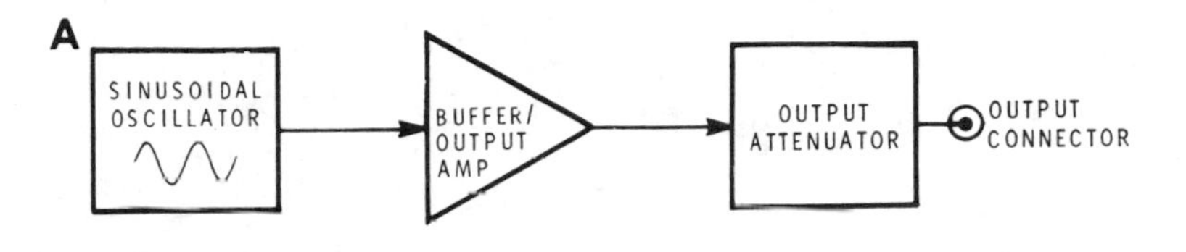

BASIC AUDIO GENERATOR-SIMPLIFIED BLOCK DIAGRAM

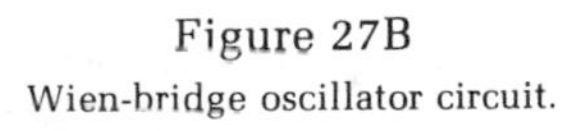

Figure 27B
Wien-bridge oscillator circuit.

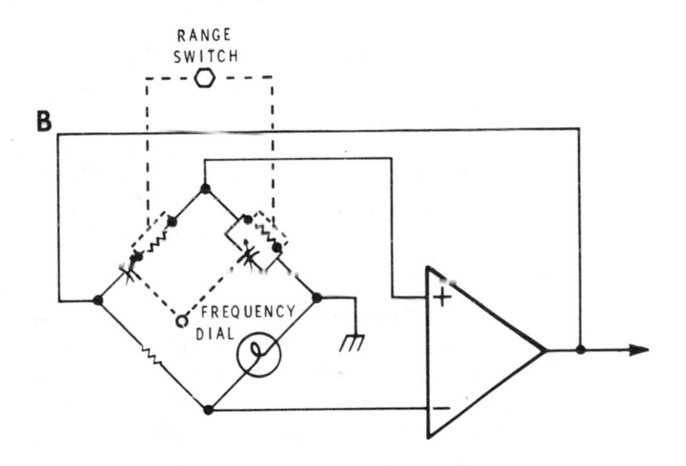

WIEN-BRIDGE OSCILLATOR CIRCUIT

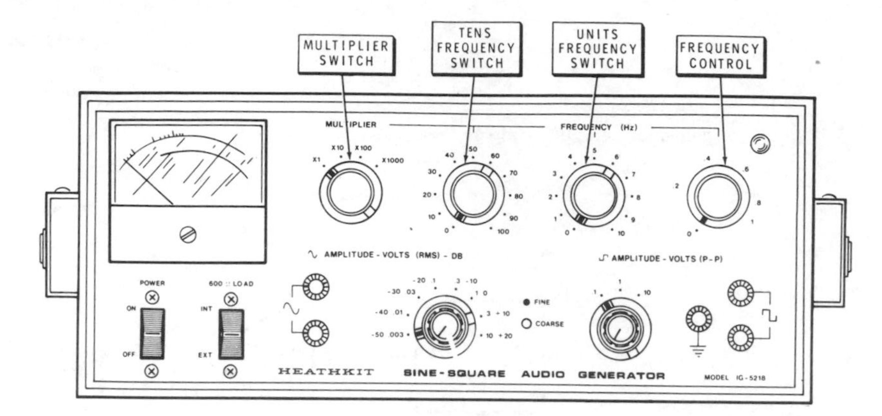

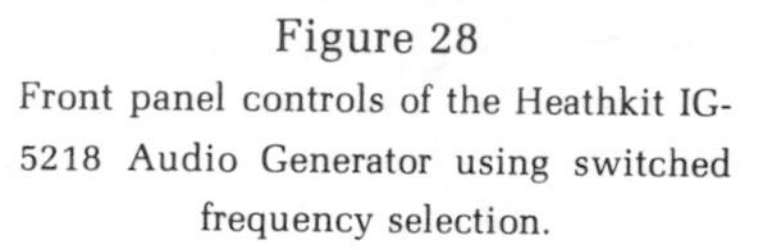

Figure 28
Front panel controls of the Heathkit IG-5218 Audio Generator using switched frequency selection.

A second tuning method that is becoming very popular is the selection of discrete frequencies by switches. These may be push-button or rotary-wafer type switches. Figure 28 shows the front panel controls of an audio generator using rotary-wafer switch frequency selection. This unit uses two decade switches (Units and Tens) and a potentiometer (Frequency Control) in conjunction with a multiplier switch. With the controls shown, you may select any frequency between 1 Hz and 111 kHz.

When switched frequency selection is used, range changing (in this case, the Multipler) is usually accomplished by changing capacitance values, while frequency is selected by selecting appropriate precision resistors. Thus the switched-tuning technique reverses the roles played by capacitance and resistance.

A typical circuit, using this type of tuning, is shown in Figure 29. If you compare this circuit with the controls shown in Figure 28, the "**multiplier**" switch controls the selection of $\mathbf{C}_x$ and $\mathbf{C}_y$, the "**Tens**" switch selects the two $\mathbf{R}_x$ resistors, the "**Units**" switch selects the two $\mathbf{R}_y$ resistors while the "**Frequency Control**" adjusts both sections of potentiometer $\mathbf{R}_z$. R_f and L_1 provide a regenerative feedback path and L_1 is used for amplitude stability. The tuned RC circuit forms a notch filter in the degenerative feedback path. As excessive degenerative feedback will prevent oscillation, the circuit will operate only at the frequency where feedback is minimum.

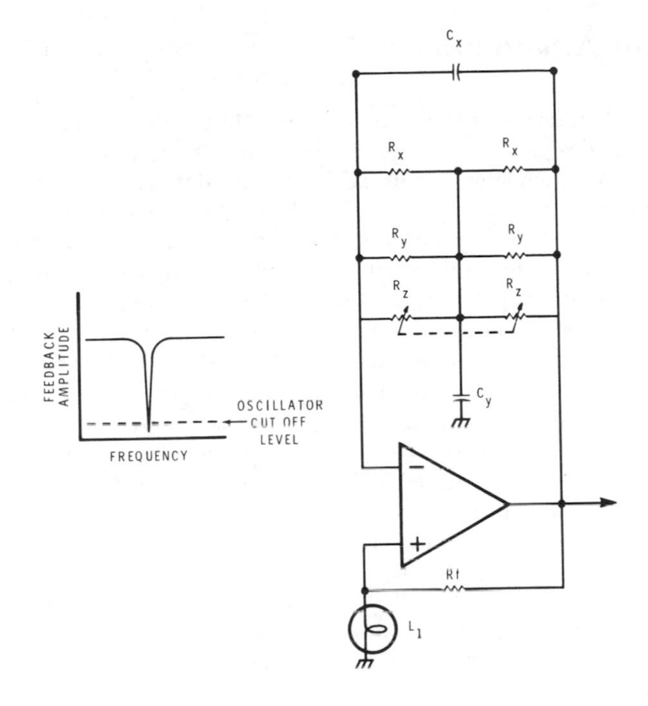

Figure 29

Buffer/Output Amplifier Circuits

In order to maintain amplitude and frequency stability, oscillators must operate into a constant load. On the other hand, you may require your audio generator to drive a wide range of load impedances. Therefore, it is very important to isolate the oscillator circuit from the output terminals. This is one major function of the amplifier circuit. It may also provide voltage and power amplification to increase the amplitude of the oscillator signal to the level required for the generator output.

The amplifiers used in this section usually consist of several wideband, low-noise, class A stages. Large amounts of negative (degenerative) feedback are used to improve frequency response and distortion characteristics. The design of these amplifiers involves great care to avoid hum, noise, and modulation. For these reasons, you should be very careful if you attempt to repair these instruments. Component placement and wire routing may be critical to proper operation.

Output Attenuators

The output attenuator allows you to control the amplitude of the output signal. The design of this circuit varies from extremely simple to quite complex. An attenuator circuit, such as you might find in a medium priced generator, is shown in Figure 30. This particular circuit is used in the Heathkit IG-5218, but it is typical of audio attenuators used throughout the industry. The only feature which might be considered unique is the meter circuit which monitors the signal voltage being applied to the step attenuators. Some generators provide this meter circuit. Others provide terminals to allow monitoring with an external voltmeter, or rely on internal calibration to establish the signal reference level. Refer to Figure 30 for the following discussion. The sine wave output of the amplifier circuit is felt across R106, which is a front panel control labeled "Amplitude, Fine." A portion of this signal, adjustable from 0% to 100%, is coupled through the isolating resistor (R107) to both the meter circuit and the sine wave "Amplitude, Course" switch.

The resistor network (R109 to R121) on two sections of this wafer switch comprises an eight-step voltage divider. This allows you to select output amplitudes in steps of 10 dB. The remaining switch section allows you to connect an internal load resistor (R122) across the output terminals in the six lowest output ranges.

The meter circuit is calibrated to indicate the voltage amplitude present at the junction of R107 and R108. Note that it does not directly measure the output voltage, as the markings of the amplitude selector switch and the meter scale might lead you to believe. The accuracy of the meter reading, as compared to the actual output, relies upon the precision of the attenuator network resistors. It also relies upon the generator operating into a 600 Ω load. On the six lowest output ranges, R122 can be used to provide this load, but on the two highest output ranges, an external 600 Ω load must be provided if the meter indications are to be considered calibrated.

Squaring Circuits

Many audio generators provide you with a square wave output in addition to sine waves. The circuit most commonly used to convert the oscillator sine wave to a square wave is the Schmitt trigger. The operation of this circuit was covered in detail during the discussion of counter signal conditioners in Unit 4. The output of the Schmitt trigger circuit is passed through an amplifier stage, usually an emitter follower. This reduces the output impedance enough to properly drive an attenuator circuit.

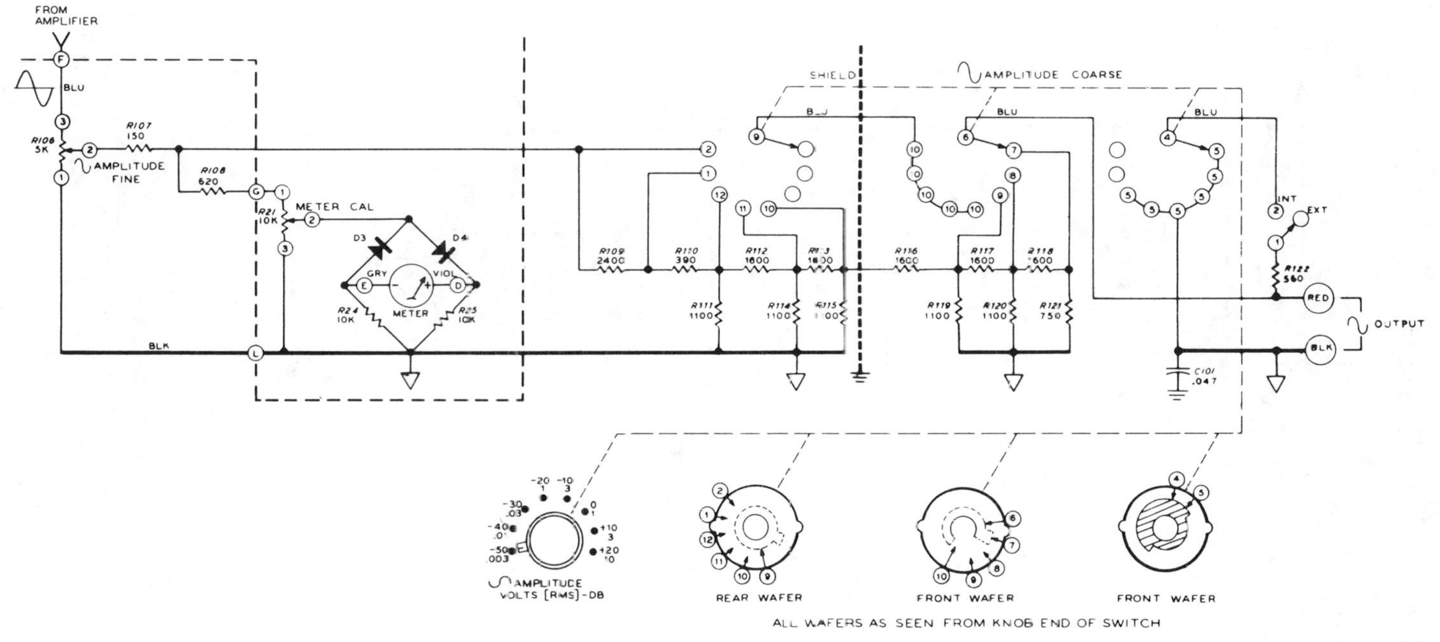
FROM AMPLIFIER
BLU
R106 5K
R107 150
AMPLITUDE FINE
R108 620
R21 10K
METER CAL
D3
D4
GRY
VIOL
METER
R24 10K
R25 10K
BLK
SHIELD
AMPLITUDE COARSE
BLU
R109 2400
R110 390
R111 1100
R112 1600
R114 1100
R116 1600
R117 1600
R118 1600
R119 1100
R120 1100
R121 750
R122 560
INT
EXT
RED
BLK
OUTPUT
C101 .047
AMPLITUDE VOLTS [RMS]-DB
REAR WAFER
FRONT WAFER
FRONT WAFER
ALL WAFERS AS SEEN FROM KNOB END OF SWITCH

Figure 30

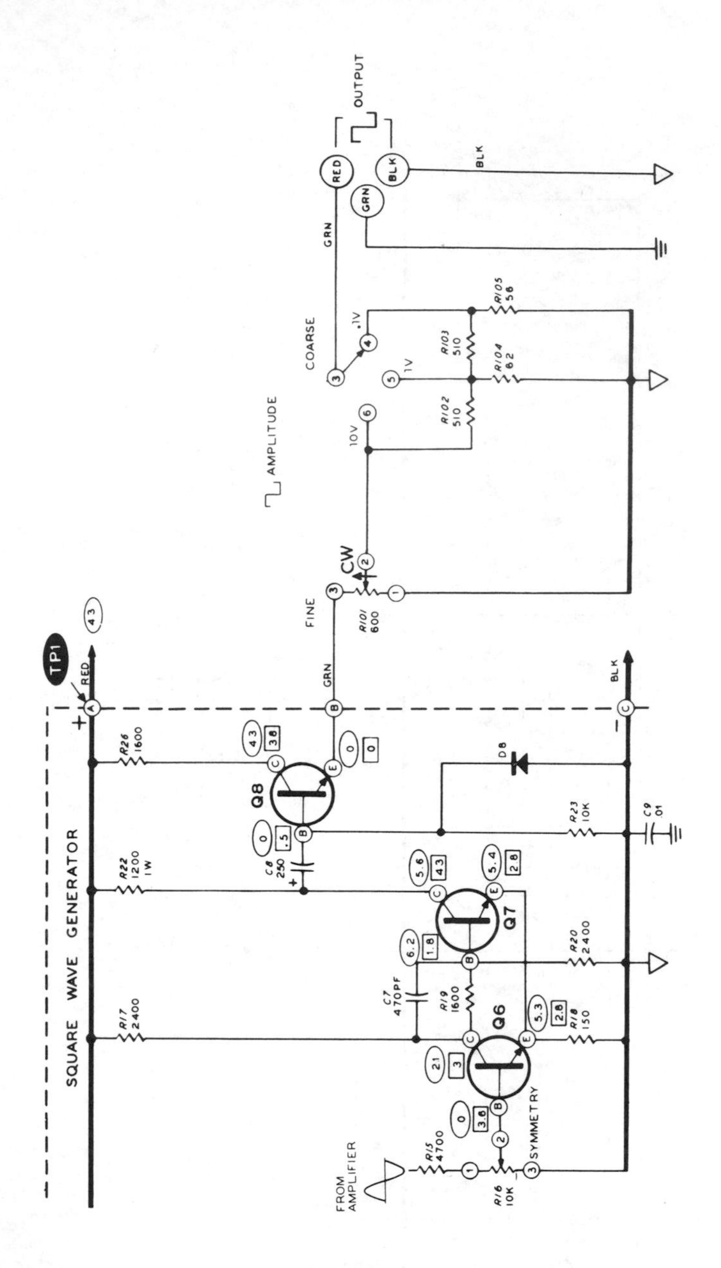

Figure 31
Typical square wave generator and output circuit (Heathkit IG-5218).

Figure 31 shows a typical square wave generator and attenuator circuit. Refer to this Figure for the following discussion.

The square wave section of this generator consists of a Schmitt trigger circuit comprised of Q6 and Q7, emitter follower Q8, and the square wave attenuator.

The square wave is produced by the Schmitt trigger circuit, which is triggered by the oscillator sine wave signal. This signal is coupled through R15 and R16 to the base of Q6. The voltage on the base of Q6, of course, varies with the rising and falling voltage of the sine wave input. Symmetry control R16 is adjusted to produce time intervals between switching-on and switching-off that are of equal length; therefore, producing a symmetrical square wave.

This square wave output, from the collector of Q7, is coupled to the base of emitter follower Q8, which provides a low impedance output with no voltage gain. The output from the emitter of Q8 passes directly to the "**Square Wave Amplitude, Fine**" control R101. From R101, a portion of the signal is applied, through the attenuator network connected to the square wave amplitude switch, to the square wave output terminals.

Except for the appropriate power supply circuits, this completes the discussion of audio generator operation. The power supplies are generally composed of standard circuits but, of course, must be well regulated and filtered.

AUDIO GENERATOR APPLICATIONS

It is impossible to discuss all possible applications of such a universal instrument as the audio generator. Any test requiring a source of relatively pure sine wave voltage within its frequency range is a likely candidate. However, as you will see, a generator is seldom used by itself. Usually it provides a test signal which is passed through the circuit under test. The output of this circuit is then monitored by a voltmeter, oscilloscope, distortion analyzer, or other indicating instrument, which displays the circuit performance.

As the name implies, the majority of audio generator applications involve measuring audio equipment parameters, such as input sensitivity, frequency response, gain, distortion and power. However, audio generator applications are not limited to testing audio equipment. There are enough other applications to warrant including a good quality audio generator in your general-purpose test equipment inventory, along with your multimeter and oscilloscope.

Next, we will examine a few of the more common audio generator applications.

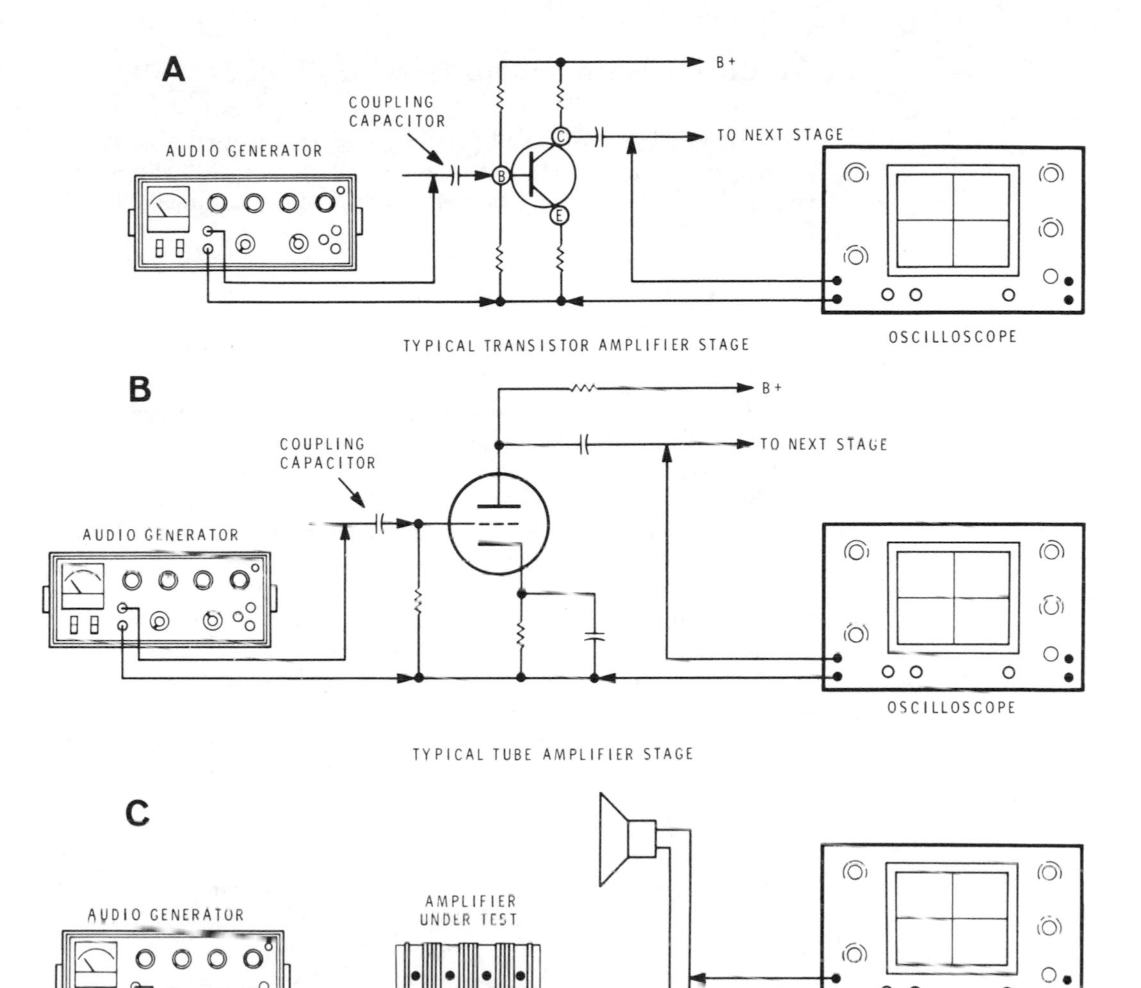

C

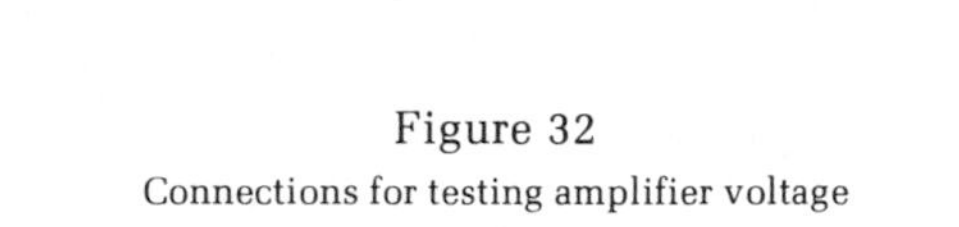

Figure 32
Connections for testing amplifier voltage gain.

Amplifier Gain (Amplification) Factor

To determine the gain, or amplification factor, of a circuit it is necessary to compare the amplitude of the input voltage to that of the output voltage. The amplification factor is equal to the output divided by the input, or:

$$\text{Gain} = \frac{E_{out}}{E_{in}}$$

Figure 32 shows how you would connect an audio generator and oscilloscope to perform this test. Figures 32A and 32B show connections for testing single amplifier stages while Figure 32C shows the connections for testing an entire unit. Note that in Figure 32A and 32B, a coupling capacitor is used in series with the signal lead. This is to prevent any DC voltage in the circuit under test from damaging the generator output circuits.

It is important that you do not introduce distortion during this test. Therefore you should be careful not to overdrive the amplifier. Select a frequency within the operating range of the amplifier and set the generator output to minimum. Next, connect the equipment as shown in Figure 32 and increase the generator output until an undistorted sine wave signal is displayed on the oscilloscope. You can now calculate the amplifier's voltage gain by dividing the input voltage amplitude into the output voltage amplitude. Of course, both readings must be expressed in the same units. The circuit in Figure 32 shows the input monitored by the generator output meter. This meter is calibrated to indicate the rms value of generator output. The oscilloscope, of course, displays the peak-to-peak value of amplifier output voltage. Therefore, you must convert either the rms input to peak-to-peak, or the output voltage to rms.

As an example, assume the rms input were 200 mV and the output read 12 V peak-to-peak. 200 mV rms equals .5656 V peak-to-peak (rms $\times$ 1.414 $\times$ 2). Therefore, gain equals 12 $\div$ .5656 or 21.2. On the other hand, 12 V peak-to-peak equals 4.242 V rms ($\frac{12 \times .707}{2}$); therefore, gain equals 4.242 $\div$ .2 or still 21.2.

These conversions are not necessary if you have a dual-trace oscilloscope. By displaying the input signal on one trace and the output signal on the other, you can read both voltages directly in peak-to-peak value.

This procedure works well with fixed gain amplifier stages. However, if you attempt to measure the gain of a high power amplifier unit, be careful. To measure maximum gain, the gain, or volume control, must be adjusted to full volume. A continuous tone at full power could result in speaker cone damage. Therefore, you should substitute a resistor, equal to the speaker impedance and of suitable power rating, for the speaker. You can then connect the oscilloscope across the resistor to measure the output voltage.

Amplifier power gain is measured in the same manner; however, it is slightly more complicated because you have to consider both input power and output power. The most convenient power formula to use is $P = \frac{E^2}{R}$. Once you have determined the input and output power levels, you can calculate the power gain. This is expressed in decibels (dB). The formula for this is:

$$dB = 10 \log \frac{P_o}{P_i}$$

Where: dB = power gain
P_o = output power
P_i = input power

You will assume some values and determine the gain of a hypothetical amplifier in the following discussion. First, assume your amplifier has an input impedance of 50 kΩ and an output impedance of 8 Ω. If you inject 0.1 V rms from your generator into your amplifier input connector, the input power is:

$$P_1 = \frac{(0.1 \text{ V})^2}{50000}$$

$$= 0.0000002 \text{ watts}$$

or

$$2 \times 10^{-7} \text{ watts}$$

Now measure the amplifier's output voltage. You can use an AC voltmeter or an oscilloscope but make sure you use the rms value of the reading. Also, the amplifier output must be terminated in its characteristic impedance, in this case, 8 Ω. If the output voltage is 10 V rms, the output power is:

$$P_o = \frac{(10)^2}{8}$$

$$= 12.5 \text{ watts}$$

Now that you know both the input and output power, the power gain can be determined as follows:

$$\text{dB} = 10 \log \frac{12.5}{2 \times 10^{-7}}$$

$$= 10 \log 6.25 \times 10^7$$

$$= 10\ (7.7959)$$

$$= 77.959$$

Input and Output Impedance

The preceding measurement required that you know the input and output impedances of your amplifier. These values are often available for commercial amplifiers in the owner's manual or specification sheet. If not, they are quite easily determined.

Figure 33 shows the equipment hookup for measuring input impedance. This method is exact if the impedance is purely resistive but only approximate if it is reactive. For measuring input impedances that are high compared to the generator output impedance, use the connections shown in Figure 30A. It may be necessary to try variable resistors of several different values from 10 kΩ to 5 MΩ in order to obtain the correct indication on the output voltmeter.

Set the generator frequency to some low value, such as 20 Hz. Short out the variable resistor, as indicated by the dashed line in Figure 33A, and set the signal level of the generator for some convenient reading on the AC voltmeter.

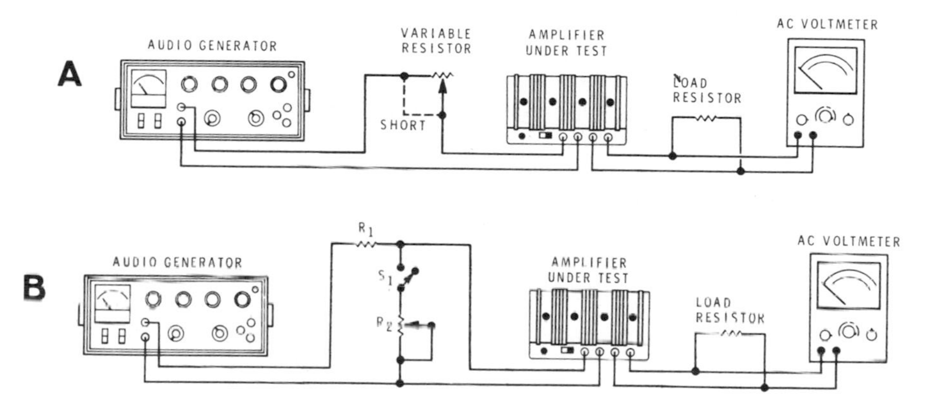

Figure 33
Equipment hookup for measuring input impedance.

Now remove the short from the variable resistor and adjust the resistor until the meter reads exactly 1/2 the value of the former reading. The resistance of the variable resistor is now equal to the input impedance of the amplifier. Remove the variable resistor from the circuit and carefully measure its resistance with an ohmmeter.

If you perform this same test using a higher generator frequency, the input impedance will probably appear to be lower. This is because of the amplifier's input capacitance which presents less impedance as frequency increases.

For low input impedances (a few hundred ohms or less) connect the test circuit as shown in Figure 33B. The fixed series resistor (R_1) should be at least ten times the input impedance to be measured. The value of this resistor is not critical as long as it is high enough. Therefore, a 10 kΩ to 47 kΩ should work nicely. The variable resistor should be about 1 kΩ.

First, with switch S_1 open, set the reference level to some convenient output voltage reading. Now, close the switch and adjust R_2 until the output meter again reads 1/2 of the reference reading. The active portion of R_2 now equals the input impedance of the amplifier.

Output impedance is determined using a similar technique. Although this procedure is presented to measure the output impedance of an amplifier, it can be used for determining this characteristic of any circuit. One note of caution; never operate a power amplifier, such as a hi-fi or stereo unit, with no load attached to the output terminals. Severe damage to the output stages can result.

Figure 34 shows the equipment hookup for this test. Set the generator frequency to 1,000 Hz and adjust the generator output amplitude for a convenient reference level on the AC voltmeter. Next, set the variable resistor near its midrange and connect it into the circuit by closing switch S_1. Adjust R_1 until the voltmeter indicates 1/2 of the reference level reading. R_1 is now equal to the output impedance value.

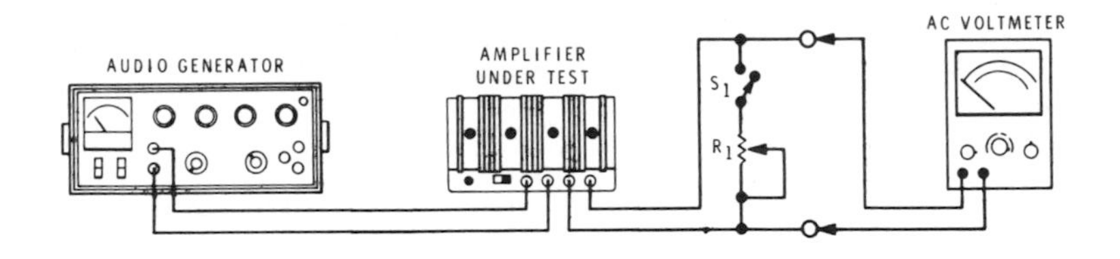

Figure 34

Input Sensitivity

Input sensitivity is another factor of amplifier performance that you may wish to measure. This specification is defined as the value of rms input voltage required to obtain full-rated output power across the rated output impedance. The equipment hookup for this measurement is shown in Figure 35. Notice that the speaker has again been replaced by a load resistor (R_L). This is done for two reasons; a resistive load will yield better results and also, as this test requires operating the amplifier at full output, eliminates any chance of speaker damage.

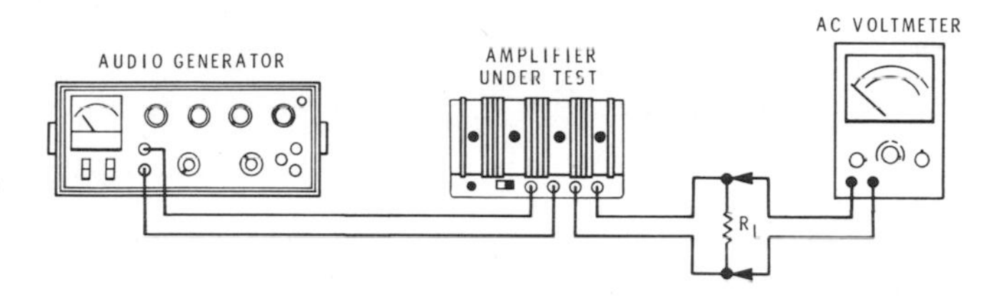

Figure 35
Equipment hookup for measuring input sensitivity.

First, you must determine the amount of AC voltage output that equals the power rating of the amplifier. For instance, for a 20-watt amplifier with an 8 Ω output, the rms voltage across R_L equals 12.65 V. This is determined by the formula $P = \frac{E^2}{R}$.

Set the generator frequency to about 400 Hz and increase the generator output until the voltage across R_L just reaches 12.65 V rms. The rms amplitude of the generator output is now equal to the amplifier input sensitivity.

The generator shown in Figure 35 has a built-in output meter. If only approximate readings are required, you can read the sensitivity directly from this meter; however, the accuracy of this metering system is usually about ±10%. If more accuracy is required you will have to measure the input voltage with a high impedance voltmeter or an oscilloscope.

Frequency Response

The final audio amplifier test we will discuss is frequency response. The frequency response specification of an amplifier states the uniformity of the output amplitude over a range of input frequencies. A typical frequency response specification might read as follows:

Frequency Response (1 watt level). . . ±1 dB from 7 Hz to 100 kHz

. . . ±3 dB from 5 Hz to 150 kHz

To perform this test, you must hold the input amplitude constant while measuring the output power at several different frequencies.

The equipment hookup for this test is shown in Figure 36A. For the test results to be valid, the amplifier must be loaded with a resistor (R_L) which is equal to the output impedance of the amplifier and of sufficient power rating to prevent it from overheating. Overheating of this resistor could cause a gross change in resistance value. This, in turn, would result in gross errors in your output power indications.

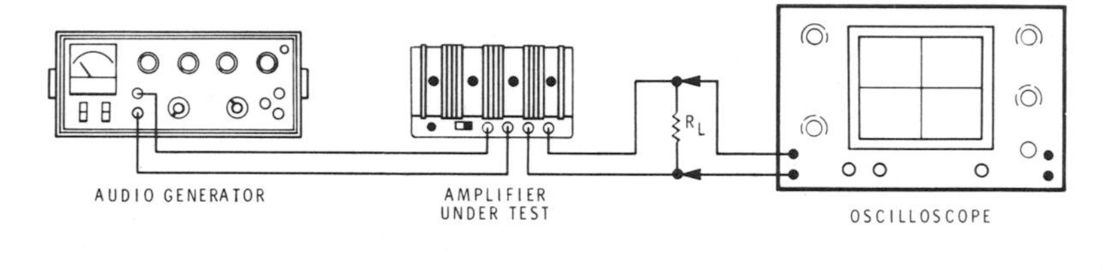

Figure 36A

From this point on, two different techniques can be used, depending upon the purpose of your test. If you are only verifying published specifications, a relatively simple procedure may be followed. If, on the other hand, you are establishing the specification from scratch, a more detailed procedure is needed.

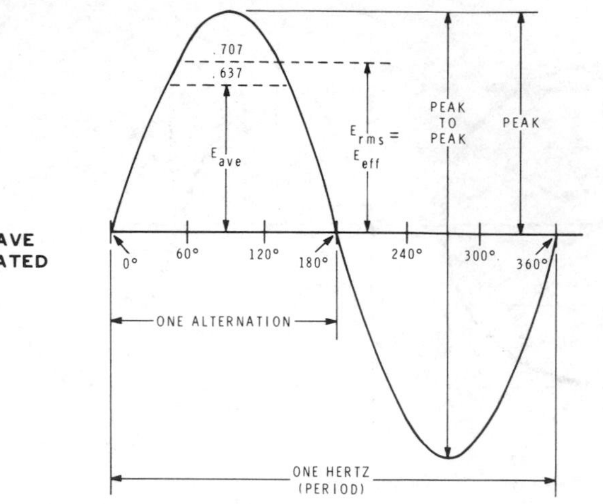

B

SINE WAVE VOLTAGE CONVERSION CHART

FROM	TO			
	EFFECTIVE	AVERAGE	PEAK	PEAK TO PEAK
EFFECTIVE (RMS)		0.900	1.414	2.828
AVERAGE	1.110		1.571	3.142
PEAK	0.707	0.637		2.000
PEAK TO PEAK	0.354	0.318	0.500	

Figure 12

Sine wave voltage values.

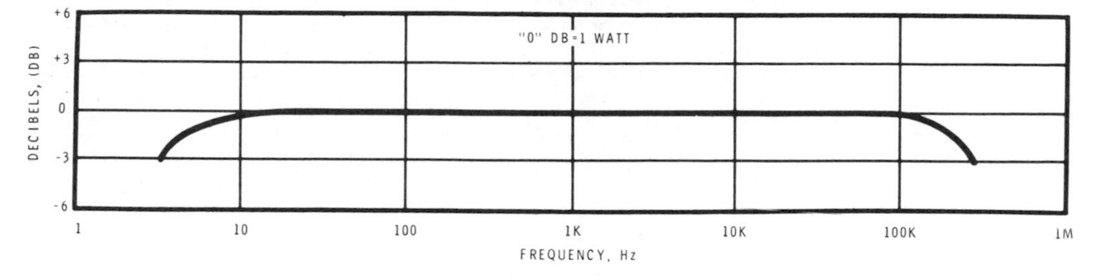

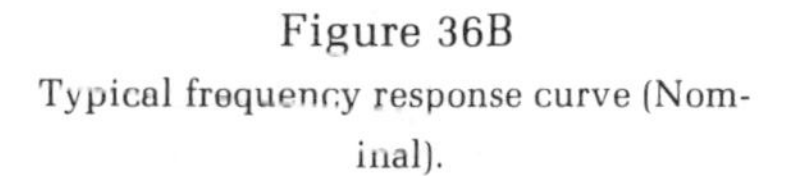

Figure 36B
Typical frequency response curve (Nominal).

First, assume that you have a commercial amplifier, complete with owner's manual. In the manual, you should find the specified frequency response, and perhaps even a graphic presentation of this specfication, such as shown in Figure 36B. The test, in this case, would consist of verifying that the amplifier meets this performance standard.

This specification is stated in dB, referenced to 1 watt, but you will be monitoring the amplifier output with an oscilloscope. Therefore, before making any measurements, you should convert dB factors into absolute voltage values. Although this can be done by using the power and dB formula, it is quite time consuming. By using the conversion factors contained in Figure 37A, it is quite easy to convert from dB to voltage or power. First, however, you must establish your reference point. In this case, the 1-watt reference is converted to an absolute voltage reading by the formula $p = \frac{E^2}{R}$. Assuming your amplifier has a typical 8 Ω output impedance, this would be $1 \text{ watt} = \frac{E^2}{8}$ or 2.83 volts rms. 2.83 volts now corresponds to the "0" dB level of the response curve. In order to know if the output meets the ±1 dB and ±3 dB tolerance limits, the absolute voltages for these levels must be established. Using Figure 37A, read down the center column (dB) until you find 1 dB. Now look directly across to the left hand column marked "Decrease (−) Voltage and Current Ratio". This value is .8913, which means if the output is reduced 1 dB, the

Decrease (−) Voltage and Current Ratio	Decrease (−) Power Ratio	Number of dB	Increase (+) Voltage and Current Ratio	Increase (+) Power Ratio
1.0000	1.0000	0	1.0000	1.0000
.9886	.9772	.1	1.0120	1.0230
.9772	.9550	.2	1.0230	1.0470
.9661	.9330	.3	1.0350	1.0720
.9550	.9120	.4	1.0470	1.0960
.9441	.8913	.5	1.0590	1.1220
.9333	.8710	.6	1.0720	1.1480
.9226	.8511	.7	1.0840	1.1750
.9120	.8318	.8	1.0960	1.2020
.9016	.8128	.9	1.1090	1.2300
.8913	.7943	1.	1.1220	1.2590
.7943	.6310	2.	1.2590	1.5850
.7079	.5012	3.	1.4130	1.9950
.6310	.3981	4.	1.5850	2.5120
.5623	.3162	5.	1.7780	3.1620
.5012	.2512	6.	1.9950	3.9810
.4467	.1995	7.	2.2390	5.0120
.3981	.1585	8.	2.5120	6.3100
.3548	.1259	9.	2.8180	7.9430
.3162	.1000	10.	3.1620	10.0000
.1000	.0100	20.	10.0000	100.000
.03162	.0010	30.	31.6200	1000.00
.0100	.0001	40.	100.000	10,000.0
.00316	.00001	50.	316.20	1×105^5
.0010	1×10^{-6}	60	1,000.0	1×10^6
.000316	1×10^{-7}	70	3,162.0	1×10^7

Figure 37A

voltage output will decrease to 89.13% of its original value. If you multiply the reference voltage, 2.83 volts rms by .8913, you find that the −1 dB level is equal to 2.52 volts rms. Since you are monitoring the amplifier output with an oscilloscope, which displays a peak-to-peak value, you must now convert these rms values to peak-to-peak values for direct read out. Thus, the 1-watt, 0 dB reference becomes 2.83 volts rms × 1.414 × 2 or 8 volts peak-to-peak, while the −1 dB point becomes 7.13 volts peak-to-peak.

You can determine the −3 dB, +1 dB and +3 dB points in the same manner. For dB voltages, use the factors found in the "Increase (+) Voltage" column. Thus, your specification check points can now be listed as follows:

dB	**volt rms (1 watt across 8 Ω)**	**oscilloscope (P-to-P) display**
+3	4 V	11.31 V
+1	3.56 V	10.07 V
0	2.83 V	8.00 V
−1	2.52 V	7.13 V
−3	2 V	5.66 V

Now that you know where the specification levels are, in terms of voltage, the rest of the test is quite simple.

Connect the equipment as shown in Figure 36A and preset the controls as follows:

Audio Generator

Power	Off
Frequency	1,000 Hz (approx)
Output Amplitude	Minimum

Amplifier

Power	Off
Volume Control	Maximum
Bass	Midrange
Treble	Midrange
Loudness Contour	Off

Turn on the equipment and allow it to stabilize. If all of the equipment is solid-state, this will only require a minute or so, but if any of the equipment uses vacuum tube circuits, you should allow thirty minutes for warmup.

After the equipment is completely warmed up, slowly increase the generator output amplitude until a signal of 8 volts peak-to-peak is displayed on the oscilloscope. Do not change the setting of this control during the remainder of the test, as it is now the input reference level.

Now that you have set up the test conditions, verify the lower frequency limits. The lowest frequency specification states that, at 5 Hz, the output should not have decreased more than 3 dB from the 1-watt level. You saw that −3 dB was equal to 5.66 volts peak-to-peak for this circuit. Therefore, by slowly decreasing the generator frequency while watching the oscilloscope, you should be able to reduce the frequency to 5 Hz without the scope signal dropping below 5.66 volts. If so, the amplifier meets this specification. To find the actual −3 dB point, continue tuning the generator downward until the 5.66 V output is observed. The generator tuning dial now indicates the amplifier's −3 dB frequency response point.

To find the −1 dB point, you should increase the generator frequency until the scope waveform is 7.13 volts peak-to-peak. To meet the stated specification, this should occur at a frequency of 7 Hz or less.

From 7 Hz to 100 kHz the output voltage should stay between the −1 dB and 1 dB points, or between 7.13 and 10.07 volts peak-to-peak. Somewhat above 100 kHz, the output should again drop to the −1 dB level of 7.13 volts. Continue upward in frequency until the scope shows an output of 5.66 V. At this point, the generator dial equals the upper 3 dB frequency response point. It should be 150 kHz or greater.

Of course, this test procedure assumes a number of things. The generator must have a tuning range greater than the amplifier's bandwidth and its output level must remain constant over this range. Also, the oscilloscope must have a calibrated vertical input and its vertical amplifier must have a bandwidth greater than that of the amplifier under test. By using a dual trace oscilloscope, you can monitor the input signal along with the output. This will allow you to make sure the input reference level remains constant. One additional factor to consider is the accuracy of the frequency readings. The tuning dials of most audio generators are not highly accurate. Therefore, if precise frequency readings are required, you should monitor the generator output with a frequency counter. By using an electronic counter in the period mode, frequencies down to 1 Hz or less can be measured with great accuracy.

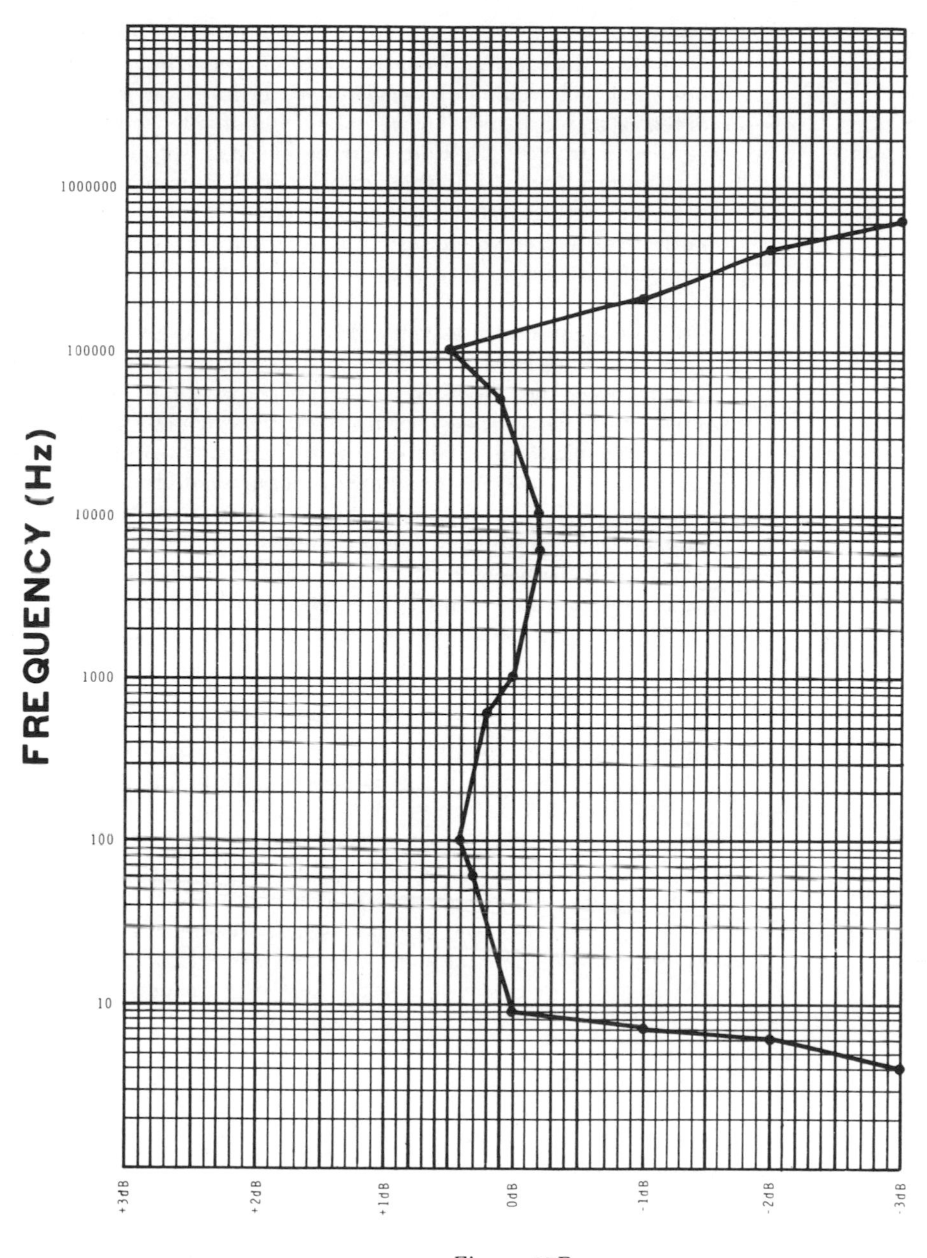

Figure 37B

Frequency response curves plotted for a typical audio amplifier.

The previous procedure showed you how to verify a specification. To determine the capabilities of an amplifier whose frequency response is unknown, you would follow a similar procedure, except you would have to record actual values of amplitude-versus-frequency instead of merely making sure the amplitude did not exceed certain minimum and maximum values. To do this, you should first prepare a chart such as shown in Figure 37B. A sheet of "log" graph paper is excellent for this purpose as it allows good resolution at low frequencies and compresses the upper frequencies, where resolution is not as important.

Set up the equipment as you did in the previous test and establish the input reference level. Now decrease the frequency until the output voltage decreases by 3 dB. Mark your graph, showing the frequency at which this occurs. You are now charting the low frequency rolloff curve; therefore, you should plot the response at least each dB until the response levels off. Figure 37B shows how the curve would look if you measured −3 dB at 3 Hz, −2 dB at 5 Hz and −1 dB at 7 Hz. At 9 Hz the output reached the 0 dB reference level. From this point on, it is a matter of slowly increasing the frequency and plotting the output amplitude. The chart in Figure 34B shows eight separate readings were taken between 10 Hz and 100 kHz. Of course, the more readings you plot, the more accurate your response curve will be.

Once the curve starts to roll off on the high end, you should once again plot the frequency in 1 dB steps. Figure 34B shows the −1 dB point at 110 kHz, the −2 dB point at 130 kHz and −3 dB point at 150 kHz.

This concludes the procedure for one method of testing amplifier frequency response. More comprehensive tests may call for you to plot a series of these curves at various gain and power settings. Also, other equipment may be used, such as a swept frequency audio generator and an X-Y plotter, to perform the test much faster. However, this procedure has shown you a way to analyze amplifier frequency response using relatively common, inexpensive test equipment.

RADIO FREQUENCY GENERATORS

General Characteristics

Radio frequency (RF) generators are similar to audio frequency generators in several respects. Both have oscillator, amplifier, buffer, and attenuation circuits. However, different design techniques must be used because of the higher frequency and greater tuning range involved. Additional features and circuits are also needed to meet the different operational requirements.

The price ranges of these units are somewhat unusual. One group of instruments are priced comparably with audio generators; about $200. However, these instruments have relatively loose specifications and only basic features. The next step up, in specifications and features, currently start around $1,000. Very little is available in between.

The low-cost generators allow the hobbyist or service shop technician to perform simple alignments and signal tracing. However, they normally lack the frequency range, stability, spectral purity, or modulation control necessary to verify the performance specifications of sophisticated communications, industrial or high-fidelity equipment.

In this unit, we will discuss the basic RF generator in detail, while touching on a few of the features found on the more sophisticated instruments.

Figure 38 shows the Heathkit IG-5242 RF generator, which is a low-cost, general-purpose generator. The following specifications for this generator point up some important differences between RF generators and the audio generators you have just studied.

RF Output

Frequency Range	100 kHz to 30 MHz in five bands.
Frequency Accuracy	±5%.
Output Impedance	50 ohms.
Output Voltage	Variable from .005 to 100 millivolts.

Internal Modulation

Frequency	400 Hz.
Depth	Variable from 0 to 50%.

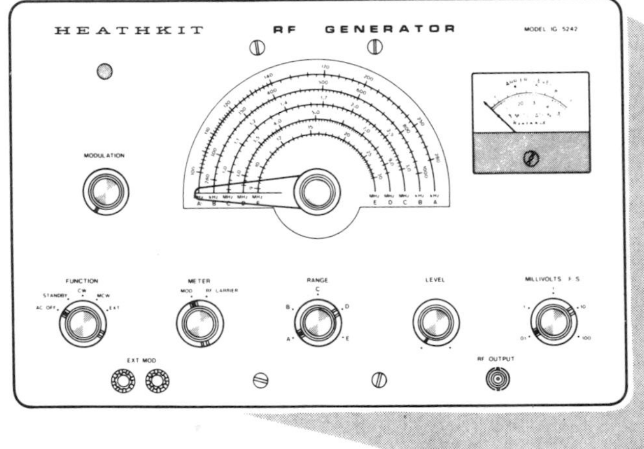

Figure 38
Heathkit IG-5242 RF generator.

General	
Front Panel Controls	Modulation control.
	Variable Frequency control.
	Level control.
	Function switch.
	Meter switch.
	Millivolts F.S. switch.

Obviously, the output frequency range is much higher, but also note the output voltage range. RF generators typically have much lower output voltage levels than audio generators. This is due, in part, to the difference in applications of the two units.

Another interesting point is the output impedance. The use of a 50 Ω output impedance is typical of all general-purpose RF equipment. At lower frequencies and higher signal levels, generator output impedance was not as critical. As frequency increases, however, proper impedance matching between the generator, the connecting cables, and the load becomes increasingly important. Mismatches cause reflected signals, reduced power transfer, signal distortion, and other undesirable effects.

The 5% frequency accuracy refers to the accuracy of the tuning dial. This means, for instance, that a frequency dial setting of 25 MHz could produce an output frequency anywhere from 23.75 MHz to 26.25 MHz without being out of tolerance. Therefore, if a more exact frequency setting is required, you will have to monitor the output with a counter, a wavemeter, or other frequency measuring instrument having the required accuracy.

One unique feature of RF generation equipment is the lingering popularity of vacuum tube circuitry. Solid-state devices began as relatively low power, low frequency components. As the technology advanced, the power and frequency capabilities increased. Today, the knowledge and hardware is available to produce high quality RF generators which operate well into the UHF region. However, a great number of RF generators currently on the market, and nearly all of these instruments more than a few years old, use vacuum tubes.

Most electronic training programs, developed within the past five to seven years, tend to ignore this area. Therefore, many persons who are relatively new to the field have a limited knowledge of how vacuum tubes operate.

For these reasons, a tube-type generator was selected as a training vehicle for this section. If you have a good working knowledge of vacuum tube operation, proceed on with this section. If, however, your training did not include these devices, it would be to your advantage to review Appendix A. This appendix provides a brief resume of common vacuum tube types, operation and characteristics.

Now that you have a general idea of the capabilities of RF generators, the following discussion will take a closer look at the internal operation.

Overall Operation

Figure 39 shows a simplified block diagram of an RF generator. As with the audio generator, it uses an oscillator, an amplifier, an attenuator, and an output amplitude metering circuit.

In addition to these basic circuits, an audio oscillator and modulator circuit have been added. In this case, a single 400 Hz modulating signal is provided; however, some generators allow you to select either a 400 Hz or 1000 Hz tone. The internal/external modulation switch also allows you to modulate the output carrier with an externally generated signal. If, for instance, you need a 20 MHz carrier modulated by 2 kHz, you can switch to external modulation and connect a 2 kHz signal from an audio generator to the external modulation input. Note that you can switch the metering circuit to measure either the RF output level or the percent of amplitude modulation.

Another important feature of the RF generator is the extensive use of shielding. This is needed for two reasons. In order to achieve a high

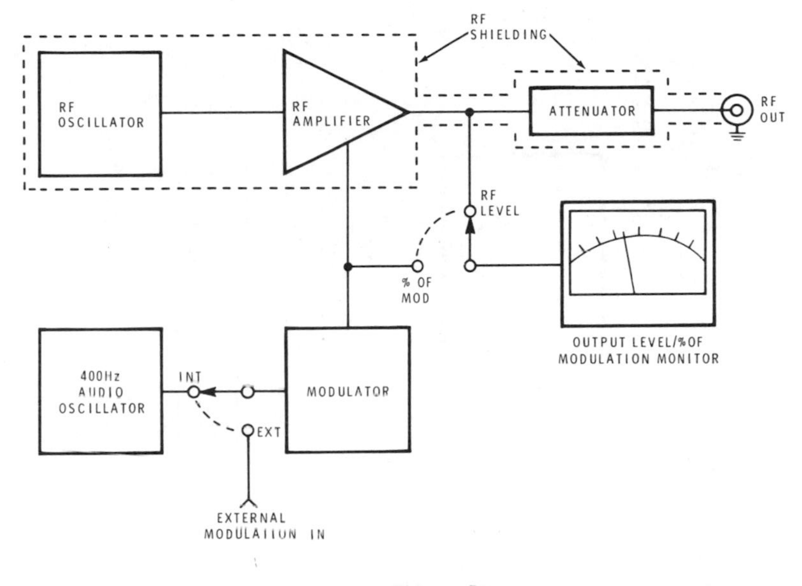

Figure 39

degree of spectral purity, the RF sections must be isolated from stray radiation. Extraneous signals, which are always present in electronic work areas, can cause reactions in RF oscillators and amplifiers, resulting in distortion and instability. Also, since high frequencies radiate much easier than lower frequencies, shielding is needed to prevent the RF generator from interfering with the operation of other equipment in the area. Shielding often involves rather complex mechanical design, such as a box within a box. All cables and connectors carrying those frequencies must also be shielded.

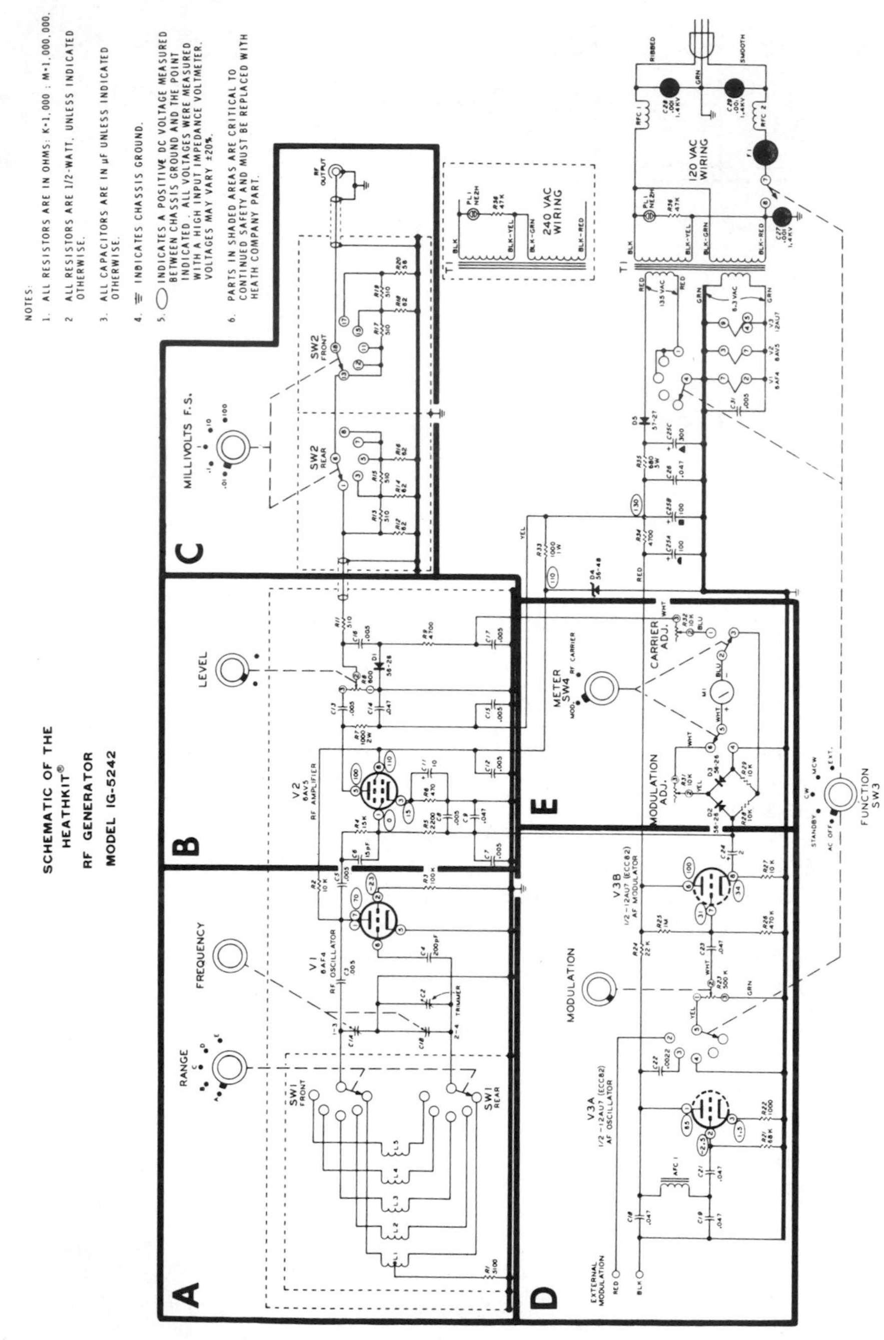

Figure 40
Schematic of the Heathkit RF generator, Model IG-5242.

Figure 40 shows the overall schematic diagram of the low-cost RF generator described earlier. This schematic has been divided into functional areas that correspond to the block diagram in Figure 39. The correlation of circuits to blocks is as follows:

A	RF oscillator circuit
B	RF amplifier circuit
C	Attenuator
D	Audio oscillator and modulator circuits
E	Metering circuit

This makes the schematic somewhat easier to follow, but, for explanation purposes, each circuit will be presented separately. In most cases, simplified schematics will be provided. In these schematics, we may reposition components for easier understanding or delete components not required for basic operation. However, they can be referenced back to actual schematics.

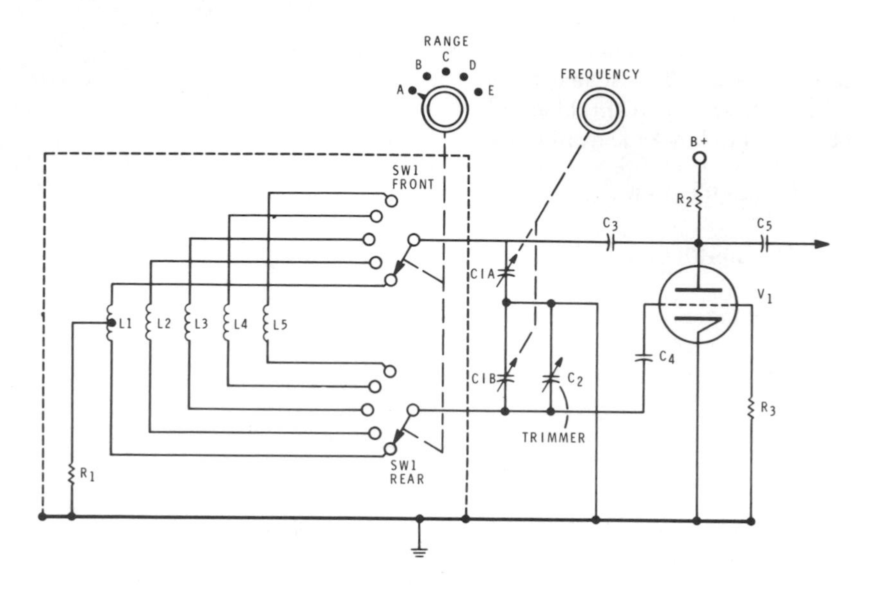

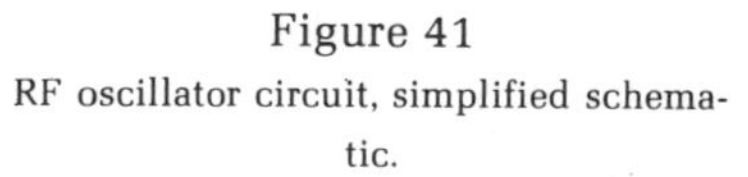

Figure 41
RF oscillator circuit, simplified schematic.

RF Oscillator

Figure 41 shows a simplified schematic of the RF oscillator circuit. This is a vacuum-tube version of the familiar Colpitts oscillator, as identified by the tapped capacitor arrangement. The frequency of oscillation is determined by a tuned tank circuit made up of C_{1A}, C_{1B}, C_2 and one of the inductors, L_1 through L_5. These inductors are connected to wafer switch SW1, which is controlled by the "Range" switch. Capacitors C_{1A} and C_{1B} are ganged, variable capacitors with a common tuning shaft connected to the "Frequency" tuning knob. Thus, when you change the frequency range, or band, you are actually switching in different fixed-value coils. Frequency changes within the bands are accomplished by varying the tank circuit capacitance. Capacitor C_2 provides a small amount of trimming capacitance that allows you to calibrate the oscillator frequency within the front panel dial markings. Triode V1 and its associated components provide the necessary amplification required for oscillator operation. The output is taken from the plate and coupled, through C_5, to the RF amplifier circuit.

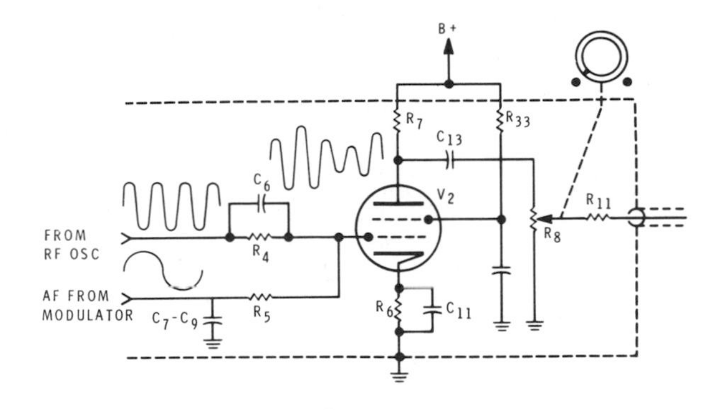

Figure 42

RF Amplifier

The RF amplifier circuit is shown in Figure 42. The oscillator signal and the modulating signal (if any) are coupled to the control grid of V2. When a modulating signal is present, modulation occurs within this tube and the amplitude-modulated carrier will be present on the plate.

V2 and its associated circuitry make up a broad band amplifier stage. The gain of this stage is sufficient to produce a signal greater than 2 V rms at the plate throughout the oscillators 100 kHz to 30 MHz tuning range. The signal is now coupled through C_{13} to the level control potentiometer, R_8. This potentiometer allows you to adjust the amplitude, or level, of RF carrier signal which is applied to the output attenuator.

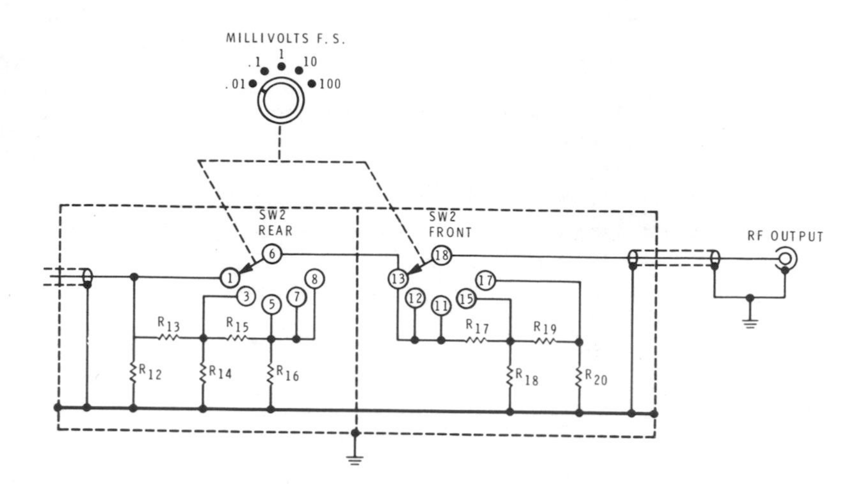

Figure 43A
Output attenuator circuit.

Output Attenuator

Figure 43A shows the output attenuator circuit. Resistors R_{12} through R_{20}, in conjunction with the "Millivolts F.S." switch (SW2), make up a five-step decade voltage attenuator. The values of these resistors and their arrangement maintain an output impedance of approximately 50 Ω regardless of the attenuator switch position.

Many generators, especially those designed for use with communications equipment, will have the output level expressed in milliwatts or dB, since most of the tests in this area are referenced to these values.

Another type attenuator, used in some higher frequency RF generators, is the piston attenuator. This assembly consists of an attenuator tube, a piston, and a drive mechanism. When the attenuator dial is turned, it drives a worm gear, which causes the piston to move within the tube.

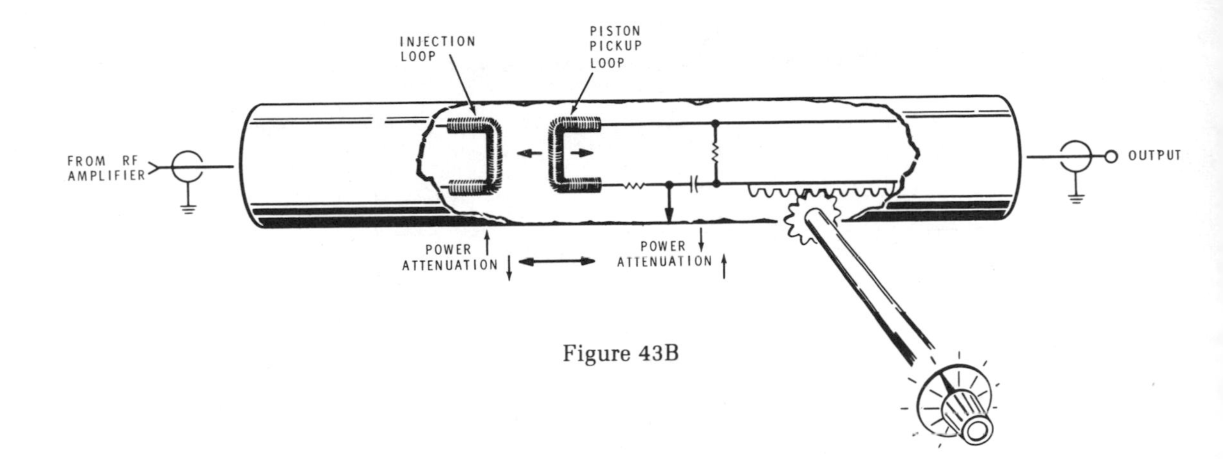

Figure 43B

This results in a change in coupling between the injection and pickup loops. Although a detailed analysis of this attenuator is quite complex, you can understand its operation by considering the injection loop to be a transmitting antenna and the pickup loop to be a receiver. As the pickup loop is moved closer to the radiation source, a stronger signal will be induced into it. Conversely, by moving it away, less signal is induced. The gear mechanism that moves the piston is linear but, as the piston is moved, the power changes at a logarithmic rate. Therefore, if the attenuator dial is calibrated in millivolts or milliwatts, the spacing between dial graduations will be unequal or nonlinear. However, if the attenuator is calibrated in dB, the dial graduations will be an equal distance apart. This is because dB varies at a logarithmic rate. These attenuators are quite expensive to manufacture due to the precision machining involved and the extensive use of silver. The interior surface of the attenuator tube and all contact surfaces are usually silver plated to improve conductivity. With good mechanical and electronic engineering, however, these attenuators can achieve excellent accuracy, even at very low power levels.

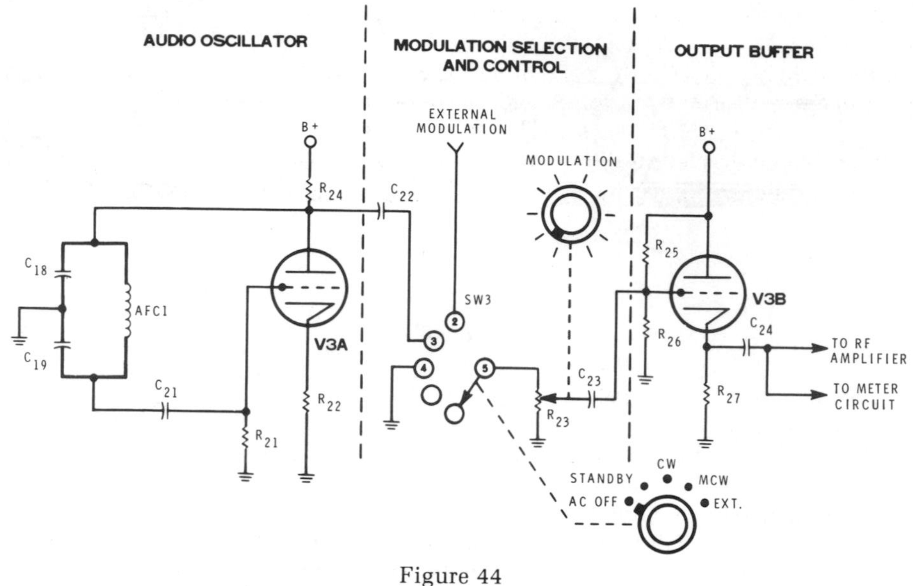

Figure 44
Audio oscillator and modulator circuit, simplified schematic.

Modulation

Now that you have learned how the RF signal is generated and processed, we will examine the circuit that allows you to amplitude modulate it.

The internal modulating signal is generated by a second Colpitts oscillator circuit designed to operate at approximately 400 Hz. A simplified schematic of this circuit is shown in Figure 44. Note that this oscillator, made up of V3A and associated components, operates at all times, producing a signal of fixed frequency and amplitude.

This signal is coupled to contact 3 of the function switch, SW3. This switch selects the source of input for V3B. In the AC OFF and STANDBY positions, the switch contacts are open. Since the instrument is not functioning in these positions, the input to V3B is unimportant.

In the CW (continuous wave) position, however, you want a constant amplitude, unmodulated, RF carrier wave output from the generator. Therefore, in this position, switch contact 5 is connected to contact 4. This places a ground on both ends of R23, preventing any stray signal from being coupled through V3B to the RF amplifier.

When you place the function switch in the MCW (modulated continuous wave) position, contact 5 of SW3 is connected to contact 3. This couples the audio oscillator output to V3B. The amplitude of this signal is controlled by the modulation control, R23. Remember, if the carrier amplitude is constant, the percent of modulation is determined by the amplitude of the modulating signal.

By placing the function switch in the EXT (external modulation) position, V3B is connected to the external modulation jack through switch contacts 2 and 5. This allows you to connect an audio generator to this input and obtain modulation at frequencies other than 400 Hz. In this case, of course, both the output amplitude of the audio generator and the setting of R_{23} will affect the percent of modulation, since both will affect the signal amplitude at V3B.

Tube V3 is a dual triode. One-half (V3A) is used as the audio oscillator, while the other half is configured as a cathode follower. This acts as a buffer stage between the audio oscillator and the RF sections. It functions much like an emitter follower circuit. The output of V3B is coupled to the control grid circuit of the RF amplifier. Here, it interacts with the RF oscillator output as you learned previously. It is also coupled to the metering circuit. This circuit is shown in Figure 45.

Metering Circuit

The front panel meter performs two separate functions which you can select with the METER switch, SW4. In the RF CARRIER position, the circuit is configured to measure the unmodulated RF carrier level being applied to the output attenuator. In the MOD position, the circuit is reconfigured to indicate the percent of amplitude modulation. This circuit is shown in Figure 45. Figure 45A shows the overall circuit as it appears on the main schematic. Figures 45B and C show the simplified circuits for both measurement modes. Switch contacts have been omitted to increase the clarity of the circuits.

The modulation measuring circuit will be considered first. With the METER switch in the MOD position, the meter is connected into a measurement bridge circuit (Figure 45B). The meter actually indicates the amplitude of the audio frequency signal. However, since this amplitude controls the amount of modulation, the meter is calibrated to read directly in percent of modulation. During the positive alternation of the AF signal, shown in dark black, current will flow within the bridge as indicated by the dark black arrows. Note that current flow through the meter is from left to right. During the negative alternation, shown in light black, the light black arrows show the direction of bridge current. Note that the meter current flow is still from left to right. The amount of current through the meter is controlled by adjusting R_{31}, Modulation Adjustment. The name of this internal adjustment may be misleading. It does not adjust the modulation; only the reading of the modulation meter.

To calibrate the meter for the correct "percent of modulation" reading, you should connect an oscilloscope to the generator output, place the function switch in the MCW position and adjust the front panel MODULATION control until an AM envelope at 33% modulation is displayed on the scope. This will occur when the maximum amplitude of the envelope is equal to twice the minimum amplitude. (% of modulation = $\frac{E_{max} - E_{min}}{E_{max} + E_{min}} \times 100$) Now, with the METER switch in the MOD position, adjust R_{31} for a meter indication of 33%.

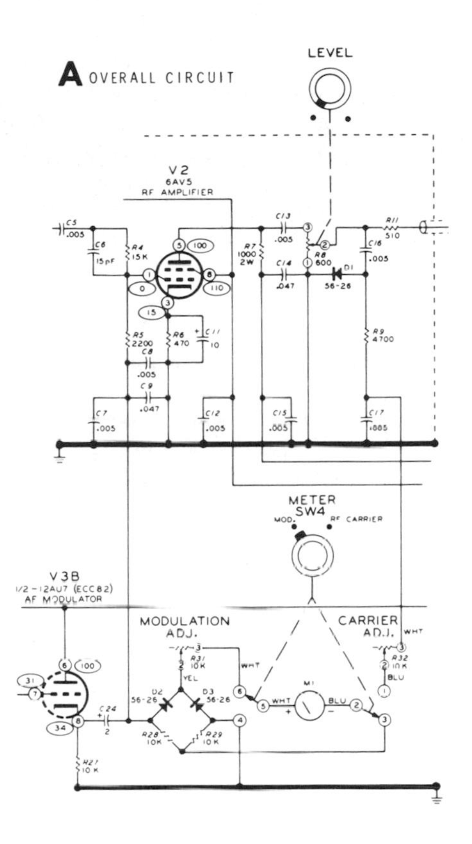
A OVERALL CIRCUIT
LEVEL
V2
6AV5
RF AMPLIFIER
METER
SW4
MOD.
RF CARRIER
V3B
1/2-12AU7 (ECC82)
AF MODULATOR
MODULATION
ADJ.
CARRIER
ADJ.

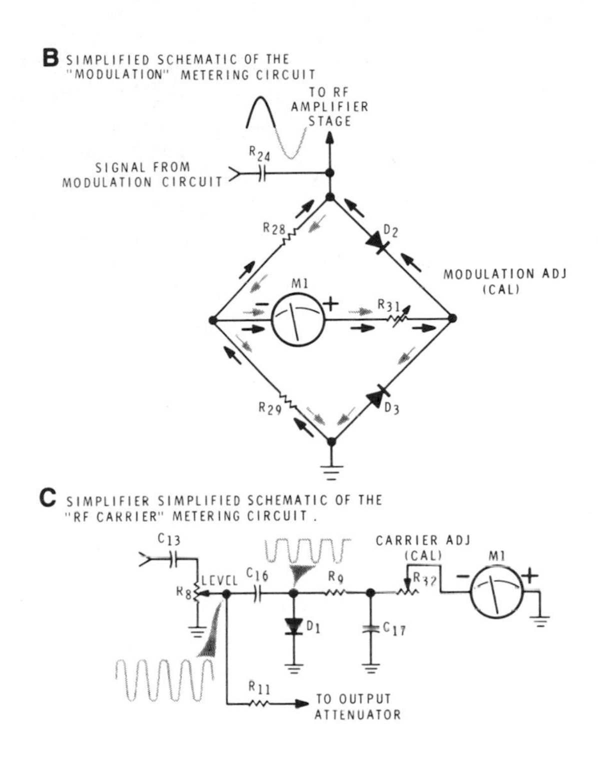
B SIMPLIFIED SCHEMATIC OF THE "MODULATION" METERING CIRCUIT
TO RF AMPLIFIER STAGE
SIGNAL FROM MODULATION CIRCUIT
R24
R28
D2
M1
R31
MODULATION ADJ (CAL)
R29
D3
C SIMPLIFIER SIMPLIFIED SCHEMATIC OF THE "RF CARRIER" METERING CIRCUIT .
C13
CARRIER ADJ (CAL)
M1
R8
LEVEL
C16
R9
R32
D1
C17
R11
TO OUTPUT ATTENUATOR

Figure 45

Moving the METER switch to the RF CARRIER position connects the meter into the circuit shown in Figure 45C. With the LEVEL control turned to maximum, the voltage at the wiper contact should be somewhat greater than 2 volts rms (2 V rms corresponds to a maximum output of 100 mV from the output attenuator). Diode D_1 acts as a half-wave rectifier while R9 and C_{17} provide some filtering. CARRIER ADJ (R_{32}) is an adjustable current limiting resistor for the meter. To calibrate this circuit, you should connect a high impedance AC voltmeter between terminal 2 of R_8 and ground. Next set the LEVEL control for a reading of exactly 2 volts on the voltmeter. Now, with the METER switch in the RF CARRIER position, adjust R_{32} for a full-scale indication (100 mV) on the panel meter.

This completes the discussion of the RF generator's operation. However, since the power supply requirements for tubes are somewhat different than for transistors, power supplies will be discussed separately.

Power Supplies

One operational characteristic of the vacuum tube is the warm-up time required for stable operation. As they usually require about 1/2 hour to stabilize, you will not want to turn the generator off after every use. However, you also will not want to leave it operating while performing other measurements because of unwanted RF radiation. Therefore, a standby function is often provided. In this mode, power is applied to the tube heaters (filaments). This maintains the tube at its operating temperature but removes all other operating voltages, keeping the generator ready for "instant-on" operation while disabling all of the oscillator and amplifier stages.

You should also be more careful when working inside these instruments. Supply voltages for solid-state equipment is usually quite low; 15 VDC to 30 VDC. Tubes, on the other hand, require much higher operating voltages. It is not uncommon to encounter 250 to 300 VDC or more in vacuum tube circuits. Therefore, an electrical shock which would be "uncomfortable" in a solid-state instrument could be much more dangerous in a vacuum-tube instrument. These higher voltages are also coupled with much higher current-capacity power supplies, which compounds the risk. There is no need to fear these circuits, but you should respect them.

RADIO FREQUENCY GENERATOR APPLICATIONS

You will find the RF generator to be an invaluable aid in designing, testing, or aligning radio circuits. It is also necessary for aligning the IF and RF stages of commercial receivers. Other applications include measuring input sensitivity, selectivity, bandwidth, image rejection, and RF gain.

Before discussing individual applications, it is necessary that you be aware of the problems involved in connecting the RF signal from the generator to the equipment under test.

Impedance Matching

An impedance match is defined as, "the condition in which the impedance of a connected load is equal to the internal impedance of the source, thereby giving maximum transfer of energy from source to load". Stated in more understandable terms, the load impedance must equal the generator impedance if reliable test results are to be obtained.

The load is seldom connected directly to the generator; therefore, coaxial cable is used to connect the generator to the load. This "coax" cable must have a characteristic impedance equal to the output impedance of the generator. Equally important, the cable must be terminated in its characteristic impedance. Any impedance mismatch will cause signal reflections, resulting in standing waves on the cable. When this occurs, the attenuator and meter settings will indicate the **average** voltage on the line. However, the voltage at any given point on the line, including the generator output connector or the load, may be considerably different. To prevent this condition, you must use some type of impedance matching device. With this in mind, we will consider some of the methods you can use to achieve a match.

Of course, the most simple condition occurs when you are using a 50 Ω generator to drive a 50 Ω load. In this case, you can connect the generator directly to the test circuit with a 50 Ω coax cable.

If you wish to connect this generator to the antenna input of a television receiver, impedance matching problems arise. TV uses an antenna impedance of either 75 Ω or 300 Ω. Therefore, an impedance matching device must be used. If you are doing enough of this work to justify the cost, the best solution is to purchase an impedance matching transformer. For occasional testing, however, you can construct a resistive pad from standard composition resistors which will effectively match the two impedances. This can be done in one of two ways. You can match the generator output to the cable/load impedance or match the generator/cable impedance to the load. In either case, some signal attenuation occurs, but it will be a known, constant loss which can be taken into consideration.

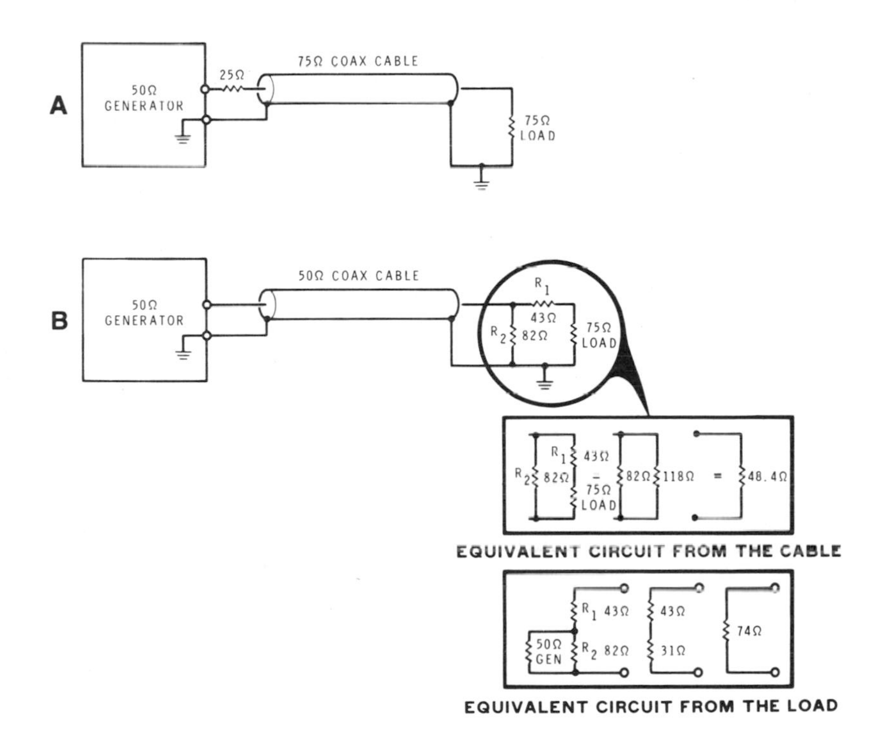

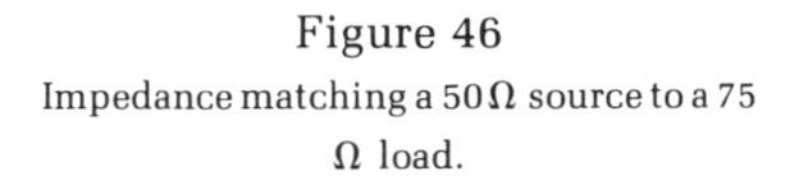

Figure 46
Impedance matching a 50 Ω source to a 75 Ω load.

Figure 46A shows how a 50 Ω generator can be matched to a 75 Ω cable/load combination. In this configuration, you have increased the apparent generator impedance to 75 Ω, thus matching the cable and the load.

Another method of accomplishing this match is shown in Figure 46B. Resistors R_1 and R_2 form a minimum loss L pad. To construct this circuit, select resistance values that will allow both the output cable and the load to be terminated in their characteristic impedances. In other words, looking toward the load, the generator should "see" a 50 Ω impedance while looking back toward the generator the load should "see" a 75 Ω impedance. Notice the resistor values in Figure 46B. The center conductor of the coax sees 82 Ω (R_2) paralleled by 43 Ω (R_1) and 75 Ω (load) to ground. The equivalent resistance of this circuit is 48.4 Ω, which is quite close to the 50 Ω impedance of the cable. Looking from the load back into the generator the 43 Ω resistor (R_1) is in series with the paralleled 82 Ω (R_2) and 50 Ω (generator impedance). This appears to be 74 Ω, which again is very close to the load's characteristic impedance.

Why do the impedances vary slightly from the optimum values? Figure 47 shows the circuit and formulas for calculating the selected resistor values. This circuit can be used to match a generator to either a higher or lower load impedance. Just remember, the higher impedance is always Z_1 and the lower is Z_2.

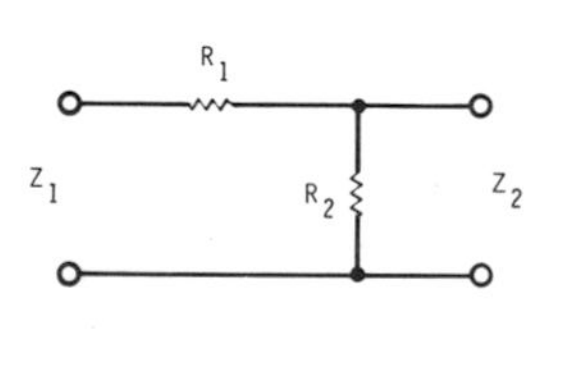

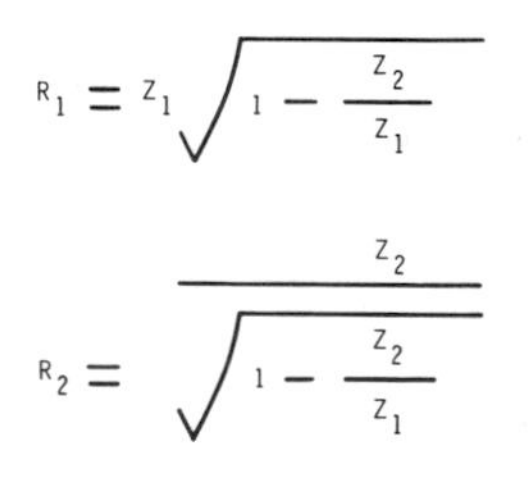

$$R_1 = Z_1\sqrt{1 - \frac{Z_2}{Z_1}}$$

$$R_2 = \frac{Z_2}{\sqrt{1 - \frac{Z_2}{Z_1}}}$$

Figure 47

To solve the problem of matching a 50 Ω generator to a 75 Ω TV receiver, you merely have to insert the impedance values into the two formulas and solve for resistance. Since Z_1 has to represent the largest impedance, it will be equal to the 75 Ω receiver, and Z_2 will be equal to the 50 Ω generator.

Thus,

$$R_1 = 75\sqrt{1 - \frac{50}{75}}$$

or

43.3 Ω

While,

$$R_2 = \frac{50}{\sqrt{1 - \frac{50}{75}}}$$

or

86.6 Ω

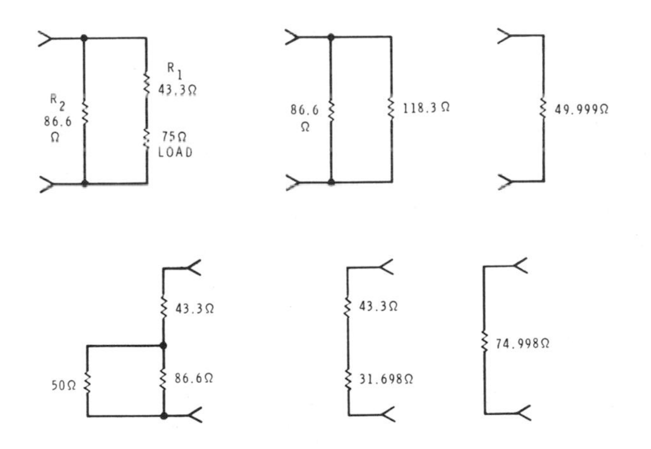

Figure 48
50 Ω to 75 Ω impedance matching pad
using ideal resistance values.

Figure 48 shows the impedance match achieved by using these theoretical resistance values of R_1 and R_2. This match would be exact if the computations were not rounded off. Therefore, an exact impedance match is possible if you can obtain the exact resistance values. However, by using the closest standard value 5% resistors (43 Ω and 82 Ω), the impedance match will be close enough for all general test procedures.

The pad described in Figure 46B will cause a 4 dB attenuation loss; therefore, the voltage applied to the load will be only 62% of the value indicated by the generator.

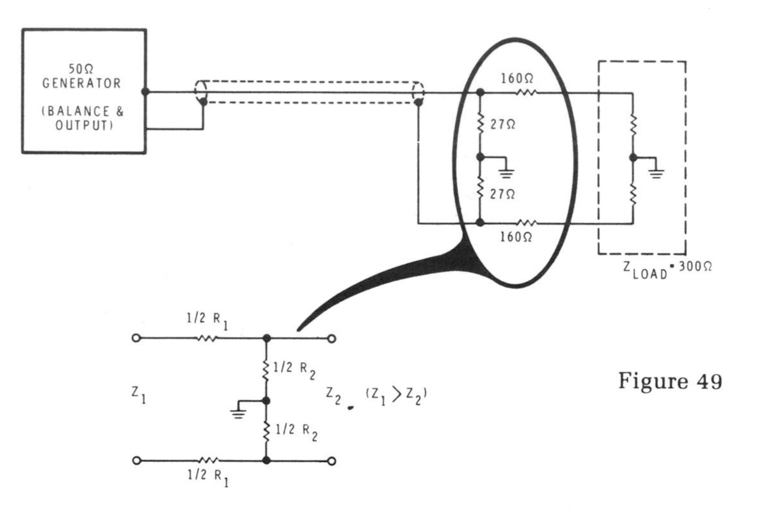

Figure 49

The pads discussed thus far are designed to drive unbalanced inputs. This means one side of the load is grounded. For a balanced input, where each side of the load has the same impedance to ground, a slightly different pad configuration is needed but the same formulas are used to find resistance values. For instance, the 300 Ω TV antenna input is usually a balanced-type input. The most efficient way to drive this load is with a balanced generator output. The pad for this impedance match is shown in Figure 49. The resistor values are calculated in the same manner as before except that each side of the pad will contain half of the resistance. If your generator is not capable of producing a balanced output, you may delete the ground connection between the two 27 Ω resistors and replace them with one 56 Ω resistor. This will provide a reasonable impedance match. The attenuation loss in this pad is about 6 dB or a 50% voltage drop.

These pads can also be used to match the generator output to a lower impedance load. Just remember that the higher of the two impedances will always be Z_1, and connect the pad accordingly.

One final impedance matching problem you may encounter is that of coupling an RF signal to a radio antenna input.

For general test work on all-wave receivers, the I.R.E. (Institute of Radio Engineers) has developed a standard dummy antenna. The schematic for this load is shown in Figure 50A.

For coupling into a standard AM broadcast receiver, a 200 pF capacitor connected in series with the antenna connection provides good coupling. This is illustrated in Figure 50B. For short-wave receivers, using the same type antenna, use a load resistor of about 400 Ω as shown in Figure 50C.

In all of these connections, note that the cable is terminated in its characteristic impedance by a 51 Ω resistor.

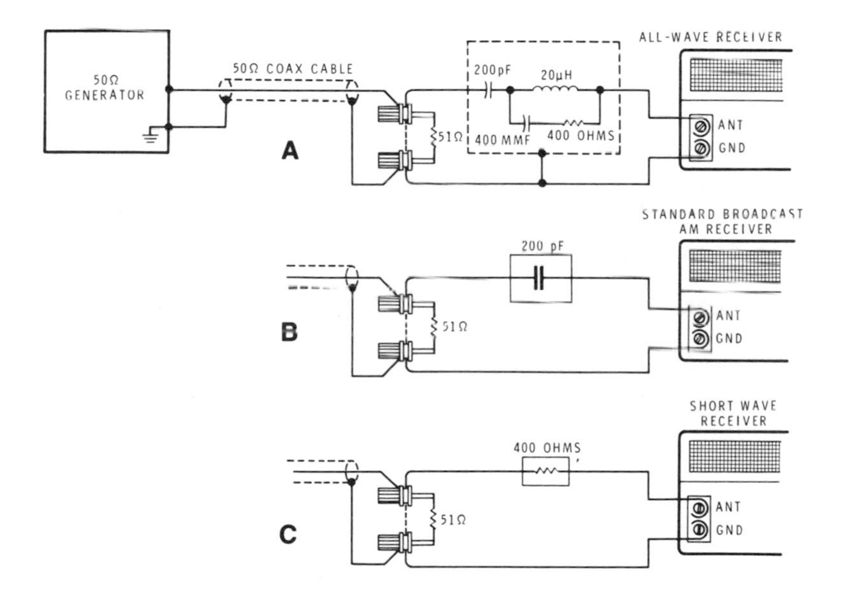

Figure 50
Generator coupling to radio receivers.

General Coupling Techniques

Proper impedance matching is very important, if your test results are based upon a known value of input amplitude. For other tests, such as tuning a resonant circuit to a specific frequency, the coupling is much less critical. In these tests, you are more concerned with the frequency of the test signal than with its absolute amplitude.

In many cases, merely placing the generator output cable close to a circuit under test will provide enough signal strength to tune the circuit. This is similar to the technique you used for loose-coupling a dipmeter in the previous unit.

When using your generator for point-to-point signal tracing, you will want to inject the signal directly into the circuit under test. For example, by injecting a test signal into the base circuit of an IF amplifier, you can use your oscilloscope to check the operation, gain, or distortion of that individual stage. To inject this signal, you should use an RF probe. This will pass the RF signal but prevent any DC potential in the test circuit from being coupled back into the generator, thus, protecting the generator output attenuator. If an RF probe is not available, you can use a small value capacitor (100 pF or so) in series with the generator lead.

Caution is necessary when working on transceivers (radio transmitter-receiver combinations). Many tests of the receiver section require that you connect the generator directly to the antenna input connector. Since these units use the same antenna for transmitting and receiving, it is extremely important that you do not inadvertantly key the transmitter during this test. To do so, will result in the entire power of the transmitter being coupled into the generator. This is almost sure to burn out a portion of the attenuator circuit. Some generators have a fuse between the attenuator and the output connector to prevent this type of damage. However, they are not always 100% effective and should not be relied upon.

AM Receiver Alignment

To align a radio receiver, means to adjust all of the circuits in which RF signals are present. These circuits must be adjusted for best receiver operation throughout its tuning range. Figure 51 shows the block diagram of a typical superheterodyne receiver. It includes all of the circuits that have to be aligned and shows the signals present in each circuit.

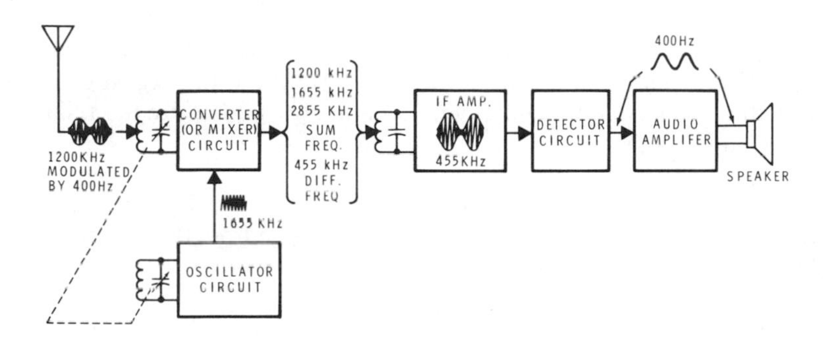

Figure 51

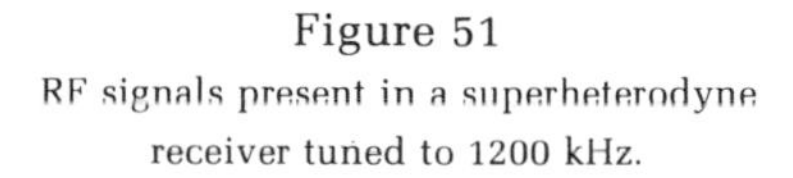
RF signals present in a superheterodyne receiver tuned to 1200 kHz.

A typical input signal of 1200 kHz is being selected by the tuned antenna input circuit and coupled to the converter stage. At the same time, a signal that is 455 kHz above the receiver frequency, or 1655 kHz, is being generated by the oscillator circuit. This 1655 kHz oscillator signal is also coupled to the converter stage where it is mixed with the incoming RF.

Remember, mixing action between two frequencies results in the presence of four frequencies at the mixer output. In this case, the two original frequencies (1200 kHz and 1655 kHz), the sum of the two (2855 kHz), and the difference between them (455 kHz) are all coupled to the IF amplifier stage.

The input to the IF amplifier is sharply tuned to 455 kHz. Thus, only the 455 kHz difference signal will be amplified. However, this signal contains all of the modulation information originally contained in the 1200 kHz RF input signal.

The difference signal is amplified and coupled to the detector stage where the audio modulating frequency is separated from the IF signal. This audio signal is amplified and applied to the speaker.

The converter and oscillator circuits must be adjusted so that the oscillator frequency "tracks" exactly 455 kHz above the antenna frequency throughout the tuning range. The converter must also tune to the frequency indicated by the tuning dial. In addition to this alignment, the IF stage must be tuned to exactly 455 kHz.

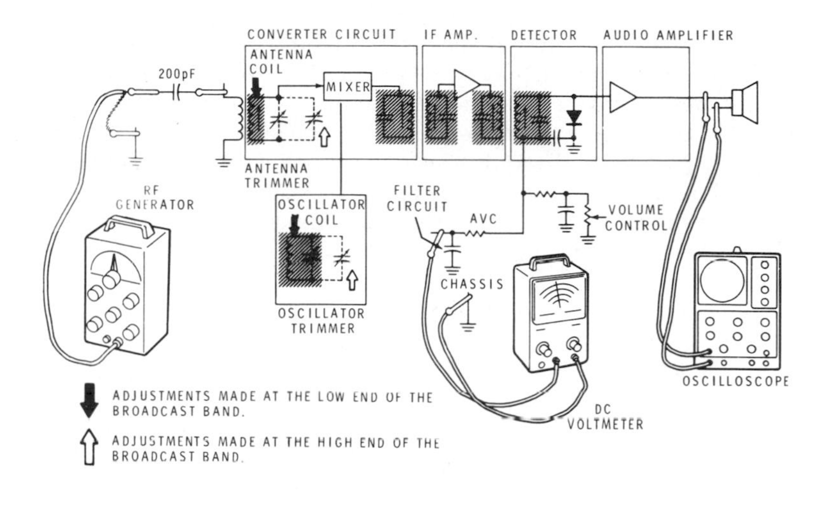

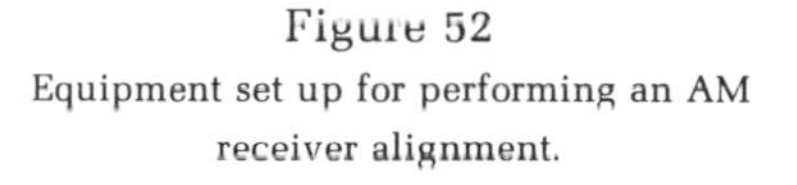

Figure 52
Equipment set up for performing an AM receiver alignment.

To perform this alignment, you will need a calibrated RF generator with AM capability and either an oscilloscope or a DC voltmeter. Figure 52 shows you where these instruments should be connected as well as the components you will be adjusting. Connect the generator to the antenna terminal of the receiver. Adjust the generator output for 455 kHz at about 30% modulation. If a counter is available, use that to monitor the generator frequency. Next, connect a voltage indicator to the audio section of of the receiver. Two separate methods are shown in Figure 52. If an oscilloscope is used, connect it across the speaker leads and set the receiver volume control to maximum volume. If a voltmeter is to be used, set it to measure negative DC volts and connect it to the output of the AVC (automatic volume control) filter, as shown. Tune the receiver to a spot on the high end of the dial where no station is heard. Once this is accomplished, disable the oscillator circuit by shorting out the oscillator tuning capacitor.

The first step in receiver alignment is to tune the IF transformers to 455 kHz. A large part of the modulated 455 kHz input will not pass through the antenna and mixer circuits, but since only a small amount is needed, enough signal will get through to align the IF stages.

Now, using a plastic or fiber alignment tool, adjust both slugs of each IF transformer. Starting with the transformer nearest the audio section and working back toward the RF section, carefully adjust the tuning slugs for maximum voltage indication on the oscilloscope or meter. You should see a sharp peak as the slugs are turned. Since these adjustments interact with each other, you should repeat this process several times for best results. When you reach the point where additional adjustment does not yield any further output amplitude, the IF stage is precisely tuned to the generator frequency.

Now, tune the receiver to 600 kHz and adjust the generator to produce a modulated 600 kHz output. It may be necessary to disable the receiver AVC circuit and reduce the generator output power. Remove the shorting lead from the oscillator tuning capacitor. Some audio output should now be indicated. The antenna coil and the oscillator coil, shown by the solid arrows in Figure 52, will be adjusted here at the low end of the broadcast band. First, adjust the oscillator coil for maximum audio voltage output. Then, do the same with the antenna coil. Repeat these adjustments until no further output increase is obtained.

To align the high end of the tuning range, tune the receiver and the generator to 1400 kHz. At the high end, you will adjust the oscillator and antenna trimmer capacitors, indicated in Figure 52 by the light arrows. Using the same sequence as before, alternately adjust the two trimmer capacitors for maximum audio output.

Retune the generator and receiver to 600 kHz. You will probably find that a slight additional adjustment is needed to compensate for the adjustment of the trimmers at 1400 kHz. If you do have to peak up the coils, you should recheck the alignment at 1400 kHz.

During this step, a counter is especially useful, because it will allow you to return to these frequency settings with greater repeatability. If, for instance, you initially peaked the low end at an actual frequence of 570 kHz and then rechecked this alignment at an actual frequency of 630 kHz, you would notice a pronounced difference in audio output level. However, both these frequencies are within tolerance for a ±5% frequency dial indication of 600 kHz.

After you are satisfied with the alignment of the receiver at the high and low ends of the band, you should make a quick check of the oscillator tracking ability. This can be done by tuning the receiver to several points on the dial, tuning the generator for maximum receiver audio output, and noting the generator frequency. It should be close to the frequency indicated by the receiver dial. Don't expect it to be exact, however, especially on inexpensive receivers, as these dials are only approximate.

Testing Wideband Amplifiers

Many pieces of electronic equipment contain wide band RF amplifiers. These include oscilloscopes, AC voltmeters, communication receivers and transmitters, television receivers, and many others. By using an RF generator and an appropriate output monitor, you can perform many tests on these amplifiers.

The simplest of these tests are frequency response and bandwidth. There are accomplished much like the frequency response tests described in the section on "Audio Generator Applications."

Connect your generator to the amplifier input, through proper coupling, and monitor the amplifier output with an appropriate indicator. This indicator can be either an RF voltmeter or an oscilloscope. Of course, your generator must be capable of producing frequencies both below and above the amplifier bandpass. Your indicator also must have a greater bandwith than the amplifier under test. If not, you will record the indicator's limitations instead of the actual amplifier response.

Starting well below the amplifier's lower frequency limit, tune the generator upward while maintaining a constant output amplitude. When the generator frequency reaches the operating range of the amplifier, you should begin recording the amplifier output-versus-frequency values at incremental frequency steps. Continue increasing the generator frequency and recording output values until you have tuned through the amplifier frequency response range. Now, by plotting these readings on graph paper, you have a visual presentation of your amplifiers bandpass, bandwidth, and frequency response.

You can also use the RF generator to determine the overload limit of the amplifier. This is the point where the amplifier starts to overdrive and distort the output signal. For this test, you should monitor the output with a oscilloscope or a spectrum analyzer.

Set the generator frequency at some value within the frequency range of the amplifier and slowly increase the output amplitude until distortion is noted on the oscilloscope display or until spurious response (random spikes) is seen on the spectrum analyzer. Reduce the generator output level until these indications just disappear. The generator output level now equals the maximum signal level the amplifier can faithfully reproduce.

You can test for minimum usable input as well as for maximum input. Using the same test setup, and employing an oscilloscope or RF voltmeter as a monitor, reduce the generator output level until the amplifier output level reaches a specified value which is determined by the rated gain of the amplifier. The generator output level now represents the input sensitivity of the amplifier.

By using your RF generator in conjunction with a spectrum analyzer, you can also determine the inherent distortion characteristics of an amplifier. To do this, merely compare the purity of the input signal to the purity of the output. Any decrease in purity (increase in distortion) can be attributed to the amplifier circuit. To obtain reliable results from this test, the purity of the input signal must be much better than the desired output purity. This limits you to one of the more sophisticated generators, as low-priced generators do not usually have a high degree of spectral purity.

FUNCTION GENERATORS

Several types of test instruments have been used in laboratories for many years before being used on service benches. These include electronic counters, dual-trace oscilloscopes, spectrum analyzers, and function generators.

When function generators were first introduced to the laboratories in the 1950's, they were large, bulky instruments costing several thousand dollars. But advancing technology and the increasing need for more powerful and versatile instruments made it possible to sell these generators to the service market around 1970 at a cost of about $250. Today, function generators are available, in the $100 price range, that are capable of producing high quality waveforms over a wide frequency range. These generators are rapidly taking the place of the traditional audio generators in general test applications. If current trends continue, it appears that within a few years, audio generator applications will be limited to audio tests requiring ultralow distortion sine waves (less than 0.05%). The function generator will become the standard bench generator providing sine, square, and triangle waves at frequencies from less than 1 Hz to over 5 MHz with very good purity.

Multiple waveforms, wide frequency range, and easy frequency control have made the function generator very popular.

One unique feature of the function generator is its method of signal generation. The traditional audio and RF generators use some type of oscillator circuit to produce a sine wave. The function generator, on the other hand, produces a triangle wave as its primary signal. All other output waveforms are derived from this triangle wave.

General Operation

Figure 53 shows the block diagram of a typical function generator. Note that the triangle generator circuit contains both the range selector switch and the frequency tuning control. The output of this stage is fed through an amplifier stage (A1) to a level detector.

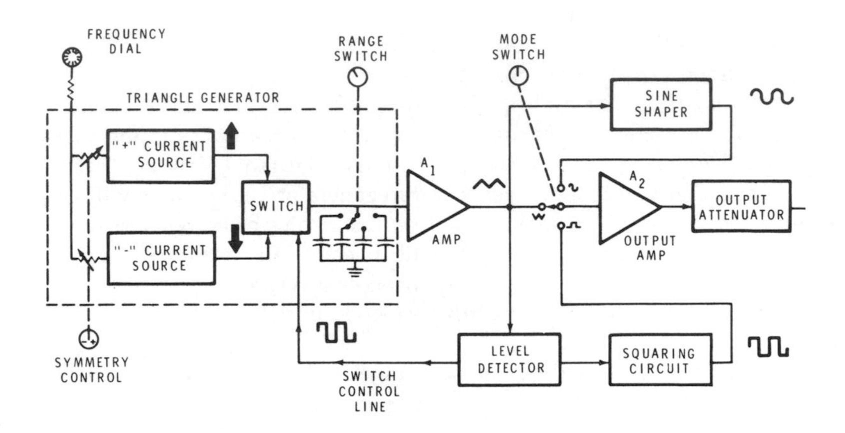

Figure 53
Typical function generator block diagram.

This level detector is actually a dual voltage comparator which detects two separate voltage levels. When the input level reaches a preset positive voltage, the detector output changes state. This causes the current switch in the triangle generator to reverse the charging current. As a result, the generator output begins its downward slope. Upon reaching the detector's preset negative voltage level, its output again changes state, switching the generator current switch and reversing the output. Each time a positive or negative peak is detected, a signal is also routed to the squaring circuit. This circuit produces a very sharp square wave at the same frequency as the triangle waveform.

The triangle output from A1 is felt at the input to the sine shaper circuit. This circuit shapes the triangle wave into a fairly low distortion sine wave.

The outputs of the triangle generator, the squaring circuit, and the sine shaping circuit are connected to separate contacts of the three-position mode selector switch.

The mode selector switch supplies a continuous series of sine, triangle, or square waves to the output amplifier. Here, the waveform is amplified to its desired level and impedance matched to the output attenuator. This stage also contains the variable output amplitude control (often labeled "variable attenuation") and DC offset control. DC offset refers to the DC reference of the output waveform. For instance, a 2-volt peak-to-peak signal referenced to "0" volts would swing from 1 volt to −1 volt. The same signal with 1 volt DC offset would swing from 2 volts to 0 volts.

The output of this amplifier circuit is routed through an output attenuator before being applied to the generator output connector. This attenuator typically attenuates the amplifier output in 10 dB steps from 0 dB to −50 dB or more. Its operation is similar to the RF output attenuator you studied previously.

Triangle Generator Circuit

The operation of a triangle generator is based on the voltage-versus-time characteristics of a capacitor, charged by a constant-current source.

As you recall, when a capacitor is charged by a constant-voltage source (a battery) through a series resistance, the capacitor charges at an exponential rate. This is illustrated in Figure 54A. However, by removing the resistance and replacing the battery with a source of constant current, as shown in Figure 54B, the capacitor charging rate becomes linear. Of course, to attain perfect linearity you must have pure capacitance. Therefore, an extremely low-leakage capacitor must be used.

To reverse the slope of the waveshape, it is only necessary to reverse the direction of the charging current. Conversely, as the capacitor is discharged, a linearly decreasing voltage is produced. This is shown in Figure 54C.

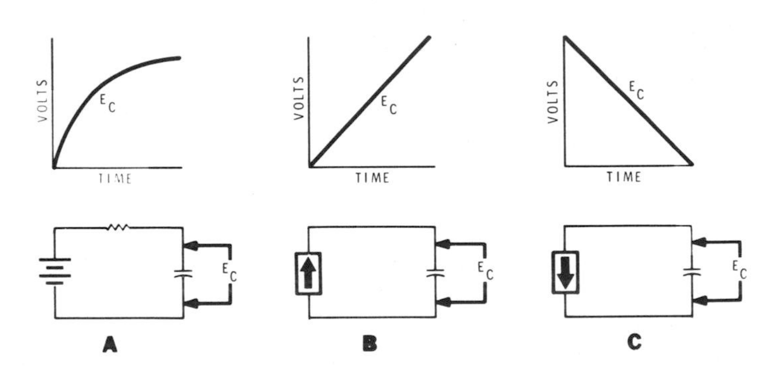

Figure 54
Capacitor charge rates, constant-voltage versus constant-current sources.

The rate of charge depends on only two factors, the size of the capacitor and amount of charging current. Thus, a triangle wave can be generated by using two constant-current sources and a switching device of some type.

First a regulated current supply is turned on, and the capacitor is allowed to charge until a selected positive peak voltage is reached. At this point, the charging current source is turned off, and a discharging source is turned on. When the capacitor voltage reaches a selected negative peak value, the current sources are again switched. Note that the switching action was initiated by the capacitor voltage level. By causing the current sources to switch whenever the capacitor voltage reaches a predetermined level, only two factors control the output frequency, the capacitor value and the amount of current supplied by the source.

The instantaneous voltage across the capacitor is found by the formula

$$E_c = \frac{(t \times I)}{C}$$

where: E_c = Capacitor voltage

t = time of the charge in seconds

I = amount of charging current in amps

C = Capacitor value in farads

When solving for time, the formula becomes:

$$t = \frac{(E \times C)}{I}$$

This gives you the time for either the positive or negative slope of the triangular waveshape, which is a half cycle. Thus,

$$\text{frequency} = \frac{1}{2t}$$

From these relationships, you can see that in order to increase frequency, it is necessary to increase the charging current or decrease the capacitance.

In practice, both parameters are varied. Different values of capacitance are switched into the circuit when changing ranges. These values are usually in decade steps. The charging current is continuously variable by means of the front panel tuning dial. A 1000:1 current range is not uncommon.

For example, assuming that the constant-current sources can be varied from 1 μA to 1 mA, the capacitor sizes range from 5 μF to 10 pF, and the voltage limit is 10 volts, determine the minimum and maximum frequencies of the generator.

To find the maximum frequency, select the smallest capacitor (10 pF) along with the largest charging current (1 mA). By inserting these values into the formula, you will find that the time for either the positive or negative slope equals:

$$\frac{10 \text{ V} \times 10 \text{ pF}}{1 \text{ mA}} = 0.0000001 \text{ sec} = 0.1 \ \mu\text{s}$$

Since this is the time for only one slope, you must multiply this by two to find the period of the waveform, or 0.1 μs $\times$ 2 = 0.2 μs. Now, to find the frequency, it is merely necessary to determine the reciprocal of the period, or:

$$f = \frac{1}{t} = \frac{1}{0.2 \ \mu\text{s}} = 5 \text{ MHz}$$

To find the minimum frequency of this generator you would follow the same procedure, this time using the largest value of capacitance with the smallest charging current.

$$t = \frac{10 \text{ V} \times 5 \ \mu\text{f}}{1 \ \mu\text{A}} = 50 \text{ sec}$$

$$\text{period} = 2t = 100 \text{ sec}$$

$$\text{frequency} = \frac{1}{100 \text{ sec}} = 0.01 \text{ Hz}$$

Thus, this hypothetical generator has a frequency range of 0.01 Hz to 5 MHz.

One other feature incorporated into some function generators is a symmetry adjustment. If the time of the positive slope is equal to the time of the negative slope, the signal is symmetrical. If, however, you change this 50:50 charge-time/discharge-time ratio, the output waveform can be made to appear as shown in Figure 55. This is done by varying the ratios between the charging and discharging current values. For instance, if the ratio of charging current to discharging current were 5:95, the output would appear as shown in Figure 55B. Reversing this ratio would produce the waveform shown in Figure 55C. As you can see, the result is either a positive or negative ramp (sawtooth) waveform. When these waveforms are passed through the squaring circuit, positive or negative going pulses will result.

The output of the triangle generator is now applied to one contact of the mode switch as well as to the inputs of both the level detector and sine shaping circuits.

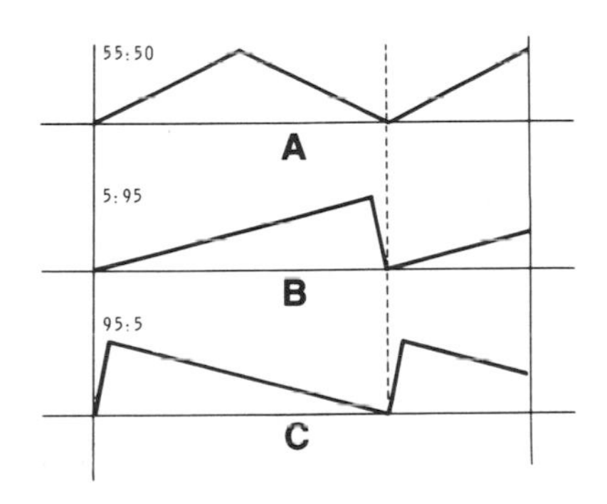

Figure 55

Level Detector

A simplified schematic of the level detector circuit is shown in Figure 56A.

The output of amplifier A1, which represents the charge voltage on the integrating capacitor, is applied simultaneously to both comparators. Comparator A3 is configured as a positive comparator, while comparator A4 is a negative comparator. The positive and negative reference voltages applied to these comparators establish the positive and negative charge limits of the integrating capacitor.

The outputs of the comparators are connected to an RS latch circuit made up of two NAND gates. Remember, this circuit functions as a flip-flop, or bistable, circuit.

The output of the RS latch circuit feeds an inverter amplifier which provides the current required to drive the generator current switch and also the squaring circuit.

As waveform A is a continuous wave, there is no real starting point, so we will pick up the operation as the voltage ramp is approaching the negative reference level. Notice that the outputs of both comparators (waveforms B and C) remain high as long as the input stays between the reference levels.

At time T1, C1 has charged to a point where the voltage of waveform A equals the negative reference voltage. This causes the output of A4 to go low as shown by waveform C. The low is felt on the input of gate 2, causing the RS latch circuit to change state. The latch output, which was high, will now go low. This switching action will activate the current source switch, reversing the charging current to the capacitor. The capacitor will now start charging in a positive direction, and the non-inverting input of A4 will once more become positive in respect to the inverting input. This will result in the output of A4 returning to the high state.

This switching action requires only a few nanoseconds to complete.

Assume there is an input signal to this circuit to see how the waveforms shown in Figure 56B are generated.

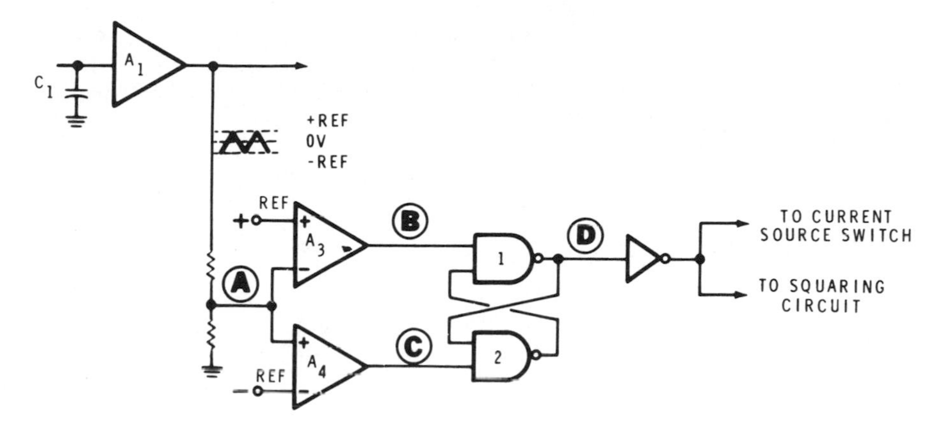

Figure 56A

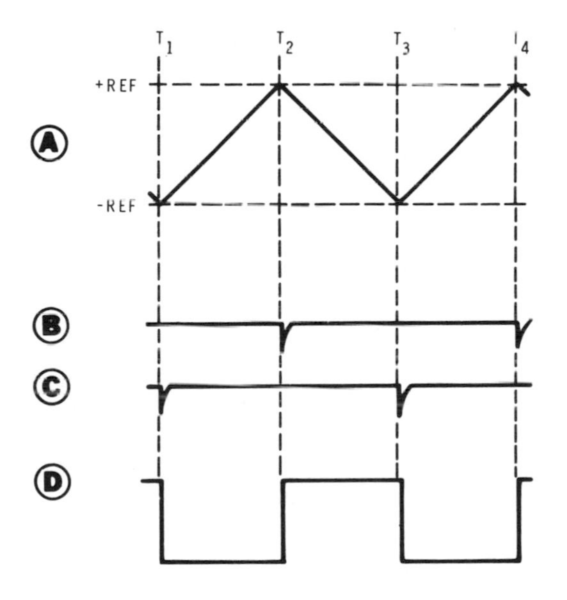

Figure 56B

The level detector circuit will remain stable as the capacitor charges in a positive direction until, at T_2, the inverting (−) input of A3 becomes positive in respect to the positive reference voltage. At this time, the output of A4 will go low, switching the RS latch and reversing the current source once more.

Each time a signal is sent to the current source switch, signaling the end of one alternation, the squaring circuit also receives a signal.

Squaring Circuit

Since the input to the squaring circuit is already a square wave (waveform D of Figure 56B), the purpose of this circuit is to sharpen the wave shape and amplify it for application to the mode selector switch.

Sine Shaping Circuit

As shown by the block diagram (Figure 53), the triangle wave is also applied to the input of the sine shaping circuit. This circuit is made up of a nonlinear load and a sine wave amplifier. A nonlinear load is illustrated in Figure 57A. The principle involved is to increase the load on the generator's high impedance output at specified amplitude points of the output waveform. This distorts the triangle wave to an approximate sine wave. Figure 57B shows the associated waveform.

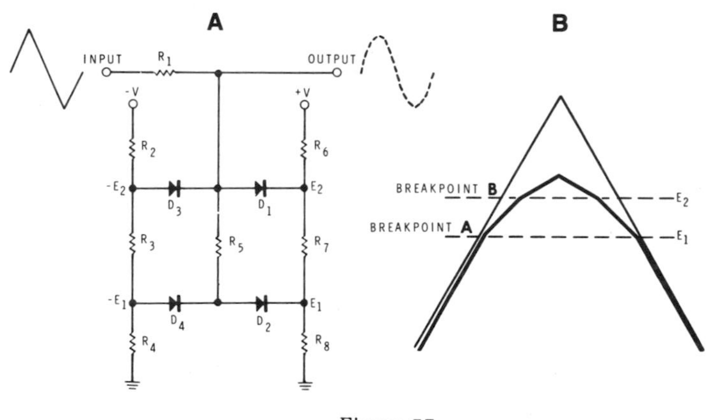

Figure 57
Sine shaper nonlinear load circuit.

Looking at this circuit, you will see that R6, R7, and R8 form a voltage divider from +V to ground. This causes D_2 to be biased at some relatively low value while D_1 is biased at a higher value. Observe what happens when a triangle wave is applied across this network.

With no input voltage, all of the diodes are reverse biased and the circuit presents an infinite resistance to ground. As the input voltage increases, it is unaffected by the circuit until its value exceeds the value of E_1. At this time, D_2 becomes forward biased. R5 and R8 are now shunted across the generator impedance, causing its output to decrease. In Figure 57B, the result of this action is shown by the decreased slope of the output waveform between points A and B. These points are known as "break-points."

At breakpoint B, the input voltage exceeds the value of E_2. Causing D_1 to conduct and placing R_7 in parallel with R_5. The resistance of the circuit is further reduced, placing an even greater load on the generator. Note that beyond breakpoint B, the slope of the output wave flattens out even more.

On the downward slope, a reverse action occurs. As the input signal drops below the level of E_2, D_1 cuts off, leaving only R5 and R8 in the circuit.

Below the level of E_1, D_2 cuts off and the circuit is once again inactive.

The negative peak is distorted in the same manner by components R_2, R_3, R_4, D_3, and D_4.

This signal is amplified and routed to the "sine" contact of the mode selector switch.

As you can see, this circuit produces only a rough approximation of a sine wave. However, by adding more diode/resistor segments to provide more breakpoints, a much closer approximation can be attained.

We have now discussed the generation of all three basic function generator waveforms and followed them to the mode selector switch. At this point, you can select any one of the three as your output test signal.

The selected signal is next applied to the output amplifier stage.

Output Amplifier and Attenuator

The output amplifier must be somewhat more sophisticated than those used in audio generators. Not only is the basic frequency range greater, harmonic frequencies are also present. Remember, a 3 MHz square wave contains significant harmonics up to 30 MHz. The amplifier must also be capable of amplifying frequencies of 1 Hz or less. This means that it must be a DC-coupled amplifier having a bandpass of at least 0 to 30 MHz.

Figure 58 shows the typical configuration of this circuit. The variable attenuator controls the signal level applied to the amplifier. This could be more accurately called an "output level control," since it will vary the output level from zero to maximum amplitude. The amplifier input circuit also contains the DC offset control, which clamps the input signal to a variable DC reference level.

This amplifier is usually capable of providing a 10V peak-to-peak signal into a 50 Ω load or about 20V peak-to-peak open circuit voltage.

The output attenuator is quite similar to the step attenuator found in other generators. It is commonly calibrated in 10 dB steps for operator convenience.

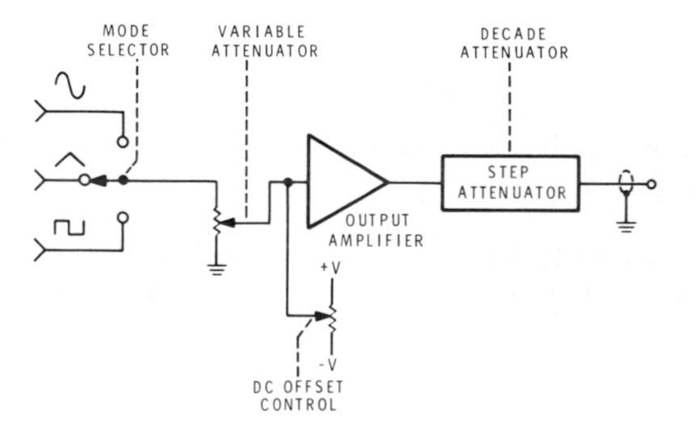

Figure 58
Output amplifier and output attenuator circuit diagram.

FUNCTION GENERATOR APPLICATIONS

A function generator is actually three or more signal generators combined into one compact instrument. As previously mentioned, if your generator has a variable symmetry feature, it can be used as a source of ramp or sawtooth voltage in the triangle mode. By using this feature in the square wave mode, you have a limited application pulse generator.

In view of this versatility, we will divide function generator applications into sub groups depending upon the output waveform.

Sine Wave Mode

In the sine wave mode, the function generator becomes a replacement for the audio generator, except for those applications requiring a very high degree of spectral purity. Therefore, all of the audio generator tests previously discussed apply to function generators, with the possible exception of critical distortion measurements. However, the function generator's expanded frequency range and controllable DC offset provide added dimensions to the tests you can perform in this mode.

For example, when measuring a circuit's frequency response, you may find that the bandwidth of the circuit under test is greater than the frequency coverage of your audio generator. In this case, you must determine the lower limit of the bandwidth with an audio generator, then switch to an RF generator in order to determine the upper frequency limit. This is complicated by several factors such as differences in output levels, output impedances, and physical connections. Many times, the added frequency range of the function generator will allow you to make the entire measurement using only the single instrument. This results in faster, less complicated test procedures. It also reduces the chances of procedural error.

The DC offset feature allows you to reference the sine wave to either a positive or negative DC level, while the audio generator signal is always referenced to zero volts DC. This is illustrated by Figure 59. Waveform A is a 10 V peak-to-peak signal referenced to 0 V. This is the signal reference you will obtain with no DC offset. It is also the only type signal you can obtain from a standard audio generator. Waveforms B and C show the same signal referenced to −5 V and 5 V respectively.

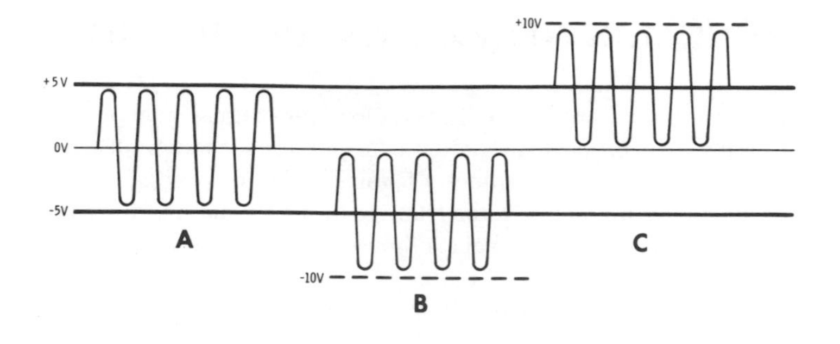

Figure 59

This feature is especially useful in testing individual circuits where the actual operating signal contains both AC and DC components. In these cases, you can simulate the operation signal very closely by adding the DC component. This is done by the DC offset control.

The DC offset feature is even more important when working with certain configurations of TTL integrated circuits, where circuit damage can occur if any portion of the input signal crosses zero volts. By using full positive offset, the negative peaks of the waveform will always be slightly above zero volts, thus protecting the IC input circuit. Referring back to Figure 59, the A and B waveforms would instantly destroy such an IC, while waveform C would be perfectly safe to use as an input signal.

Square Wave Mode

This mode also closely approximates the applications of the audio generator, assuming your audio generator has square wave capability. General test applications include frequency response testing and evaluating many digital timing and counter circuits. This output also has the same advantages of expanded frequency range and variable DC reference level.

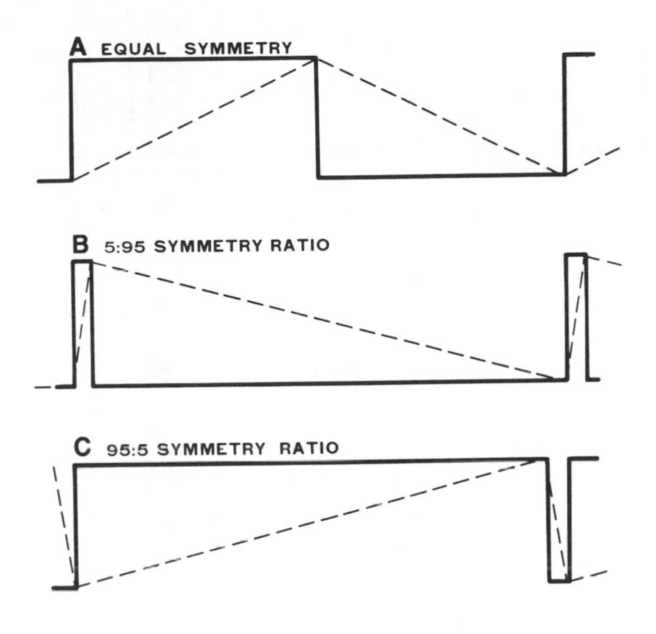

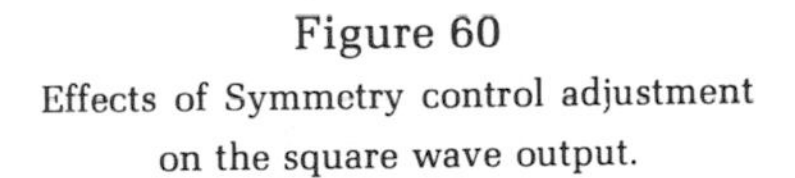
Figure 60
Effects of Symmetry control adjustment on the square wave output.

The function generator's ability to test a circuit's high frequency response is much greater than that of the audio generator. This increase is due, in part, to its greater frequency range, but a greater increase is possible with the symmetry control. Figure 60 shows the effects of varying the symmetry of a square wave to obtain a pulse output. The dashed line shows the triangle generator output that produces the positive or negative pulses. Recall from the discussion of waveform characteristics, that a square wave is composed of the fundamental frequency plus odd numbered harmonics up to at least the ninth harmonic. A pulse waveform, on the other hand, contains both odd and even numbered harmonics up to the fifteenth harmonic or higher.

Therefore, by applying a 1 MHz square wave, you could obtain an indication of a circuit's high frequency capabilities up to about 9 MHz. However, by adjusting the symmetry control to produce a pulse, frequency response indications of 15 MHz and beyond are possible.

These pulses are also useful in testing some digital circuits. By adjusting

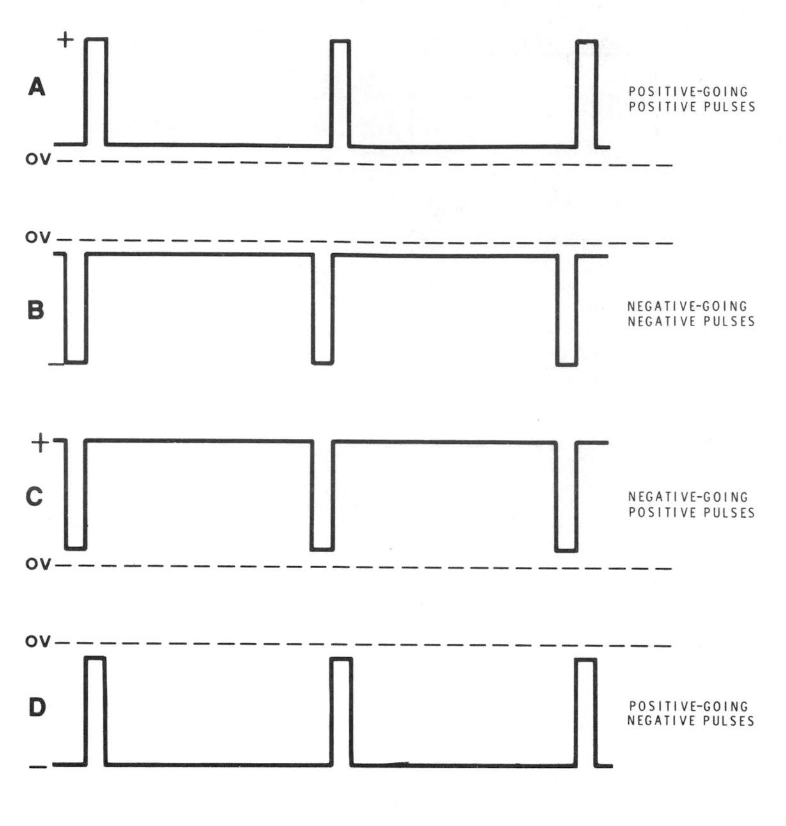

Figure 61

the symmetry control for positive-going pulses and adding positive DC offset, you can produce a test signal such as shown in Figure 61A. This can be used to check the operation of counting or timing circuits requiring a positive input. If a negative input is required, the test signal can be inverted by adjusting the symmetry for negative pulses and the DC offset for full negative offset as shown in Figure 61B. Of course, it is also possible to obtain positive-going negative pulses and negative-going positive pulses if these are desired. These are illustrated in Figure 61C and D.

Another characteristic of the function generator which increases its value as a test signal source is its low output impedance. Logic circuits of the TTL, RTL and DTL families require low impedance drive sources. Thus, the 50 Ω generator impedance is quite compatible with these circuits.

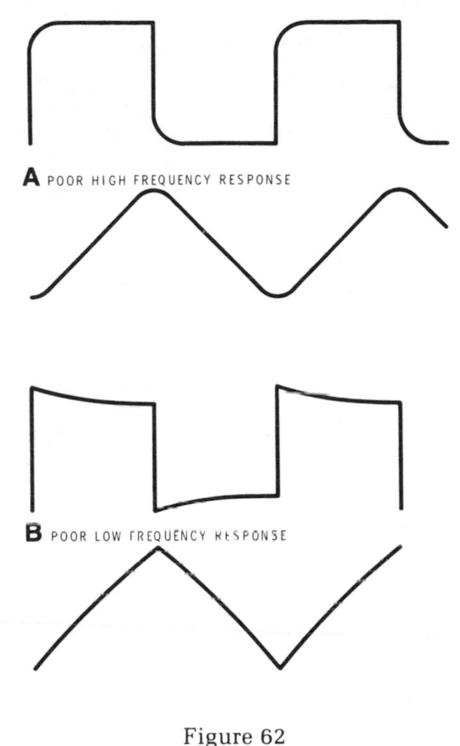

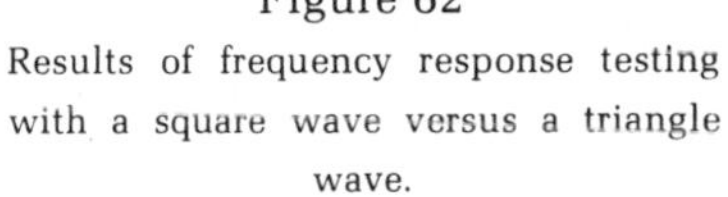

Figure 62
Results of frequency response testing with a square wave versus a triangle wave.

Triangle Wave Mode

The triangle wave is unique to the function generator. This waveform, with its linear slopes and sharp, well defined peaks, is a natural choice for time domain analysis of amplifiers, Schmitt triggers, comparators and other circuits.

As you saw earlier, during the discussion of general waveform characteristics, it is much easier to see a slight aberration on a straight line than on a curved line. It also is easier to measure the exact amplitude of a sharp peak than the rounded crest of a sine wave.

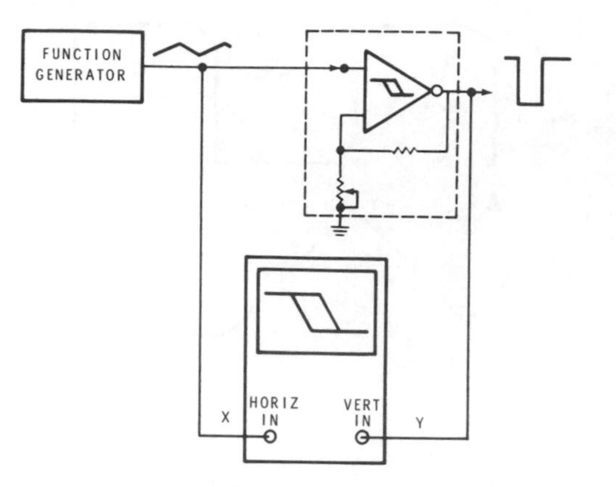

Figure 63A

A triangle wave can also be used to detect poor high and low frequency response in a circuit. Figure 62 shows the results of these conditions upon a triangle wave as compared to a square wave.

Figure 62A shows the loss of high frequency response. Note that the peaks of the triangle wave lose their sharpness, while the slopes between the peaks retain their linearity. However, poor low frequency response is evidenced by a loss of linearity while maintaining sharp peaks. This is shown in Figure 62B.

The triangle waveform is also useful in adjusting the hysteresis level of Schmitt trigger circuits.

As you will recall from your study of electronic counters, a Schmitt trigger converts a slowly changing signal, such as a sine wave, into a square or rectangular waveform. The hysteresis built into this circuit causes the output level to change states at a higher input level for the positive-going signal slope than for the negative-going signal slope. In many applications, such as the signal processing section of the counter, improperly adjusted sensitivity or an insufficient hysteresis gap can result in erratic triggering. You can easily check and adjust this circuit using a function generator and an oscilloscope. The equipment set up for this test is shown in Figure 63A.

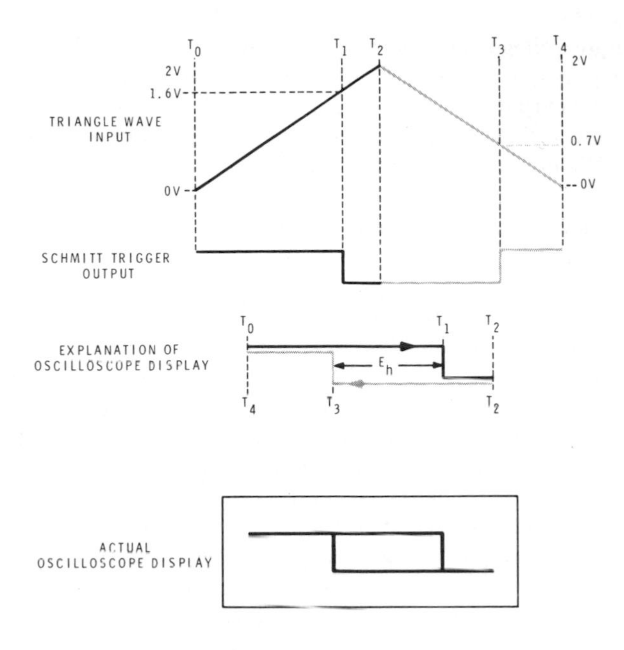

Figure 63B
Waveforms for hysteresis check.

Connect your function generator so that you are using a triangle wave to drive both the Schmitt trigger input and the oscilloscope horizontal deflection, The triangle wave is now the time base of the oscilloscope display. On the positive slope of the signal, the beam will be driven from left to right across the screen. The negative slope of the wave will cause the beam to retrace from right to left. Use the vertical and horizontal centering controls to obtain a centered presentation. Now, by applying the Schmitt trigger output to the oscilloscopo vertical input, you can compare the two signals. Refer to the waveforms in Figure 63B to see how the circuit operation corresponds to the oscilloscope display.

Note the triangle wave voltage in Figure 63B, with respect to time. The positive slope rises, at a linear rate, from 0 volts at T_0 to 2V at T_2. From T_2 to T_4, the voltage decreases back to 0 volts at the same linear rate. This signal drives the beam back and forth across the screen and, at the same time, is applied to the input to the Schmitt trigger.

Assuming that our Schmitt trigger circuit has a trigger level of 1.6 V and a hysteresis voltage of 0.9 V, look at the circuit output. At time T_0, the input voltage is below the trigger level, thus the output will be at a high level.

When the input voltage exceeds the trigger level at time T_1, the output will go to its low state. The input must drop below the lower hysteresis level before the output will again change states. Therefore, the output will remain low until time T_3. At this time, the input sinks below 0.7 V (1.6 V − 0.9 V), and the output reverts to its high state.

Looking at the simulated oscilloscope display of this action, (Figure 63B) note that the dark trace, moving from left to right, remains at a high level from T_0 to T_1. At T_1, the trace drops to a lower level and remains there. Remember, at time T_2, the triangle wave reaches its maximum positive value and begins to drop negative again. This marks the maximum oscilloscope deflection to the right. The beam now reverses direction and retraces from right to left. This is shown by the light trace. The traces are shown separately so that you can observe this action. In practice, these two traces will be superimposed as shown in the actual display illustration. The retrace will stay at a low level until time T_3, at which time, the Schmitt trigger output will drive it high once more. The resultant display will take the appearance of a box with a high level trace at one end and a low level trace at the other. The horizontal length of the box is proportional to the difference in hysteresis levels (E_h), while the distance from T_0 to T_1 is proportional to the circuit's trigger level. If you take the time to carefully adjust the oscilloscope horizontal gain to the triangle wave amplitude (for instance, make 1 volt equal 1 cm), you will be able to measure these values directly on the screen.

To set the trigger level of this circuit, merely set the amplitude of the generator output to the desired trigger level using your oscilloscope to measure the absolute peak amplitude. Now adjust the circuit until it just triggers. If you wish to have the circuit trigger on the smallest possible signal, reduce the generator amplitude until the circuit barely stops triggering, the adjust the circuit until it starts triggering once more. Repeat this procedure until you reach the point where no further improvement in sensitivity is possible.

Relays also exhibit a hysteresis-type characteristic in their operation. The voltage required to energize them is greater than the voltage at which they de-energize. These levels are known as the pull-in and drop-out voltages.

Most relays are classified by their general operating voltage and, in many cases, this data is sufficient. However, you may encounter an application that requires you to know the precise pull-in and drop-out values of the relay. In other cases, you may need to match the characteristics of two or more relays fairly closely.

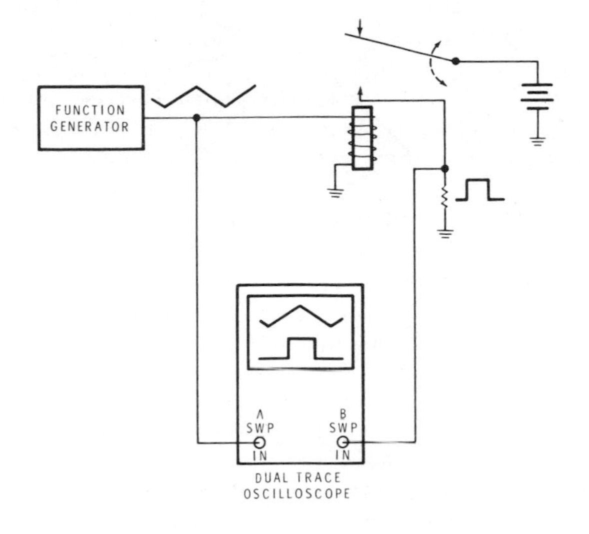

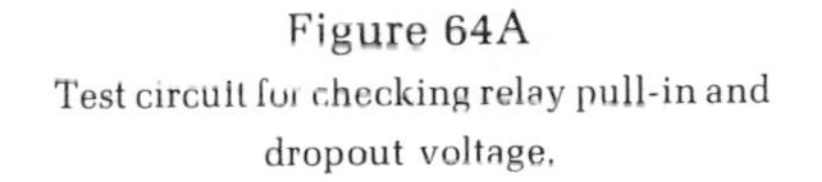
Figure 64A
Test circuit for checking relay pull-in and dropout voltage.

The equipment setup for this test is shown in Figure 64A. Note, in this case, a dual-trace oscilloscope is used. Actually, you could use the same single-trace oscilloscope used in the previous Schmitt trigger test, but the dual-trace oscilloscope will present a second, and perhaps more understandable, method of viewing hysteresis action.

Assuming your function generator is capable of driving the relay under test, you should connect the equipment as shown and adjust the generator output for a low-frequency triangle wave. Increase the output amplitude until the relay begins to operate. At this point, you will be able to adjust the oscilloscope to obtain a display similar to that shown in Figure 64B.

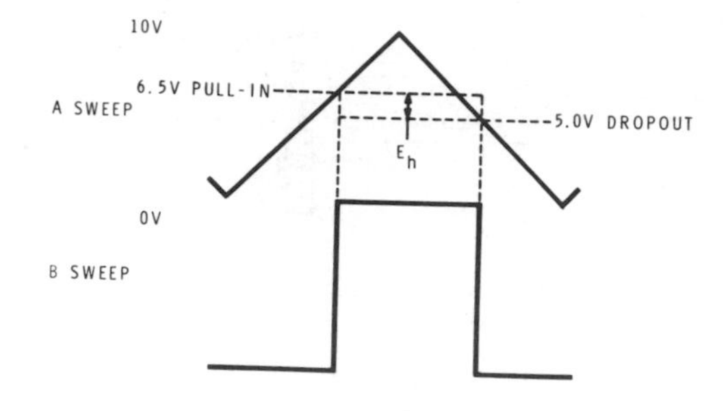

Figure 64B

Select an "A" sweep volts/division scale that will allow good resolution for voltage readings. The amplitude of the "B" sweep is not important, as you are only concerned with the timing of the relay waveform. Once you have a stable presentation on the screen, position the "B" sweep so that the leading edge of the relay signal just touches the positive slope of the A sweep signal. This is the pull-in point. Read and record the amplitude of the triangle wave at this point. Now, move the "B" sweep downward until the trailing edge of the relay waveform just meets the negative slope of the triangle wave. This is the dropout point. Read and record the amplitude of the triangle wave at this point.

You have now measured the voltage levels at which the relay pulls in and drops out. To determine the hysteresis voltage, subtract the drop out voltage from the pull-in voltage, in this case 1.5 V. While the dual-trace method may take slightly longer to set up, it does provide a more graphic presentation of the circuit operation.

In other testing situations, a sawtooth waveform may be advantageous. Figure 65 shows the waveform you can obtain using the symmetry control in the triangle mode. Again, these waveforms can be referenced to any DC level within range of your generator's DC offset control.

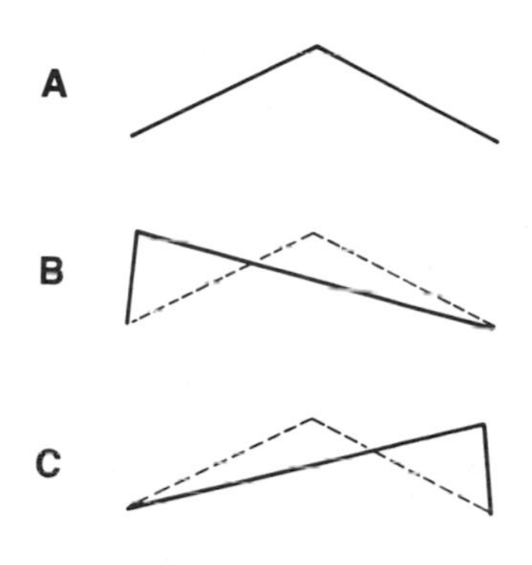

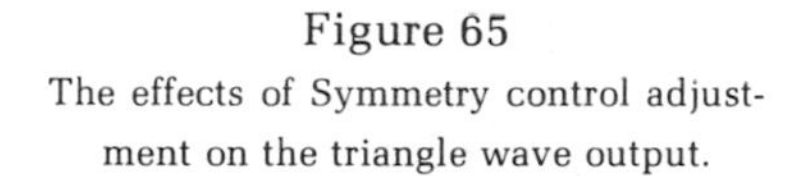

Figure 65
The effects of Symmetry control adjustment on the triangle wave output.

These ramps can be used as external sweep voltages for oscilloscope, X-Y plotters, pen recorders, and similar devices. They can also be used to drive voltage-controlled oscillator stages (VCOs). This converts the VCO to a swept-frequency oscillator, whose output frequency changes at the rate of the ramp voltage. If you wish the frequency to sweep back and forth at a linear rate, select a symmetrical triangle wave output from the function generator. However, if you require a frequency source which sweeps only upward or downward at a linear rate and returns quickly to its starting frequency, select the appropriate sawtooth waveform.

While these are by no means all of the possible applications of this versatile instrument, they do provide you with a representative sample of its uses. The knowledge you have gained, coupled with your test requirements and imagination, will allow you to find many additional uses for your function generator.

APPENDIX A

Vacuum Tubes

Diodes

The first electron or vacuum tube diode was constructed by Thomas Edison in 1885. However, it wasn't until the early part of this century that the vacuum tube diode was fully developed and put to use as a rectifier.

DIODE OPERATION

Figure C1 shows the basic structure of an electron tube diode. The diode consists of a cathode and an anode or plate. The cathode emits or gives off electrons, while the plate attracts and collects electrons.

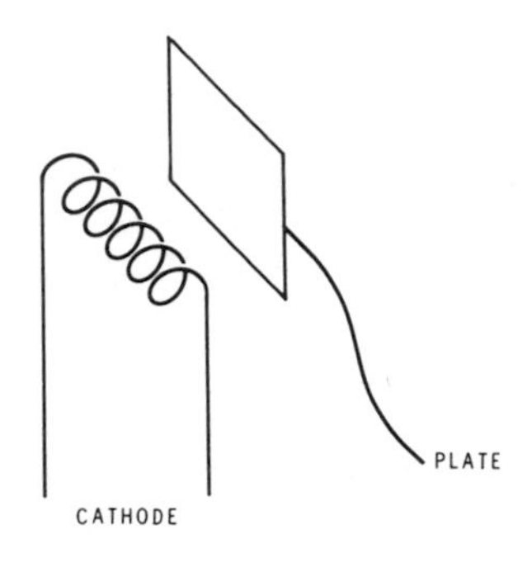

Figure C1

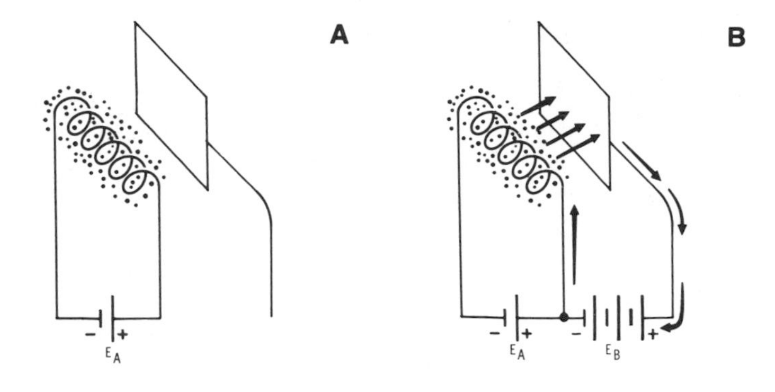

Figure C2
Diode operation. (4-2)

To emit electrons, the cathode is heated in much the same way as a light bulb filament. That is, a DC or AC voltage is applied and the resulting current heats the cathode. The high temperature of the cathode agitates the electrons so violently that many electrons actually escape the cathode surface. As the electrons are "boiled" off, a cloud of them forms around the cathode as shown in Figure C2A.

In Figure C2B an additional battery (E) is connected between the cathode and plate. Its positive terminal is connected to the plate and its negative terminal to the cathode. Now that a positive voltage is applied to the plate, the electrons "boiled" off the cathode are attracted toward the plate. These electrons travel to the plate and continue to the positive terminal of the battery.

Note that if the terminals of E_B are reversed (negative on the plate, positive on the cathode), current can no longer flow. Since the plate is not heated, it cannot emit electrons. Thus electrons cannot flow from plate to cathode. This is an important characteristic of a diode: current flows only from cathode to plate, and then only if the plate is positive with respect to the cathode.

DIODE CONSTRUCTION

Figure C3 shows the physical construction of two types of cathodes. Figure C3A shows the directly heated cathode which we have discussed. Here, the cathode is heated by passing a current directly through it. Figure C3B shows the indirectly heated cathode. It consists of a cathode sleeve which is heated by a separate heater or filament. The indirectly heated cathode gives a more uniform electron emission. However, it is suitable only for low to medium power applications. The directly heated cathode is used in high power tubes which require large amounts of current.

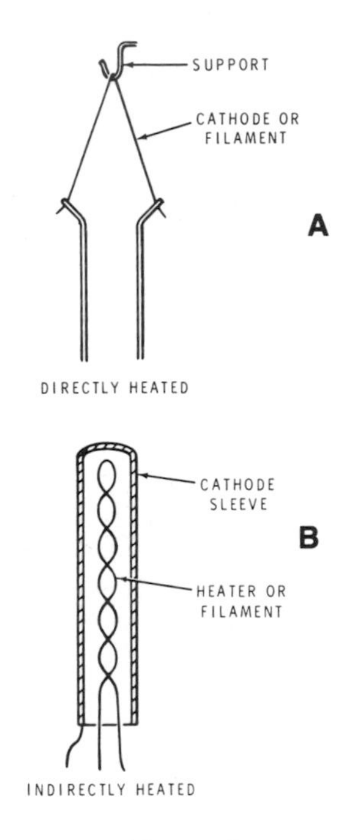

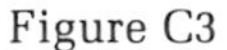

Figure C3

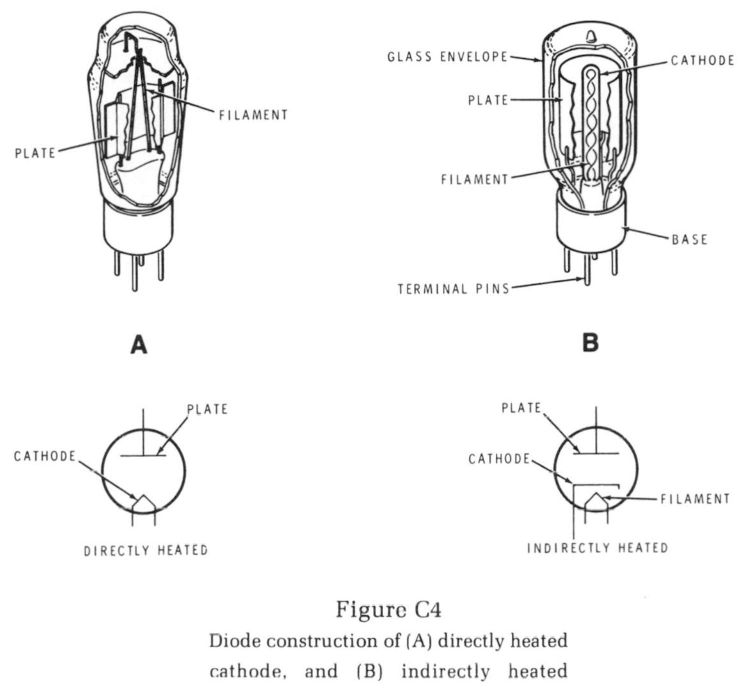

Figure C4
Diode construction of (A) directly heated cathode, and (B) indirectly heated cathode.

Figure C4 shows cutaway drawing and schematic symbols for both types of diodes. Note that the tubes are enclosed by a glass envelope. All the air is evacuated from this envelope, leaving a vacuum. This is why they are often called "vacuum tubes." The vacuum allows the cathode or filament to get extremely hot without danger of burning up. If the cathode were exposed to the oxygen in the air, it would instantly burn out.

DIODE RECTIFIER

The diode's most important application is as a rectifier. A rectifier is used to change alternating current to direct current. It does this by allowing current to flow in only one direction. Figure C5 shows the comparison between this diode and a solid-state rectifier diode.

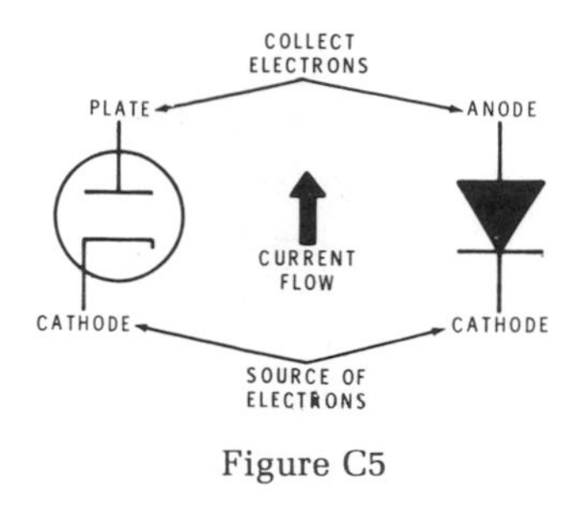

Figure C5

A diode half-wave rectifier circuit is shown in Figure C6. It is called a half-wave rectifier because it allows only one-half of the input sine wave to appear at the output. This is due to the diode's ability to pass current in only one direction. When the diode's plate is positive, current flows as shown in Figure C6A. At this time, voltage appears across the load due to the circuit current. When the plate is negative, as shown in Figure C6B, the diode does not conduct. Therefore, circuit current is reduced to zero and no voltage is developed across the load.

Although the output is not a steady voltage, it is DC, since it has only one polarity. It is known as pulsating DC, since it consists of positive pulses of DC voltage.

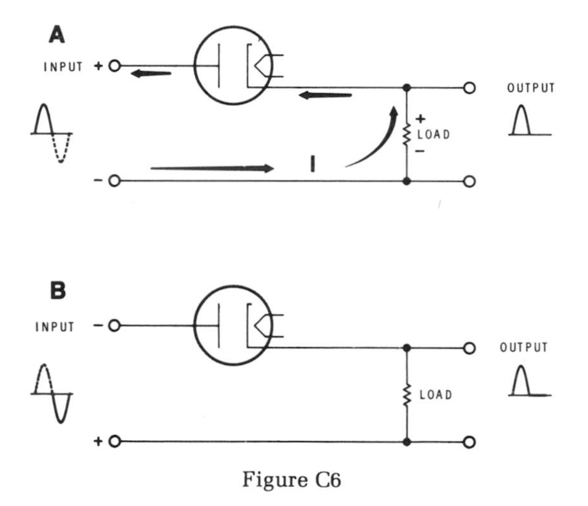

Figure C6

Triodes

The invention of the triode revolutionized the field of electronics. The triode was the first device which could amplify signals. This allowed the invention of many electronic devices including the sensitive radio receiver.

TRIODE CONSTRUCTION

The triode has three elements: cathode, plate and control grid. This is shown in Figure C7A. The control grid is a fine metal wire wrapped around two supports placed in the space between the cathode and plate. All electrons which travel from cathode to plate must pass through the openings in the grid. By applying a voltage to the grid, the flow of electrons within the tube can be controlled. This means that plate current now depends on two factors: the plate voltage and the control grid voltage.

CONTROL GRID OPERATION

The function of the control grid is to control plate current. It does this by controlling the number of electrons which travel to the plate. When a high negative voltage is applied to the grid, all the electrons from the cathode are repelled and prevented from reaching the plate. This condition is called cutoff, and it reduces plate current to zero. At the other extreme, if the grid voltage is zero or slightly positive, plate current will be maximum. This condition is called saturation, and it occurs when the plate attracts all the electrons emitted by the cathod. Once into saturation, plate current cannot increase further because no additional electrons are available.

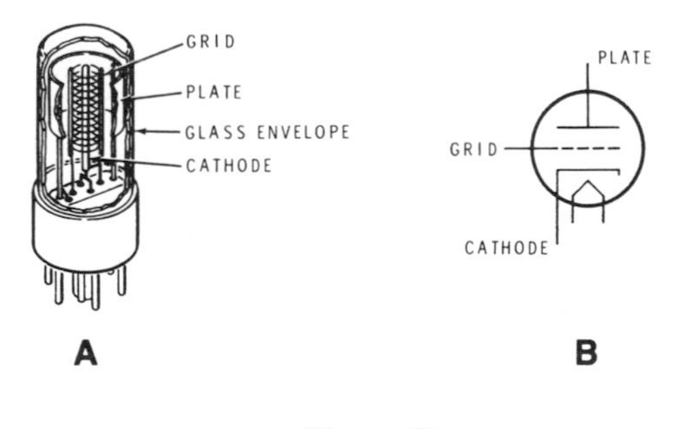

Figure C7
Triode construction is shown in (A). Its schematic symbol is shown in (B).

In normal operation, a DC bias voltage is applied to the grid. The bias voltage establishes the steady-state plate current or operating point. Usually, it sets plate current midway between cutoff and saturation. In Figure C8, the bias voltage is supplied by E_C.

Let's see what happens to the plate voltage (E_P) when grid voltage (E_C) is varied in the circuit of Figure C8. For example, let's assume that when E_C is -2 volts, plate current is 5 mA. Since the tube is in series with R_L, the current through R_L is also 5 mA. Therefore, E_{RL} is: $I_X R_L = 5\text{ mA} \times 20\text{ k}\Omega =$ 100 V. E_P is 200 V since the sum of the voltage drops must equal the supply voltage, 300 V.

Now, suppose the grid voltage (E_C) is reduced to -1 V. This allows more electrons to reach the plate, increasing plate current. In this case, plate current increases to 6 mA. The voltage across R_L is: $I_X R_L = 6\text{ mA} \times 20\text{ k}\Omega$ 120 V. This means the plate voltage (E_P) is now 180 V.

Next, if grid voltage is increased to -3 V, the plate current will decrease. In this example, let's assume that plate current decreases to 4 mA. The voltage across R_L becomes 80 V, while the plate voltage increases to 220 V.

Notice that a voltage change of ± 1 V on the grid causes a change of ± 20 V on the plate. This is amplification, a small change in grid voltage causes a large change in plate voltage. The grid voltage variation is amplified.

Figure C9 shows a comparison of a triode to the more familiar NPN transistor.

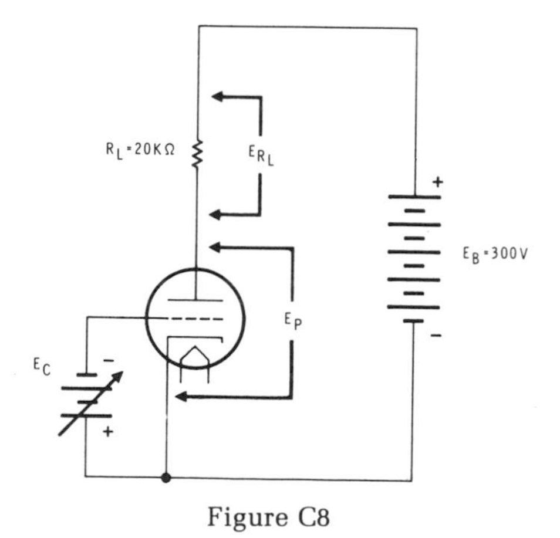

Figure C8

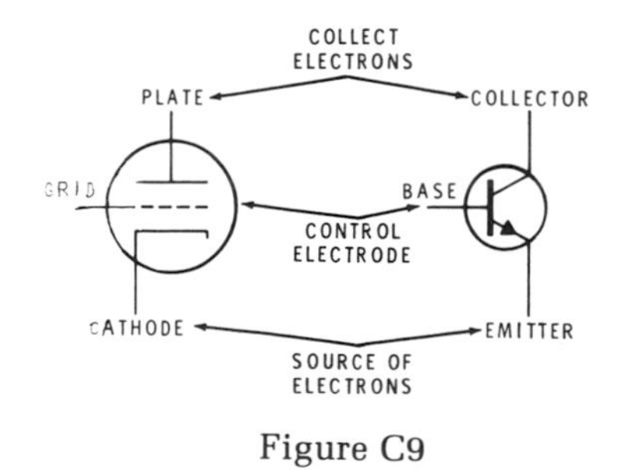

Figure C9

TRIODE AMPLIFIER

An amplifier is a circuit that increases the amplitude of a small input signal to provide a much larger output signal.

Figure C10 shows a triode amplifier. In this amplifier, E_C supplies the grid bias voltage and E_B supplies the plate voltage. C_C and R_G are the input coupling network. C_C prevents any DC voltages from reaching the grid, while allowing the input AC signal to pass. R_G prevents the input signal from being shorted out by the low resistance of E_C. R_L and C_O provide coupling for the output.

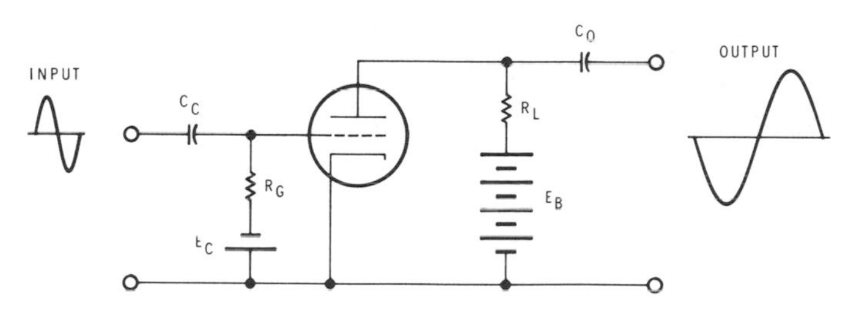

Figure C10

A triode amplifier.

Let's analyze the operation of the triode amplifier of Figure C10. The AC input signal, since it is varying in polarity, alternately adds to, and subtracts from, the bias voltage E_C. On its positive alternation, it reduces the negative grid bias, causing an increase in plate current. This causes a decrease in plate voltage and the output voltage will drop as shown. The negative alternation of the input voltage adds to the grid bias, causing a decrease in plate current. This increases the plate voltage and the output voltage rises as shown.

It is important to note that the output voltage is an amplified duplicate of the input voltage. Note also that there is a 180° phase inversion between input and output.

Amplifier voltage gain is a measure of how much amplification a circuit provides. It is the ratio of input voltage to output voltage and is expressed mathematically as:

$$A_V = \frac{E_{out}}{E_{in}}$$

where A_V = voltage gain
E_{out} = output voltage
E_{in} = input voltage

If the input to an amplifier is 100 mV and the output is 5 V, the gain will be:

$$A_V = \frac{E_{out}}{E_{in}}$$

$$A_V = \frac{5\ \text{V}}{100\ \text{mV}}$$

We must use the same units for both values, therefore:

$$A_V = \frac{5\ \text{V}}{0.1\ \text{V}} \quad \textbf{or} \quad \frac{5000\ \text{mV}}{100\ \text{mV}}$$

$$A_v = 50$$

The gain of the amplifier is 50. This means that the input signal is increased 50 times.

AMPLIFIER CONFIGURATIONS

The circuit shown in Figure C11 is a conventional triode amplifier and is called the grounded-cathode configuration. The input is applied to the control grid and the output is taken from the plate. The cathode is common to both input and output and is connected to ground. An important characteristic of the grounded-cathode amplifier is the 180° phase inversion between input and output. This amplifier is used in almost all vacuum tube equipment.

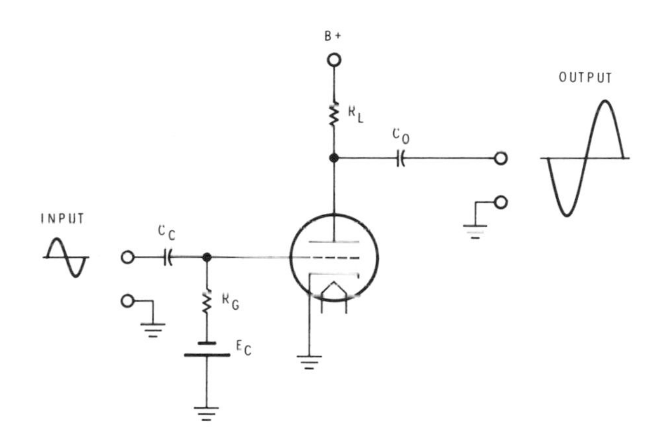

Figure C11
A grounded-cathode amplifier.

The grounded-grid amplifier is shown in Figure C12. Here, the input is applied to the cathode, output is taken from the plate, and the grid is grounded. With this amplifier, there is no phase inversion between input and output. The grounded-grid amplifier is used primarily in RF power amplifiers.

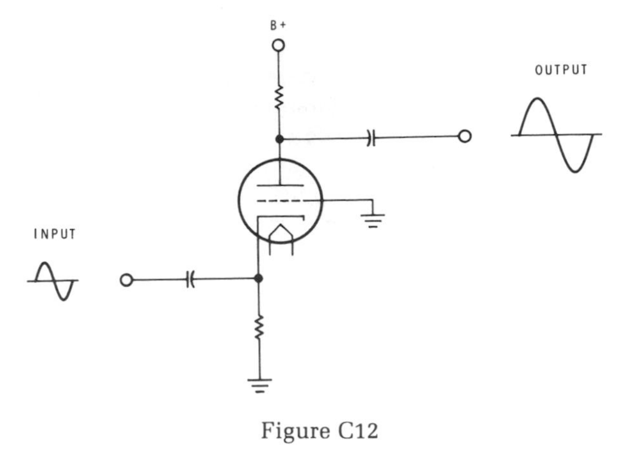

Figure C12

The grounded-plate or cathode-follower amplifier is shown in Figure C13. With this configuration, the input is applied to the grid, output is taken from the cathode, and the plate is connected to AC ground by capacitor C_1. The voltage gain of this amplifier is less than 1. Therefore, its primary use is impedance matching. It has a high input impedance and low output impedance. Note that there is no phase difference between input and output.

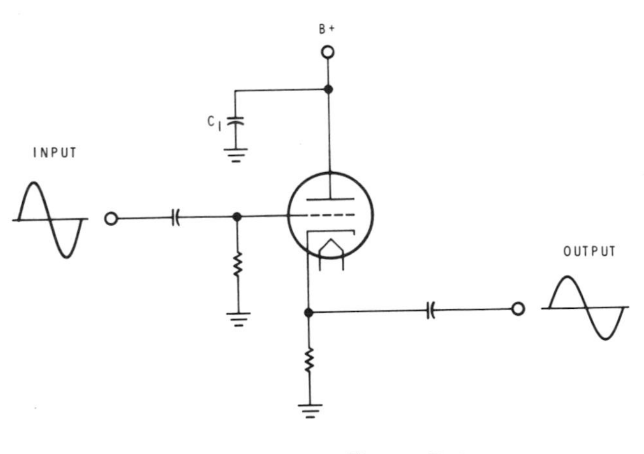

Figure C13

Tetrodes and Pentodes

Tetrodes and pentodes add extra grids to the basic triode structure. The tetrode adds a screen grid while the pentode adds both a screen grid and a suppressor grid.

TETRODE

The tetrode is a special type of vacuum tube that has four electrodes. Like the triode, it has a cathode, grid, and plate. In addition, the tetrode has a fourth element called a screen grid.

Screen-Grid

The screen-grid is an additional electrode placed between the plate and control grid. As shown in Figure C14 the control grid is grid number 1 and is placed close to the cathode. It still controls the electron flow from the cathode. The screen-grid is grid number 2 and is placed close to the plate. It is not used to control plate current, however, it has a positive voltage which accelerates the electrons toward the plate.

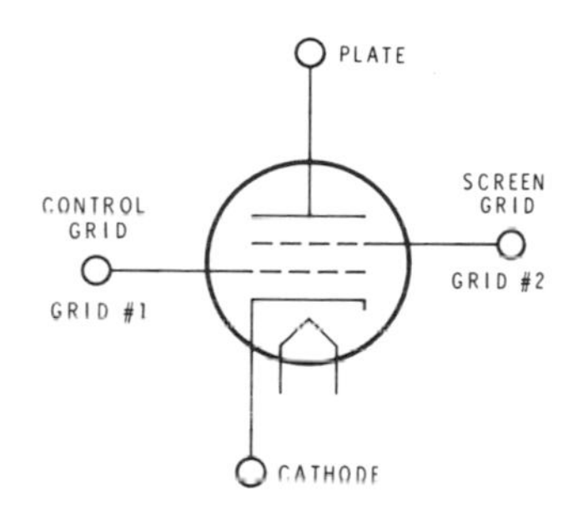

Figure C14
The tetrode.

Figure C15 shows the basic circuit of a tetrode amplifier. The screen-grid receives its positive voltage from the B+ supply through the screen-dropping resistor R_S. The screen voltage must be slightly lower than the plate voltage or it will attract more electrons than the plate. The screen-dropping resistor, R_S, accomplishes this. The screen-grid will still attract a few electrons but it will be a minor amount. Capacitor C_S is used to pass any AC variations to ground. This keeps the screen-grid voltage at a steady value.

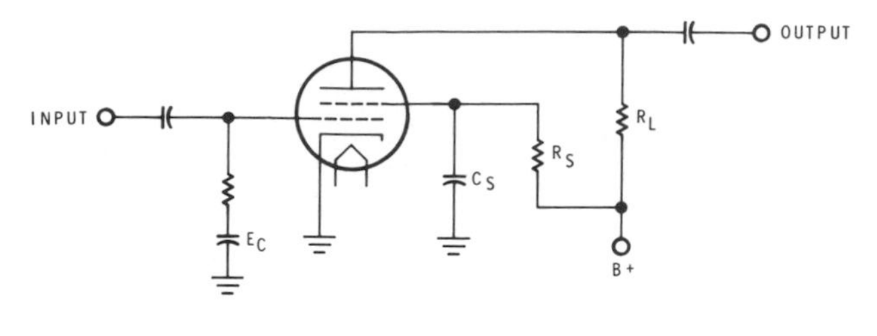

Figure C15

The major purpose of the screen grid is to reduce the influence of the plate voltage on plate current. Since the screen grid is closer to the cathode than the plate is, its positive voltage has a greater influence over electron flow. This means that the control grid is now the major controlling element. Therefore, screen-grid tubes have a higher gain than triodes, due to the increased effectiveness of the control grid when compared to the plate.

Another factor in favor of the screen grid is that it reduces interelectrode capacitance. This capacitance is shown in Figure C16 as C_{gp}. Since there are two metal electrodes (plate and control grid) separated by an insulator, capacitance must exist. A typical value of control grid to plate capacitance (C_{gp}) in a triode is 1.8 pF. When a screen grid is added, it reduces C_{gp} to a value as low as 0.006 pF. The reason for the reduction in C_{gp} is that the screen grid acts as a shield for AC signals between the plate and control grid. Low values of interelectrode capacitance are required when amplifying radio frequencies.

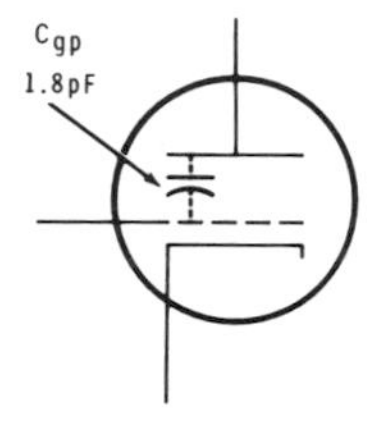

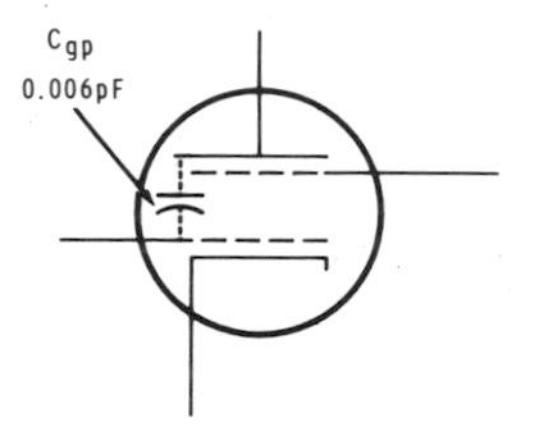

Figure C16

PENTODE

The final type of vacuum tube that we will be talking about is the pentode. It is very similar to the tetrode, however, it has one additional electrode. This extra element is called the suppressor grid. To see why this grid is used, we must discuss a phenomenon called secondary emission.

Suppressor Grid

One problem that arises in electron tubes is secondary emission. This occurs due to electrons "bombarding" the plate. When the electrons hit the plate, additional electrons are dislodged and released. In a diode or triode, secondary emission is not a problem because any electrons emitted are soon collected by the plate. However, in a tetrode, the positive screen grid may collect these electrons, reducing plate current.

To eliminate the problem of secondary emission, the suppressor grid was developed. It is placed between the screen grid and plate, as shown in Figure C17. Notice that it is internally connected to the cathode. This means it is negative with respect to the plate. Therefore, when secondary emission occurs, the negative suppressor grid will repel the electrons back toward the plate and prevent them from getting to the screen grid.

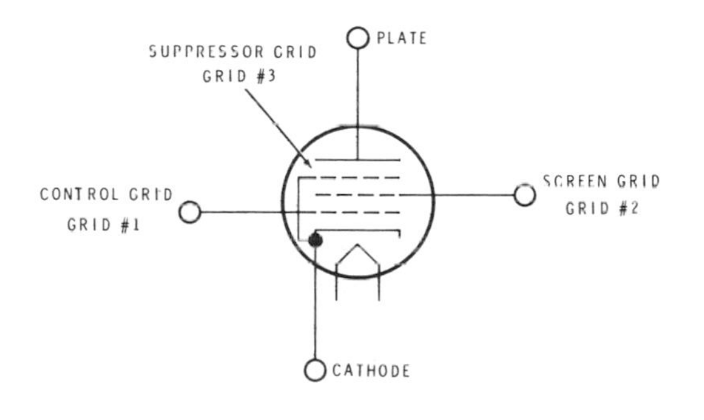

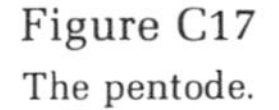

Figure C17
The pentode.

Summary of Vacuum Tubes

The chart that follows summarizes the important points of each electron tube we have discussed.

ELECTRON TUBE	SCHEMATIC SYMBOL	ELECTRODES	CHARACTERISTICS
DIODE	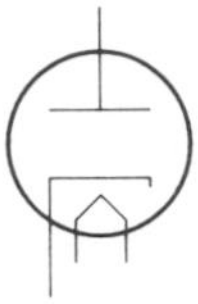	PLATE CATHODE	Conducts only when plate is positive with respect to cathode. Used as a rectifier.
TRIODE	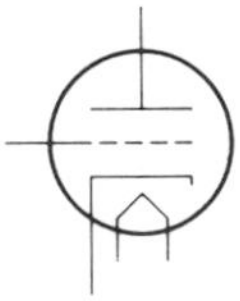	PLATE CONTROL GRID CATHODE	Grid controls plate current. Used as an amplifier.
TETRODE	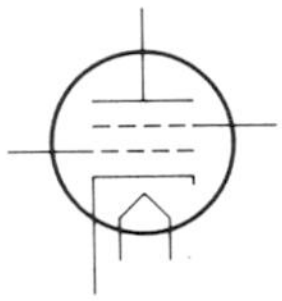	PLATE SCREEN GRID CONTROL GRID CATHODE	Screen grid reduces grid to plate capacitance, increases gain.
PENTODE	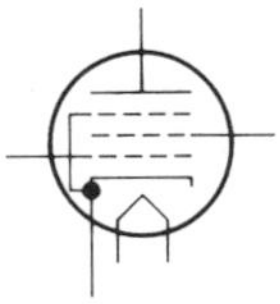	PLATE SUPPRESSOR GRID SCREEN GRID CONTROL GRID CATHODE	Suppressor grid eliminates secondary emission.

NOTE: Indirectly heated cathodes shown.

Figure C18

Unit 6

SPECIAL MEASURING INSTRUMENTS

INTRODUCTION

Previous units of this course discussed general purpose test equipment and its use in maintaining electronic equipment. While a large portion of electronic testing can be performed with this general purpose equipment, there are several places where equipment designed for a special purpose is needed.

In this unit, you will learn about some of that equipment. You will see how bridge circuits can be used to measure resistance, inductance, and capacitance much more accurately than a meter. You will learn how transistor testers and curve tracers check practically all the characteristics of a number of solid-state devices. You will also see how the spectrum analyzer, which has long been a valuable tool in radar maintenance, can be used in AM and FM circuits.

Digital circuits have not been neglected. The ever popular logic probe and its companion, the logic pulser, are both covered here.

You will learn how this equipment works, what it can measure, and what it can't. You will also learn some basic techniques for using and interpreting the measurements.

BRIDGE CIRCUITS

Bridge circuits are ideal for making precise measurements of component values. They can be used to measure almost any parameter, such as resistance, inductance, capacitance or impedance. We will discuss some basic bridge types and how various components are checked.

Resistance Bridges

While the ohmmeter is an ideal instrument for measuring resistance, it has limited accurancy. Typical accuracy of a good digital ohmmeter is ±2%. Therefore, the ohmmeter is not suitable for measuring precision resistance values.

A bridge circuit is often used to measure precision resistance values. There are many types of bridge circuits in use; however, most are a variation of the basic Wheatstone bridge which will be discussed here. A schematic of the basic Wheatstone bridge is shown in Figure 1. Resistors R_1 and R_2 are fixed. R_s is a variable resistor with provisions for accurately reading its value at any setting. R_x is the unknown value of resistance being measured.

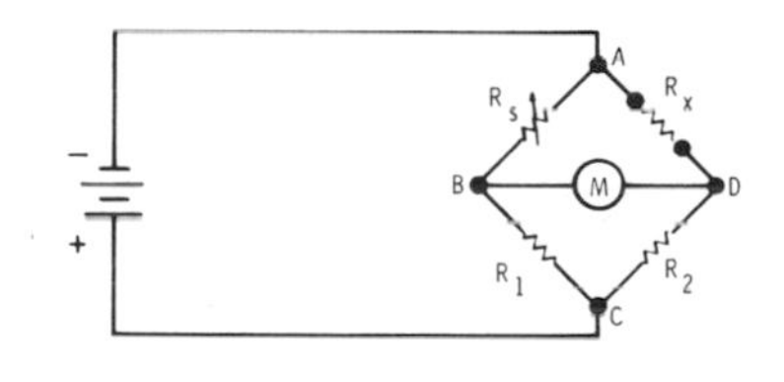

Figure 1

Operation of the bridge is based on the fact that if the voltage at point B is equal to the voltage at point D, no current will flow through the meter. This condition exists any time the ratio of R_s/R_1 equals the ratio of R_x/R_2. Thus, if R_1 equals R_2, zero current through the meter will occur when $R_s = R_2$. Obviously, if R_s does not equal R_x, then a potential difference exists between B and D and some current will flow.

If R_s in Figure 1 is larger than R_x, B will be more positive than D and current will flow from D to B. On the other hand, if R_x is larger than R_s, current will flow from B to D. Since current can flow in either direction, a zero center meter movement must be used.

Measurement is made by connecting the unknown resistance at R_x, adjusting R_s for zero current, and reading the value directly from the calibrated dial of R_s. This works great when R_1 and R_2 are equal. However, it is frequently necessary to measure a resistance that is either higher or lower than the range of R_s. This can easily be accomplished because the actual values of resistance are unimportant as long as the proportion $\frac{R_s}{R_1} = \frac{R_x}{R_2}$ is maintained.

When R_1 and R_2 are unequal, the equation $R_x = \frac{R_2}{R_1} R_s$ is used. If R_2 is

made ten times as large as R_1, the R_x will be ten times R_s when the bridge is balanced. Thus, you can find the value of R_x by multiplying the indicated value of R_s times ten.

A value of R_x that is less than R_s can be measured by making R_1 larger than R_2. If R_1 is ten times the value of R_2, then R_x will be 0.1 times the indicated value of R_s.

Wheatstone bridges are available that can measure from one ohm to one megohm with an accuracy of $\pm 0.25\%$. Different types of bridges are available for measuring larger or smaller resistances at reduced accuracy.

Capacitance Bridges

Bridge circuits are also used for measuring capacitance. There are many varieties of the capacitance bridge in use; however, the Wien bridge shown in Figure 2 is one of the most common. The capacitance bridge is similar to the resistance bridge. An unknown value of capacitance is connected at C_x, and C_s is adjusted for zero meter current. The value of capacitance is then read directly from a calibrated scale. Obviously, an AC source must be used when measuring capacitance.

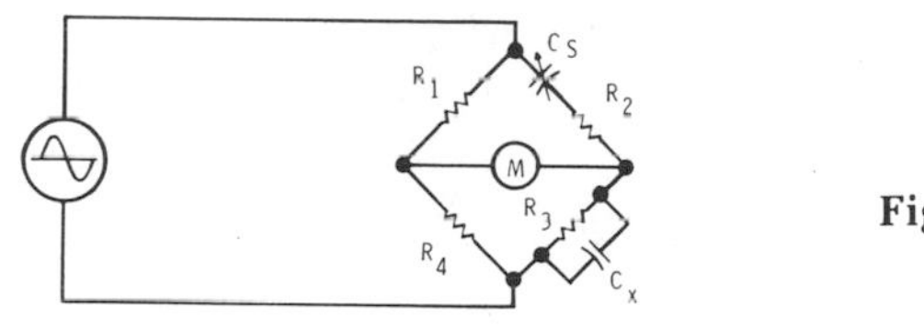

Figure 2

Inductance Bridges

Measuring inductance is another function of bridge circuits. As with resistance and capacitance, there are several different configurations of inductance bridges. The Maxwell bridge shown in Figure 3 is one of the most common. The inductance bridge is similar to the resistance bridge in that the inductor in question is connected in the place of L_x and R_x, and the bridge adjusted for zero current through the meter. L_x and R_x are not two different components, but are the two properties of any inductor. Both of these properties can be determined with the Maxwell bridge. Inductance is calculated by the formula $L_x = R_1 R_3 C_1$, while resistance is calculated by $R_x = \frac{R_1}{R_2} R_3$.

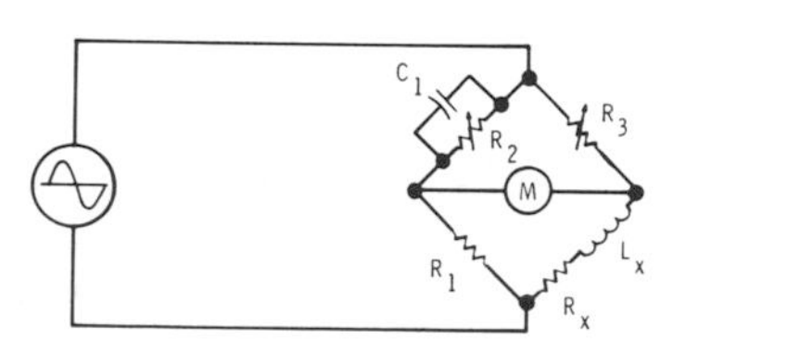

Figure 3

In practice, no calculation is normally required. R_3 is calibrated to read out the correct value of inductance, and R_2 is calibrated for resistance. Thus, when the current through the meter has been reduced to minimum, the value of both L and R can be read directly.

RLC Bridges

Frequently, inductance, capacitance and resistance measurement capabilities are combined in one instrument. The low-cost RLC bridge shown in Figure 4 is an example of such an instrument.

Figure 4

This instrument can measure resistances from ten ohms to ten megohms, inductance from ten microHenrys to ten Henrys and capacitance from ten picoFarads to ten microFarads.

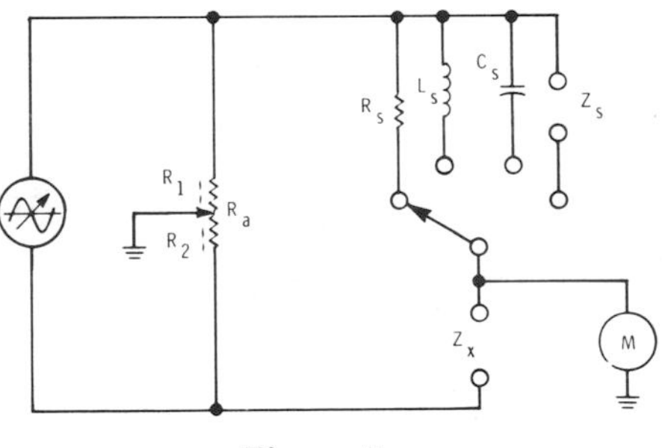

Figure 5

A conceptual drawing is shown in Figure 5. If, as shown here, the switch is placed in the resistance position, an unknown value of resistance is connected at Z_x. R_a is adjusted until the ratio of R_1 to R_2 is equal to the ratio of R_s to Z_x, then at that time, the current through the meter will be minimum.

To measure inductance or capacitance, merely connect the unknown quantity at Z_x and adjust R_a until $R_1/R_2 = X_s/Z_x$. A dial connected to R_a is calibrated to indicate the proper value of R, L or C. In the case of L, the measurement is based on Z_1 and no provision is made for separating the inductive component from the resistive component. Also, since the current source is always AC, any resistor must be noninductive or the resistance readout will be equal to the impedance of the component under test at the frequency of the current source.

Of course, the actual instrument is more complex than shown. It contains an oscillator with frequency selection and stabilization circuits. Obviously, the alternating current must be rectified before it reaches the meter. In addition, the meter signal is amplified for better accuracy and stability.

It is also possible to compare two components by inserting the known value, or the one to be used as a standard, into Z_s. The unknown component is connected to Z_x, and the balance control is adjusted for a null. Conversion formulas are written on the front of the case that allow you to calculate the value of Z_x from the dial reading and the value of Z_s. Any two values can be compared as long as one is no more than ten times the value of the other.

When using this bridge, or any other, it is usually best to keep the lead length to a minimum. Extra lead length introduces additional resistance and inductance. There is also the possiblity of picking up stray electrical fields which may cause a wrong measurement. As the accuracy of the bridge and the precision of the component under test increase, this precaution becomes even more important.

The LCR bridge is an excellent tool for measuring component values. The more sophisticated models are capable of separating the quantities of a given component. For instance, it should be possible to determine the resistive component and the inductive component of an inductor as well as the total impedance.

TRANSISTOR TESTER

You have seen how the LCR bridge can be used to test passive components such as resistors, capacitors, and inductors. Obviously there must be some means of testing active components, such as transistors, FET's, and IC's. In earlier units of this course, we discussed how certain transistor tests could be made with meters and oscilloscopes.

When working with transistors, there are four essential tests that must be made. These are gain, leakage, breakdown voltage, and, on occasion, switching time. To a limited degree, some of these tests can be made with meters and oscilloscopes. However, a transistor tester like the one shown in Figure 6 can provide more accurate information. A tester of this type measures the DC beta (β) of transistors and the Gm of FET's as well as leakage current.

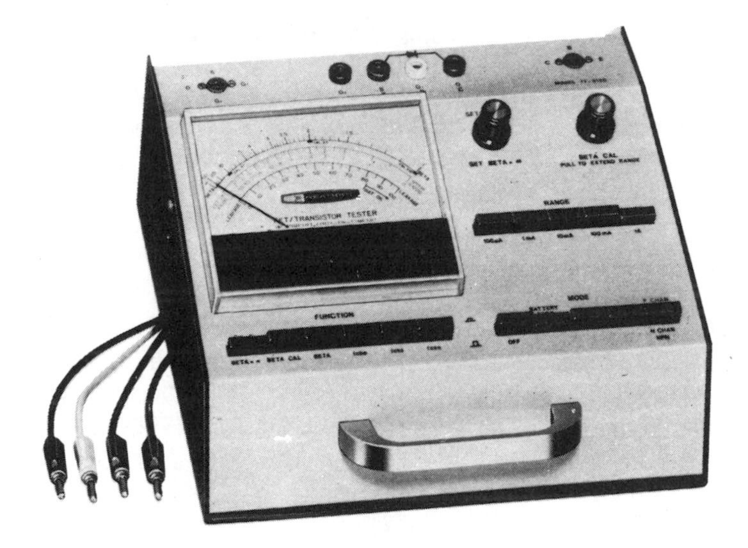

Figure 6

Beta

Basically, beta is measured with a circuit similar to the one in Figure 7. Since $\beta = \frac{I_C}{I_b}$, it can be calculated from the measurement shown here.

However, most testers have the controls arranged such that beta can be read directly from the meter scale. This is accomplished by applying a known value of I_b, then calibrating the I_c meter to indicate β.

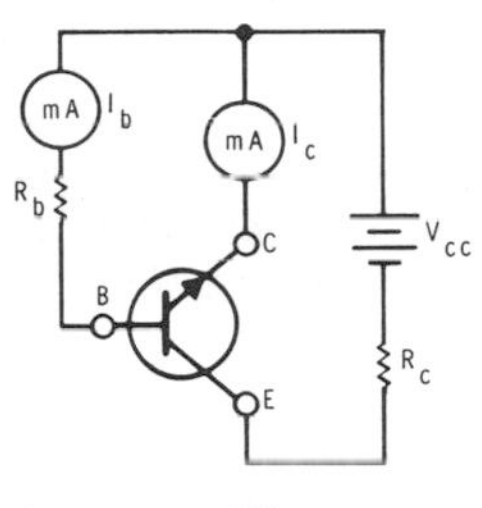

Figure 7

Transistors may be tested "in-circuit" or "out-of-circuit." Out-of-circuit beta tests are usually quite accurate, but the accuracy of an in-circuit test depends on the shunting resistances of the circuit in which the transistor is being used. If the transistor indicates a significant gain in-circuit, it can normally be considered good. However, if it shows no gain, the cause could be a bad transistor, a bad circuit component, or circuit design. If, for example, a circuit is designed to have a gain of 20 and the tester shows little or no gain, the transistor would be removed from the circuit and tested. If it shows a proper gain in the out-of-circuit test, the problem is most likely one of other circuit components.

Testers of this type measure collector current up to 1 ampere and beta up to 1000, which means they can test a wide range of transistors. They are, however, for the most part, DC devices. This means that as the frequency of operation increases, the value of the transistor tester decreases. Transistors from high frequency or switching circuits are usually tested by subsitution or with a curve tracer, since a transistor may test good at DC but not work in a critical high frequency application.

Remember, when making in-circuit tests, never connect the tester with power applied to the circuit. This can damage the tester or the circuit, or at best, will result in an erroneous reading.

Leakage

Leakage tests must always be made with the transistor out of the circuit, since the resistance of the circuit could cause erroneous readings. Three leakage measurements (I_{cbo}, I_{ces}, and I_{ceo}) are normally made. Leakage current is measured by placing a battery and a meter in series with two of the transistor leads as shown in Figure 8. This circuit is connected for measuring I_{cbo}. I_{cbo} is the measurement of the current that flows between the collector and the base of the transistor with the emitter open and the collector-base junction reverse biased. This measurement should always be the lowest of the leakage measurements.

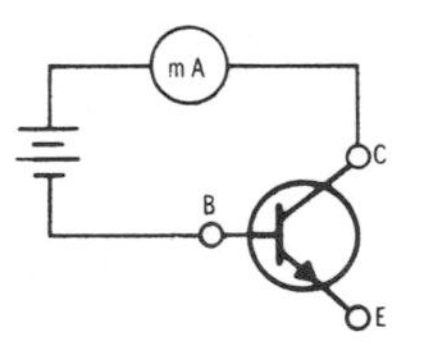

Figure 8

I_{ces} is the measurement of the current that flows between the collector and the emitter with the base connected to the emitter. Figure 9 shows this connection. It should be between I_{cbo} and I_{ceo}.

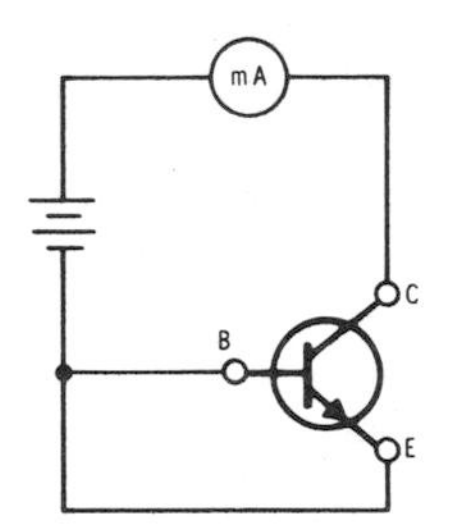

Figure 9

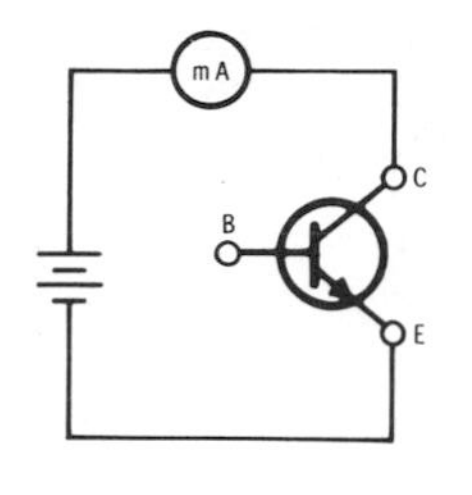

Figure 10

The third measurement, I_{ceo}, is the measurement of the current that flows between the collector and the emitter with the base open. Figure 10 shows how this measurement is made. It should always be the highest of the three leakage measurements.

For good transistors, I_{cbo} will always be less than I_{ces}, which will always be less than I_{ceo}.

On most testers, you will not obtain any measurable leakage current when testing silicon transistors. You can usually expect I_{cbo} to be less than 1 μA for low power transistors. High power transistors have higher leakage values with some having an I_{ceo} up to 50 μA.

Germanium transistors have much higher leakage currents with an I_{cbo} from several microamperes up to as much as 5 mA.

When testing transistors, it is helpful, although not essential, to know the type of transistor-NPN or PNP-and the current and voltage ratings. With most testers, you can discover some of these factors, but not all.

FET Tests

In addition to transistors, both N channel and P channel FET's (field effect transistors) can be tested. The primary characteristic measured by the tester is transconductance (Gm). Gm is the reciprocal of drain-to-source resistance and is a measure of how well the FET conducts.

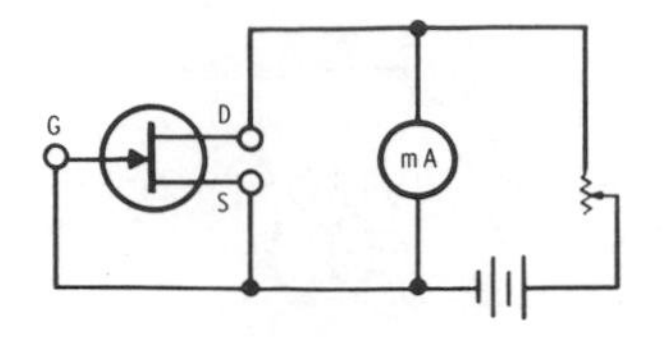

Figure 11

The circuit for measuring transconductance is shown in Figure 11. This circuit is basically a shunt type ohmmeter that measures the source-to-drain resistance of the FET. Transconductance values are usually expressed in micromhos and usually range between 500 and 10,000 micromhos.

Notice, in Figure 11, that the gate and source loads are connected together. Thus, this test does not measure the ability of the gate to control Gm. Therefore, a "gate" test must be added as shown in Figure 12. This "gate" test places a reverse bias on the gate of the FET through R_1. This causes the channel of the FET to become narrower, increasing resistance, and decreasing Gm. A noticeable decrease in Gm indicates a normal gate circuit.

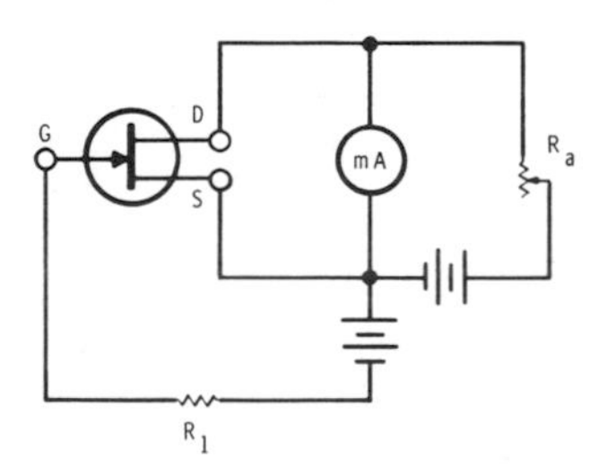

Figure 12

As with transistors, FET transconductance and gate tests may be made either in or out of the circuit. Just remember that the circuit components may change the expected value of Gm. Once more, be sure the circuit power is off before you connect the tester.

LEAKAGE

There are two basic leakage tests which are commonly made. They are I_{gss} and I_{dss}. As with transistors, leakage tests must be made with the FET removed from the circuit.

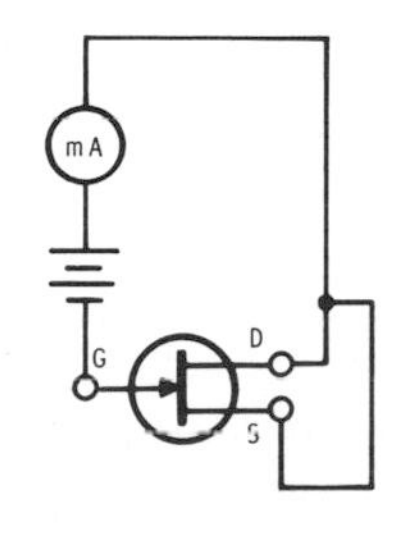

Figure 13

I_{gss} is the measurement of the current between the gate and the source. The circuit for this test is shown in Figure 13. Since the FET has a very high impedance input, this leakage current should be in the nanoampere range. Therefore, any measurable leakage current indicates a defective FET.

I_{dss}

I_{dss}, is the measurement of the current between the drain and the source with the gate shorted to the source. The circuit for this measurement is shown in Figure 14. This is a measure of forward current. The value will normally fall between 100 μA and 10 mA and is an excellent measurement when matching FET's.

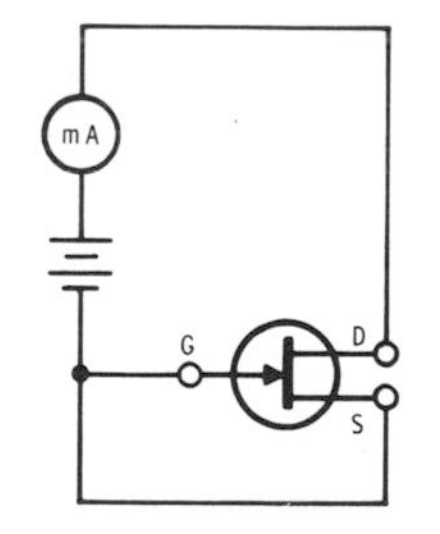

Figure 14

Diode Tests

The transistor tester can measure the forward or conducting current and the leakage of a diode. As with transistors, silicon diodes have very little leakage, while germanium diodes have a leakage current of several microamperes.

A good signal diode should have a forward current of about 1 mA or more while a good power diode or rectifier should indicate a conducting current of 1A or more.

Most transistor testers do not have the ability to test the zener voltage of a zener diode. However, the zener can be tested as a conventional signal diode. If this test is good, you can usually assume that the zener diode is good.

Diodes may be tested either in-circuit or out-of-circuit. Just remember that the circuit components may cause variations in the current readings, especially the reverse current readings.

Other Components

SCR's and triacs can be tested in or out of the circuit. However, the only test which can be performed is the turn-on or turn-off test.

While the transistor tester is quite a versatile instrument providing tests on a variety of components, it performs these tests with a fixed voltage and bias. Thus, the results of the test are really valid only under those conditions. If a more detailed analysis of the device or component under test is desired, a transistor curve tracer or similar test equipment is required.

CURVE TRACER

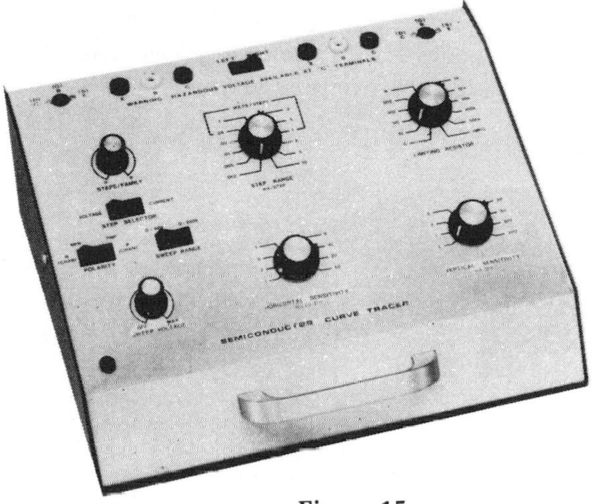

Figure 15

The curve tracer, shown in Figure 15, draws a graph of output voltage or current versus input voltage or current. This graph is drawn on a separate oscilloscope. With the curve tracer, devices such as transistors, FETs, SCRs, triacs, diodes, etc. can be tested for a variety of characteristics. For example, transistors can be tested for:

1. Current gain

2. Collector-to-emitter breakdown

3. Collector-to-base breakdown

4. Output admittance

5. Saturation voltage

6. Saturation resistance

7. Cutoff current

8. Leakage current

9. Linearity and distortion

10. Temperature effects

11. Identifying NPN or PNP

12. Matching

13. Sorting and substitution

In short, the curve tracer does everything the transistor tester does and more.

Block Diagram

A basic block diagram of the curve tracer is shown in Figure 16. The sweep supply applies a pulsating DC to the collector of the transistor under test and to the horizontal deflection circuits of the CRT. Thus, the electron beam will move horizontally across the CRT as the collector voltage increases and decreases.

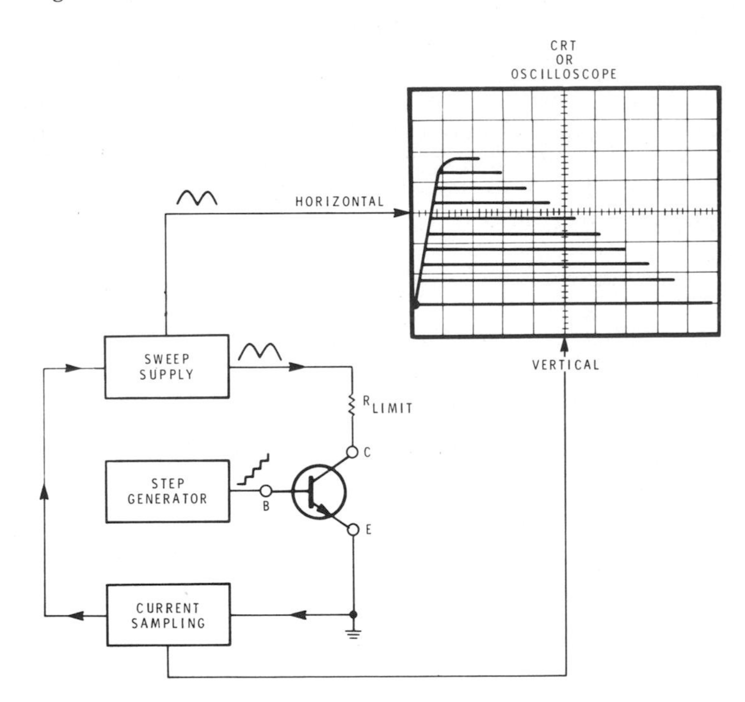

Figure 16

The step generator produces a staircase of voltage or current which is applied to the base of the transistor and controls the bias. Timing in the circuit is such that one complete horizontal sweep is produced at each level or step. Obviously, emitter current is controlled by the combination of base and collector voltage. The current sampling block converts this current to a voltage that is then applied to the vertical deflection circuits. A family of curves is thus plotted for the transistor under test.

Developing the Presentation

For this explanation, assume that a good transistor is being plotted by the curve tracer. The test starts with the electron beam located in the lower left corner of the CRT, and the step generator output at zero volts. The sweep supply starts at zero. As the sweep voltage increases, the spot moves across the CRT in a horizontal direction and the collector voltage increases on the transistor. However, since there is no emitter-base bias, no collector current will flow. Therefore, the current sampling circuit applies no vertical deflection signal to the CRT, and a straight line is drawn across the CRT as shown in Figure 17A.

When the sweep supply voltage returns to zero, the step generator increases the voltage applied to the base by one level. This level is controlled by the operator. Now, when the sweep supply voltage starts to increase, collector current will flow as the trace moves across the scope. At first, collector current increases with collector voltage; however, it soon reaches a level where it is limited by the beta of the transistor under test. Any further increases in collector current will have essentially no effect on collector current unless breakdown is reached. Thus, after two complete cycles the presentation will appear as in Figure 17B.

At the end of sweep two, the step generator raises the bias to the next level and the sweep cycle is repeated. Obviously, with the increased bias, collector current will be higher. So, after three complete sweep cycles, the presentation will appear as in Figure 17C. A complete series of curves might appear as in Figure 17D.

When operating the curve tracer, the operator selects the size and number of the base current steps and the maximum sweep voltage applied to the collector. Base current can be varied from 2 μA to 10 mA per step, and up to ten steps are available. Sweep ranges are 0-40V and 0-200V. These figures are taken from the Heathkit IT-3121 curve tracer shown in Figure 15; however, they are typical of the specifications for other tracers in the same price range. More sophisticated tracers perform essentially the same function; however, more steps may be provided and sweep voltage may go as high as 1500V.

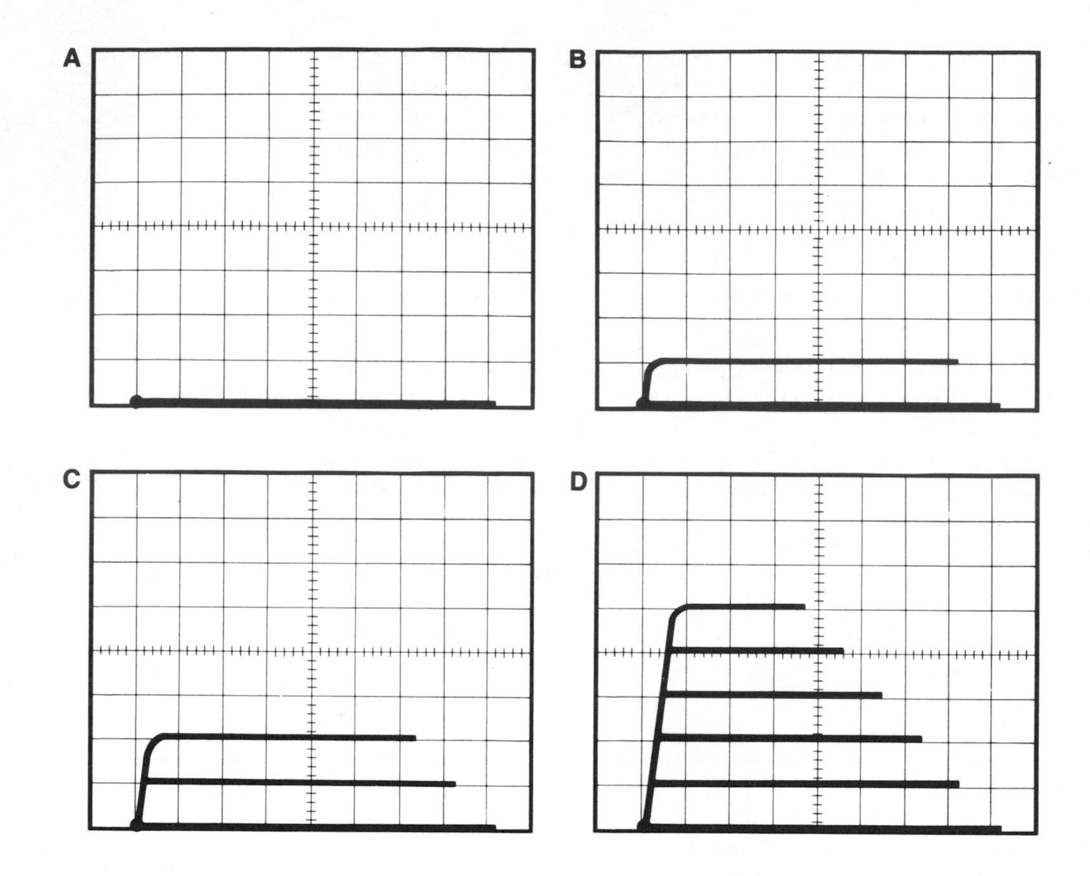

Figure 17

Testing Bipolar Transistors

Perhaps the most common use of the curve tracer is testing bipolar transistors. Since the sweep supply must provide a positive voltage for NPN transistors and a negative voltage for PNP transistors, it is helpful if you know the type of transistor you are testing. You will also need to know the power class (signal, intermediate, or power) of the transistor. The power class can usually be determined by the size of the transistor. Figure 18 shows some common sizes and power classes.

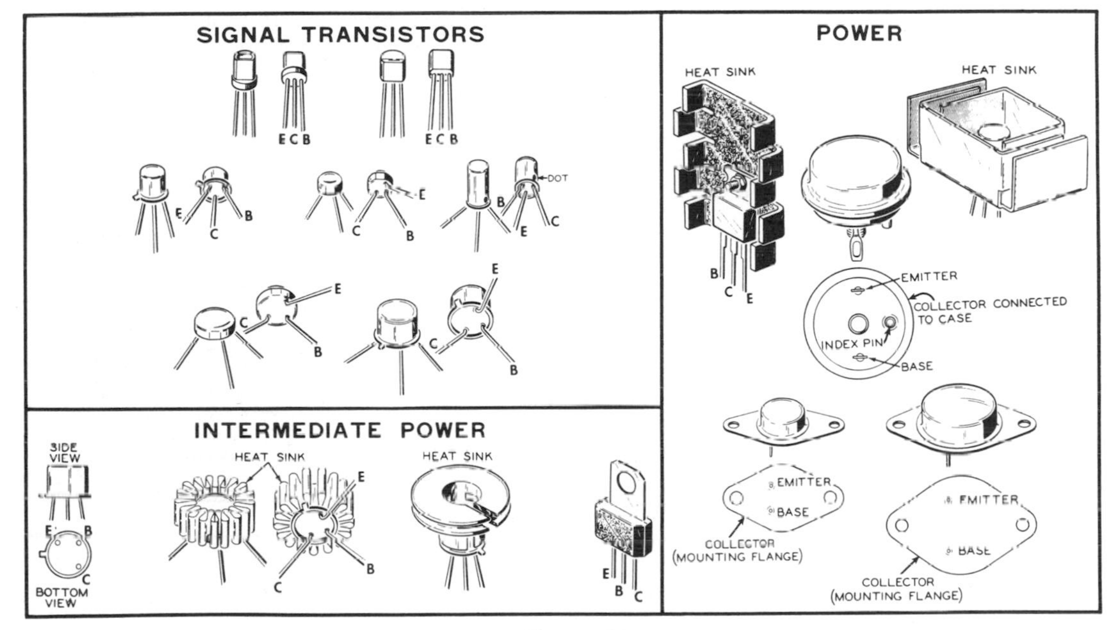

Figure 18

For signal transistors, start the test with a base step current of 2 μA/step, an R limit of 5 kΩ and a sweep range of 0-40 V. If you are using a brand X curve tracer, the operation manual will tell you what to use. As you go up in power class, move everything except sweep voltage by a factor of 10. Keep the sweep voltage low for the first part of the tests.

Assume that you have a transistor you know nothing about. By its size, you can usually tell its power rating. If you are in doubt, place it in a lower rating and start the tests.

We know that an NPN transistor requires a positive collector voltage; therefore, the sweep waveform for NPN transistors must be positive-going. We also know that a positive voltage will cause scope deflection upward and to the right. Thus, for an NPN transistor, the presentation will be like the one in Figure 19A. Since a PNP transistor requires a negative collector voltage, the display will be inverted as shown in Figure 19B.

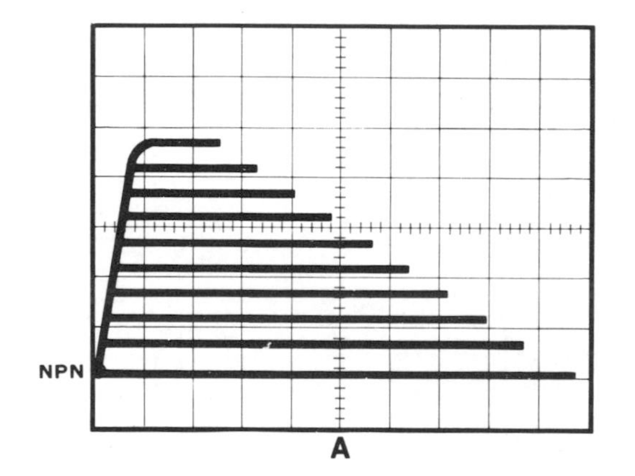

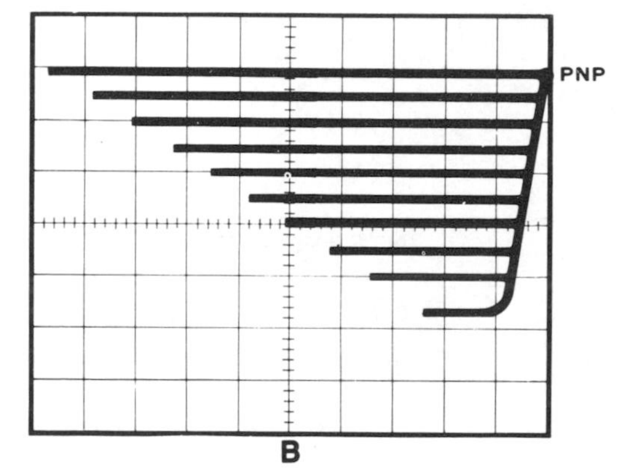

Figure 19

So, if you don't know what type of transistor is being tested, start the measurement in the center of the CRT instead of at a corner and set everything else up as if it were NPN. If it is an NPN, a presentation like the one in Figure 20A will appear as you increase the sweep voltage. If it is not an NPN, a straight line will appear as in Figure 20B. If that happens, switch the controls to PNP, and the presentation will look like the one in Figure 20C. When you know what type of transistor you have, continue with the test.

BETA

There are two types of beta that can be determined with the curve tracer. They are AC beta and DC beta. DC beta (h_{FE}) is the ratio of collector current to base current at a given operating point and is equal to current gain. The actual value of h_{FE} depends on what value of I_C and V_C are selected. Therefore, beta can vary within a given device. A group of I_C/I_B curves is shown in Figure 21A. To find DC beta, pick a value of I_C and V_C that intersect on one of the I_B lines. In Figure 21A, we have picked an I_C of 20 mA and V_C of 5 V. The intersection is on the 0.1 mA I_B line. Thus, 0.1 mA of base bias will cause 20 mA of current collector with a collector voltage of 5 V. Note that this ratio of I_C/I_B is true for a wide range of collector voltage. You can now find DC beta by the formula:

$$\beta = \frac{I_c}{I_B}$$

$$= \frac{20 \text{ mA}}{.1 \text{ mA}}$$

$$= 200$$

AC beta (h_{fe}) is the ratio of a **change** in collector current (ΔI_c) to a **change** in base current (ΔI_B). This is a more useful measurement as it is closer to the actual operating conditions.

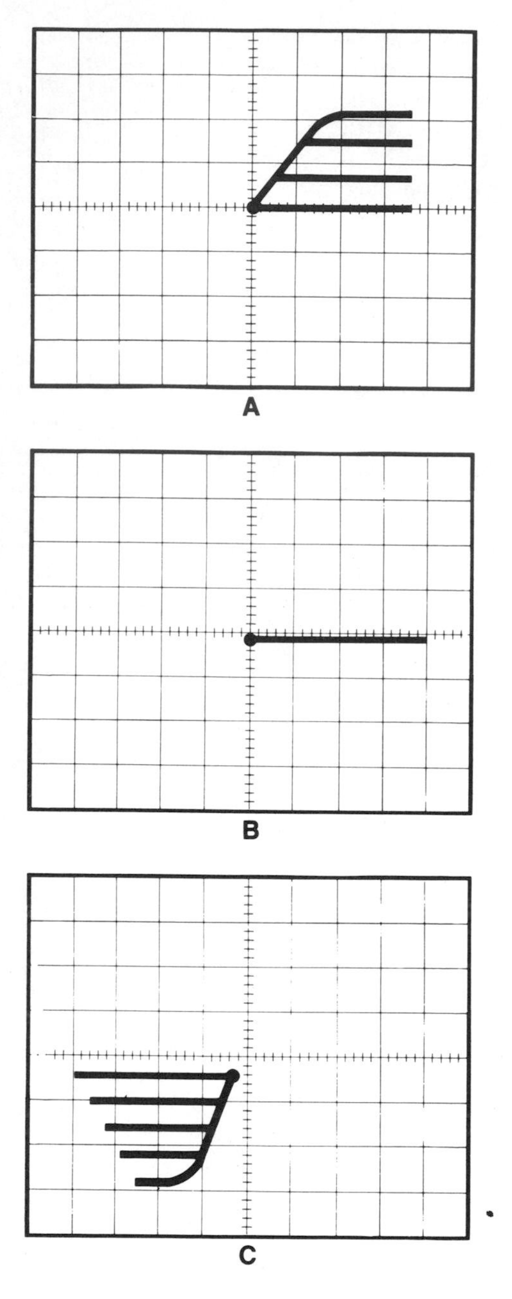

Figure 20

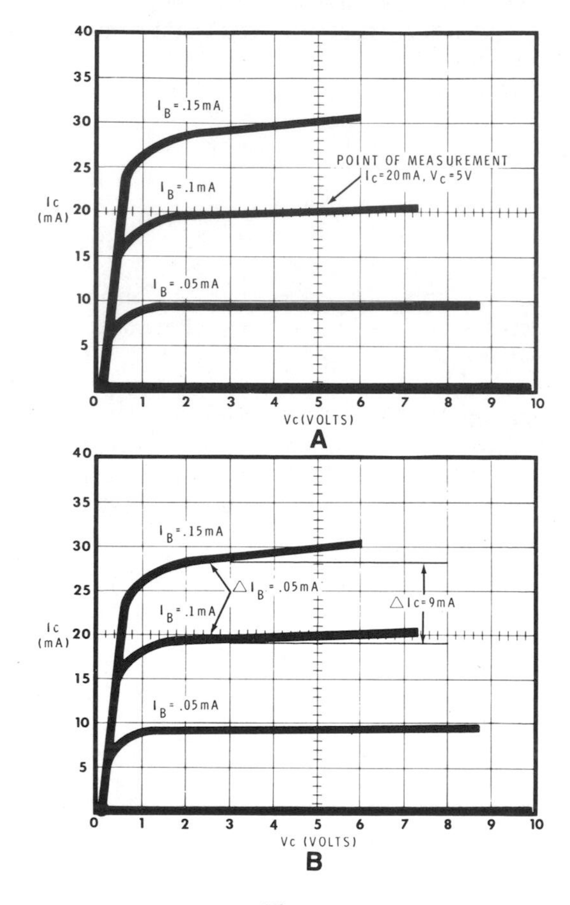

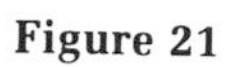

Figure 21

To find h_{fe}, adjust the **STEP RANGE** for evenly and widely spaced curves on the CRT. Take the difference between two adjacent curves or steps as shown in Figure 21B. Then find the change in I_c caused by the change in I_B. For example, in Figure 21B, if V_c is 4 volts and I_B is 0.1 mA, then I_c is 20 mA. When I_B is increased to 0.15 mA, I_c increases to 20 mA. Thus, for a ΔI_B of 0.05 mA, we have a ΔI_C of 9 mA. The AC beta can be found by:

$$\beta = \frac{\Delta I_c}{\Delta I_B}$$

$$\beta = \frac{9 \text{ mA}}{.05 \text{ mA}}$$

$$\beta - 180$$

LINEARITY

AC beta should be higher at the normal operating region of the transistor and lower above and below that region. Therefore, the normal operating region can usually be found by locating the point of highest AC beta.

Linearity is a measure of a transistor's ability to amplify, without distortion, a signal that appears on its base. The linearity of a device can be determined from the same family of curves used to determine beta. If the device is linear, a change in base current should produce a proportional change in collector current. Therefore, the spacing between the curves should be equal. If not, the device is nonlinear.

Linearity should be checked along a load line instead of at a specific V_{CE} because V_{CE} is not constant but varies with I_C in actual operating conditions. To measure linearity from the I_C I_B, first, plot a test load line. To plot a load line, two things must be known — V_{CC} and R_L. With these conditions known, two points can be located on the curves. The first, V_{CE}, is equal to V_{CC} when bias is zero and no current flows. In Figure 22, this is 10 V. The second point, maximum I_C, can be found by the formula

$$I_C = \frac{V_{CC}}{R_L}$$

If an R_L of 125 Ω is used, the maximum I_C will be 80 mA. Since this value is not shown on the curve tracer, its position must be extrapolated by mentally extending the I_C scale. A line connecting these two points is the "Test Load Line".

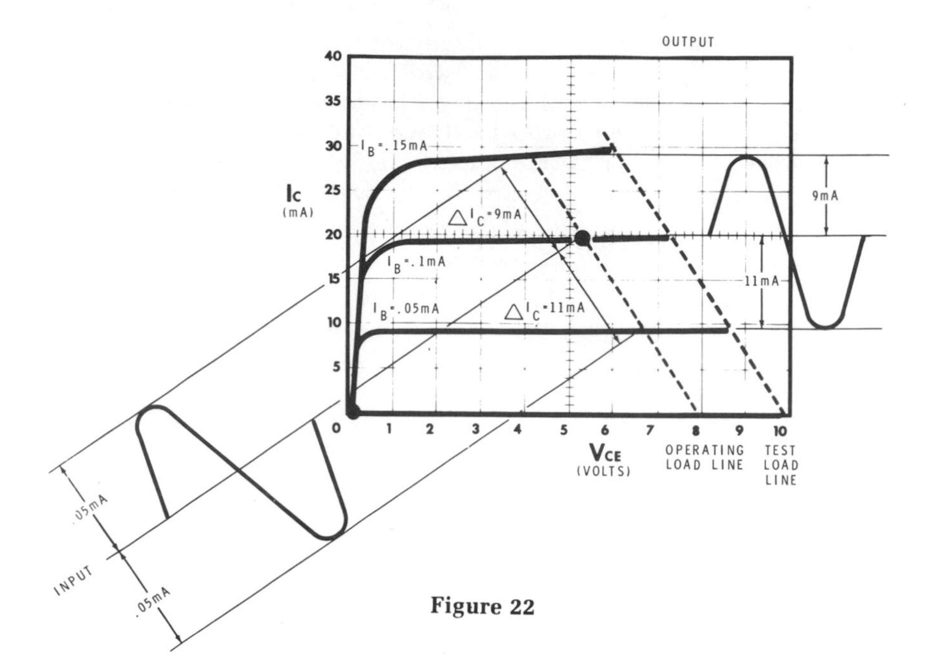

Figure 22

Plotting the load line is even easier with the curve tracer. The ends of the curves indicate the maximum V_{CE} that can be obtained with the selected V_{CC} and the various values of I_B. Therefore, the test load line can be constructed by merely connecting the ends of the curves as shown in Figure 22. Next, locate the desired V_{CE} on the zero I_C line. From that point, draw an operating load line parallel with the test load line.

Now, pick a base current for reference, in this case .1 mA. Increase and decrease the base current from this reference, and see what happens to collector current. In Figure 22, when I_B increases from 0.1 mA to 0.15 mA, I_c increases from 20 mA to 29 MA. Thus, a change (ΔI_B) of .05 mA causes a change in collector current (ΔI_c) of 9 mA. (The symbol Δ means "a change in" and will be used in this discussion.) When the input swings in the other direction from 0.1 mA to 0.05 mA, the collector current decreases from 20 mA to 9 mA. In this case, a ΔI_B of 0.05 mA causes a ΔI_c of 11 mA. Obviously, this device has some nonlinearity because a 9 mA change does not equal an 11 mA change.

Any time nonlinearity exists, the output will be distorted or different from the input. This may or may not be undesirable depending on the application. In any case, the curve tracer can show potential distortion problems.

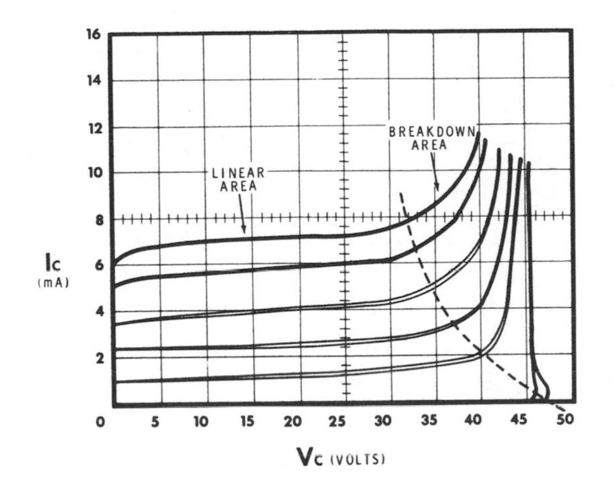

Figure 23

BREAKDOWN VOLTAGE

In the normal operating range of a transistor, collector current is affected very little by collector voltage. However, a point will be reached where collector current becomes independent of base current and rises sharply until limited by external resistance. This breakdown area is shown in Figure 23. Keep this test short, as the high collector current generates a lot of heat and the transistor could be destroyed.

In addition to the collector-emitter breakdown just discussed, there is the possibility of base-emitter and base-collector breakdown. These tests are made by treating the respective junctions as diodes. Diode testing is covered a little later in this unit.

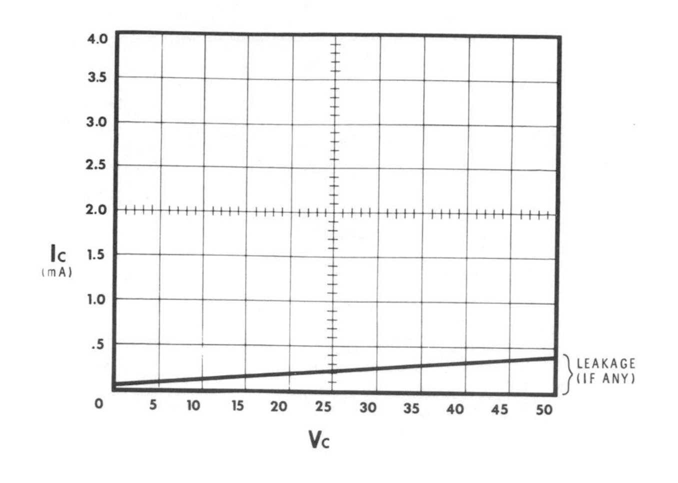

Figure 24

LEAKAGE CURRENT

Two types of leakage current can be measured with the curve tracer. They are I_{ceo} and I_{ces}. I_{ceo} is the collector to emitter current with the base open or disconnected. This test is performed just like the beta, except the base lead is left disconnected. If any significant leakage occurs, it will appear as in Figure 24.

I_{ces} is the collector to emitter current that flows when the base is shorted to the emitter. The test is performed the same as before except that both base and emitter leads are connected to the emitter terminal. Once more, leakage will appear as in Figure 24.

In most good transistors, leakage will be only a few nanoamperes; and probably will not even be visible on the screen. Therefore, any detectable leakage probably indicates a bad transistor.

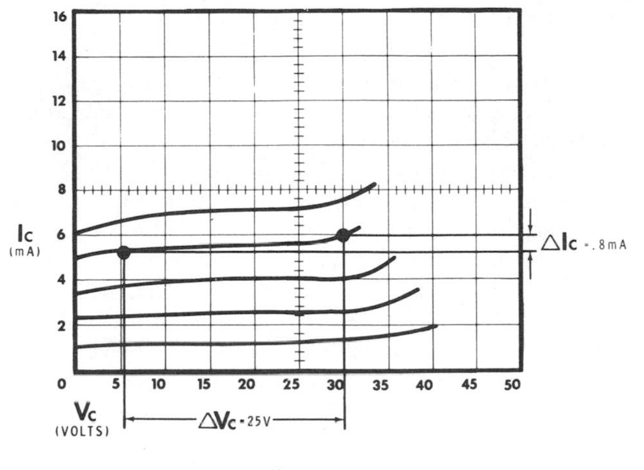

Figure 25

ADMITTANCE

The output admittance (h_{OE}) of a transistor is the change in collector current that results from a change in collector voltage with the base current held constant. Admittance is measured from a set of I_C I_B curves as shown in Figure 25. Any one of the base current curves can be selected for the test. In this case, we are measuring the change in I_C with a change in V_C from 5 V to 30 V or a $\Delta V_C = 25$ V. This causes a ΔI_C of 0.8 mA. Admittance can then be calculated by:

$$h_{oe} = \frac{\Delta I_c}{\Delta V_c}$$

$$= \frac{.8 \text{ mA}}{25 \text{ V}}$$

$$= 32\ \mu\text{mhos}$$

Output admittance is measured in micromhos and is the reciprocal impedance. Therefore, the output impedance can be found by:

$$Z_{out} = \frac{\Delta V_c}{\Delta I_c}$$

$$= \frac{25}{.8 \text{ mA}}$$

$$= 31{,}250\ \Omega$$

SATURATION VOLTAGE

Collector saturation in a transistor is the point where an increase in collector voltage no longer causes an increase in collector current. Knowing the saturation voltage is necessary if the transistor is to be used as a linear amplifier or as a switch. In digital circuits, the conduction voltage of the transistor must often be held below a given amount. For instance, in TTL logic circuits, the conducting or saturation voltage must be less than 0.4V. You can see in Figure 26A that it is less than 0.4 for any given amount of I_c. However, for switching situations it is usually more desireable to use a high value of I_B and drive the transistor to saturation as quickly as possible. Figure 26B shows what happens when I_B is increased to 1 mA. Notice that saturation is reached almost instantaneously and is limited by the external circuit, not by I_B. Here the saturation voltage is only about .04V for any value of I_c.

Saturation resistance can be calculated by:

$$r_{ce(sat)} = \frac{V_c}{I_c}$$

$$= \frac{.04}{10 \text{ mA}}$$

$$= 4\ \Omega$$

This resistance can be calculated for any given value of I_c.

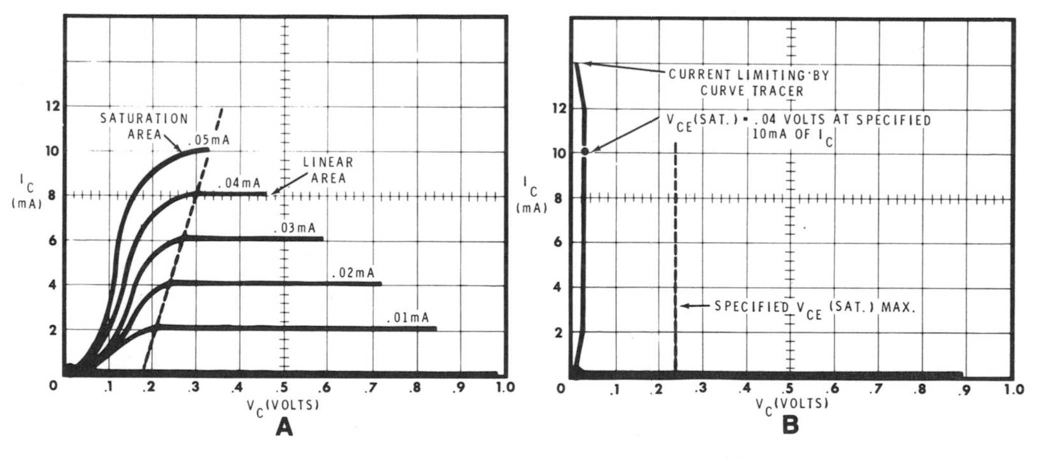

Figure 26

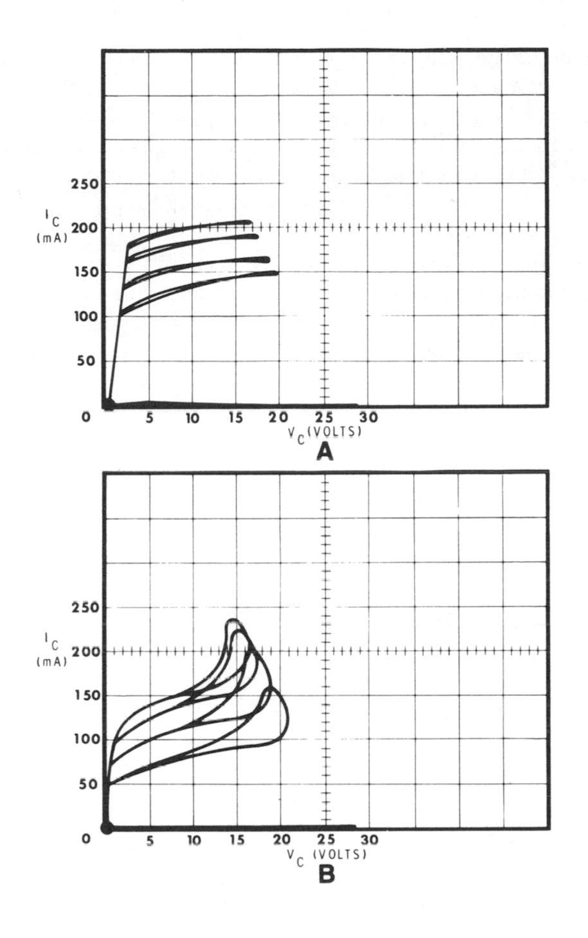

Figure 27

THERMAL EFFECTS

Transistors are greatly affected by heat. This shows up on the curve tracer in the form of loops. Loops are produced when a device heats at a different rate than it cools, thus, causing I_c to increase at a different rate than it decreases. Figure 27A shows how the display will appear with a moderate amount of heat.

Small amounts of looping like those shown here can be caused by a number of factors, such as collector capacitance and inductance, or, in certain cases, by the curve tracer itself. However, severe looping like that shown in Figure 27B indicates excessive heat and the approach of thermal runaway. If this is allowed to continue, the curves, or loops, will roll toward the top of the screen until the device is destroyed.

When conducting the thermal test, apply heat with your finger or, if necessary, a soldering iron and observe the effects. Remove the heat as the loops approach the maximum allowable I_c. This test should not be conducted unless it is necessary, as there is always the danger of destroying the transistor.

IN-CIRCUIT TESTS

Many of the tests discussed here can be made with the transistor in the circuit. Of course, you will be testing the circuit, not just the transistor. Often, only a comparison will be possible, not an actual test.

MATCHING TRANSISTORS

Frequently a "matched pair" of transistors is required. Most curve tracers have provisions for plugging in two devices at the same time so that curves can be compared. If the curves are identical, the devices are well matched.

Diode Testing

Although the curve tracer is primarily designed to test transistors and other 3-element devices, it is capable of showing the characteristics of 2-element devices. There are two basic tests to be conducted on the PN junction diode; forward conduction and reverse breakdown.

FORWARD CONDUCTION

The **forward conduction** test is made by connecting the diode into the transistor socket with the anode to the collector terminal and the cathode to the emitter terminal. A positive-going sweep is then selected. Since no connection is made to the base terminal, the step generator is not used.

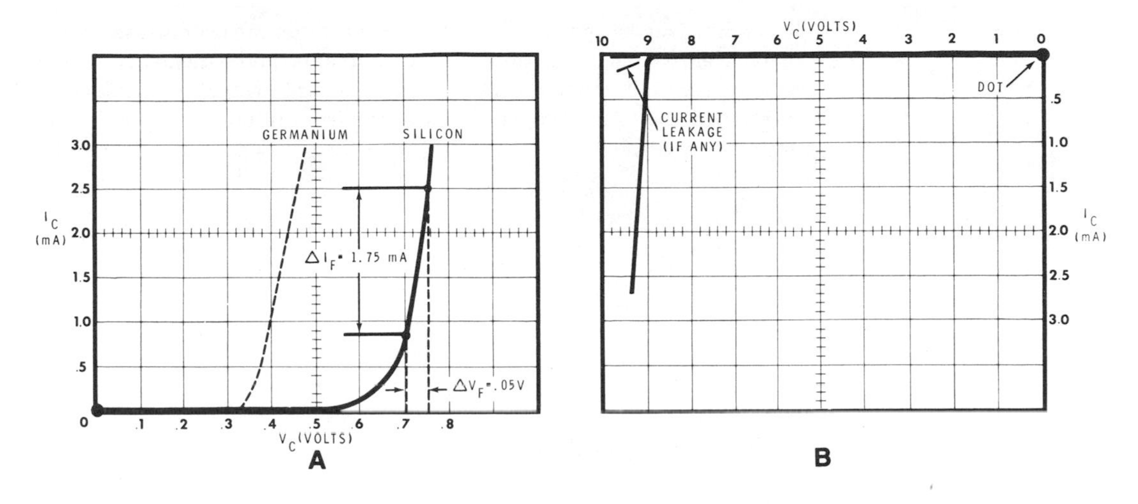

Figure 28

The resulting curve, and there is only one, looks like the one in Figure 28A. Notice that no current flows until the forward conducting potential of the diode is reached. For a silicon diode, this is about 0.7V, and for a germanium diode, about 0.3V. Once the conduction potential is reached, current increases rapidly with voltage.

The conducting or dynamic resistance of the diode can be found from the curves. Select a change in forward current that is on the linear portion of the curve, ΔI_F in Figure 28A. Then, find the change in voltage that caused the change, ΔV_F. Dynamic resistance (R_D) can then be found by:

$$R_D = \frac{\Delta V_F}{\Delta I_F}$$

$$= \frac{.05}{1.75 \text{ mA}}$$

$$= 28.57 \ \Omega$$

REVERSE BREAKDOWN

The **reverse breakdown** test measures the potential where current will begin to flow in a reverse direction, disregarding leakage current. Some diodes, especially germaniums, may show a small amount of leakage.

To measure the reverse breakdown potential, connect the diode the same as for forward conduction, but select a negative sweep. As sweep voltage increases, no current will flow until breakdown is reached. Once it is reached, current increases rapidly with any increase in voltage as shown in Figure 28B. To avoid destruction of the diode, be sure you use an adequate limiting resistance in series.

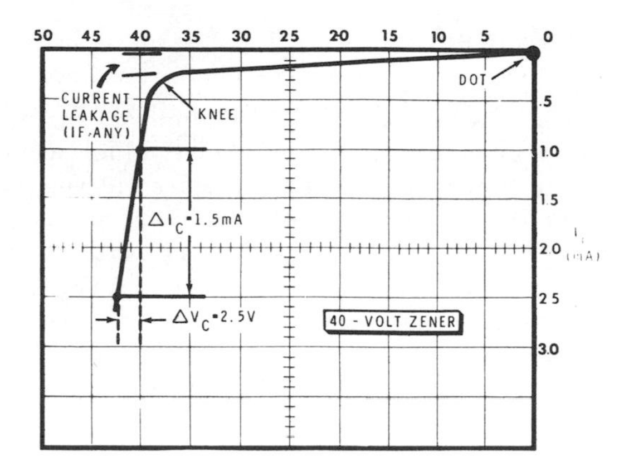

Figure 29

Zener Diodes

The curve tracer is an ideal instrument for checking zener diodes. Zener diodes are similar to regular PN junction diodes except that the reverse current is carefully controlled. Three characteristics of zener diodes are easily measured; leakage current, zener or breakdown voltage, and dynamic impedance.

Breakdown voltage is measured at the point where the current just passes the knee and enters the linear region as shown in Figure 29. Leakage current is measured just before the knee is reached. The dynamic resistance is found for PN junction diodes:

$$Z = \frac{\Delta V_c}{\Delta I_c}$$

$$= \frac{2.5 \text{ V}}{1.5 \text{ mA}}$$

$$= 1667\ \Omega$$

A lower dynamic impedance means a better zener. Obviously the breakdown voltage should be as specified, and leakage should be minimum.

Field-Effect Transistors

There is a lot of similarity between the FET and bipolar transistor: however, there is also a lot of difference. The primary difference is in the biasing arrangement. A transistor with zero base bias will have no collector current, but an FET with zero gate bias will have maximum drain current. When the bias on a bipolar transistor is increased, collector current increases; however, when bias is increased on an FET, the drain current decreases. So, the polarity of the step voltage on the curve tracer must be reversed in relation to the sweep voltage. Because of this, the curves are "backward" with zero gate volts (V_G) at the top of the presentation as shown in Figure 30A. Notice that, as V_G increases, drain current (I_D) decreases.

If sufficient bias is applied, drain current will stop completely and "pinch-off" will be reached. The value of pinch-off bias can be determined on the curve tracer by observing which step causes I_D to equal zero as shown in Figure 30B.

TRANSCONDUCTANCE

The gain or transconductance (Gm) of the FET can also be found by the curve tracer. This is the ratio of **change** in drain current to the **change** in gate voltage at a given drain voltage. It may be calculated from a set of V_D I_D curves as follows:

Refer to Figure 30B and note the difference in ΔI_D between the two adjacent curves at the same drain voltage (V_D). Next, find the difference in V_G between the two curves. Transconductance can then be calculated by:

$$Gm = \frac{\Delta I_D}{\Delta V_G}$$

$$= \frac{1.1 \text{ mA}}{.5 \text{ V}}$$

$$= 2200 \text{ micromhos}$$

Because the curves are nonlinear, Gm will vary depending on the value of V_D and the values of V_G selected.

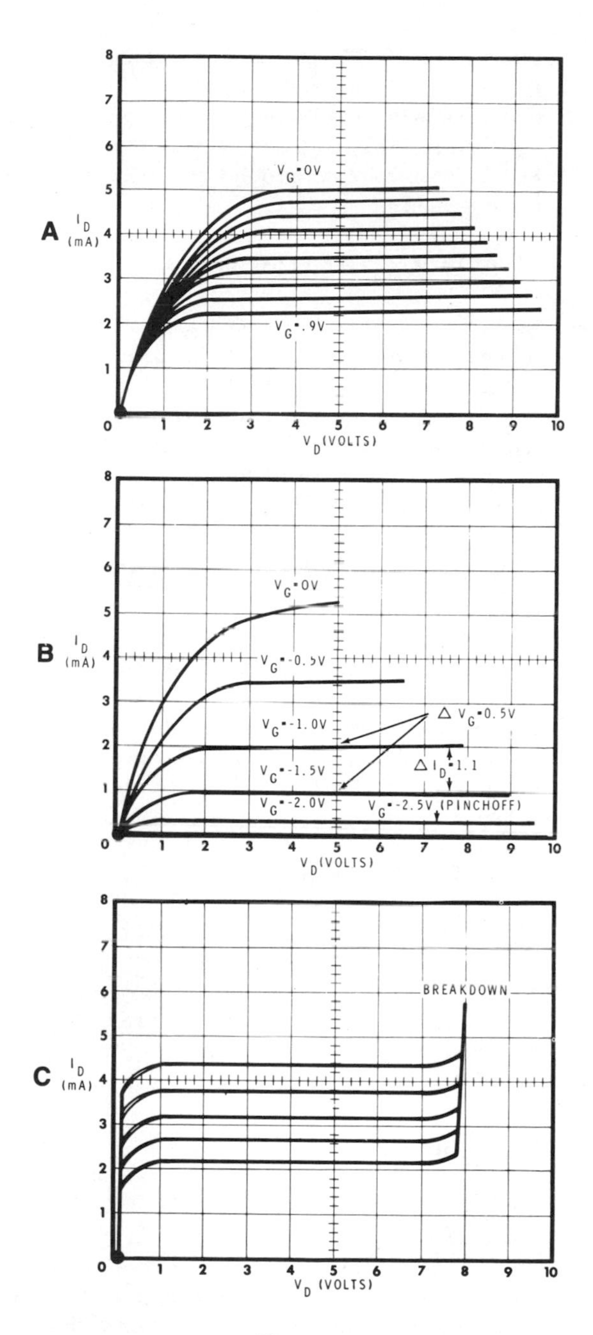
A
B
C
I_D (mA)
V_D (VOLTS)
V_G = 0V
V_G = .9V
V_G = -0.5V
V_G = -1.0V
V_G = -1.5V
V_G = -2.0V
V_G = -2.5V (PINCHOFF)
ΔV_G = 0.5V
ΔI_D = 1.1
BREAKDOWN

Figure 30

When testing FET's remember that some FET's, particularly MOSFET's can be damaged by the static electricity on your body. Therefore, be sure to discharge this electricity by touching ground before and while handling the MOSFET. Another point to remember when testing dual-gate MOSFET's is to either ground or bias the gate not being tested. Do not leave it open.

BREAKDOWN POTENTIAL

As with all devices, FET's have a breakdown potential whereas sweep voltage is increased further, drain current becomes independent of gate voltage and rises sharply until limited by the external circuit. The breakdown point is shown in Figure 30C. The curve tracer during this test should limit current to an amount that will not harm the device under test. Even so, this test should be made very quickly because the FET is more easily damaged by high voltage than is the bipolar transistor.

Silicon Controlled Rectifier (SCR)

A silicon controlled rectifier is basically a diode with a built-in switch or gate. It will not conduct in the forward direction unless the gate is turned on or the blocking voltage is exceeded. However, once this happens, the SCR will continue to conduct until the cathode-to-anode current drops below the holding current of the device. There are three tests to be made on the SCR. They are **forward blocking voltage, holding current** and **gate trigger current**.

BLOCKING VOLTAGE

Figure 31 shows the presentation for measuring blocking voltage. Notice that no current flows until the blocking voltage is exceeded, in this case, about 200 V. However, as soon as this occurs, there is a surge of current, and voltage drops rapidly to the forward voltage drop of the SCR. Current then varies with voltage until the current drops below the holding current. This figure shows a holding current of 5 mA.

GATE TRIGGER CURRENT

The test for gate trigger current is shown in Figure 32. The step generator is connected to the gate. As step current is increased, no current will flow until the trigger level is reached. However, at the trigger level, current will be controlled by anode-cathode potential. If the trigger generator is adjusted for two steps and the step level varied, the exact trigger current can be found.

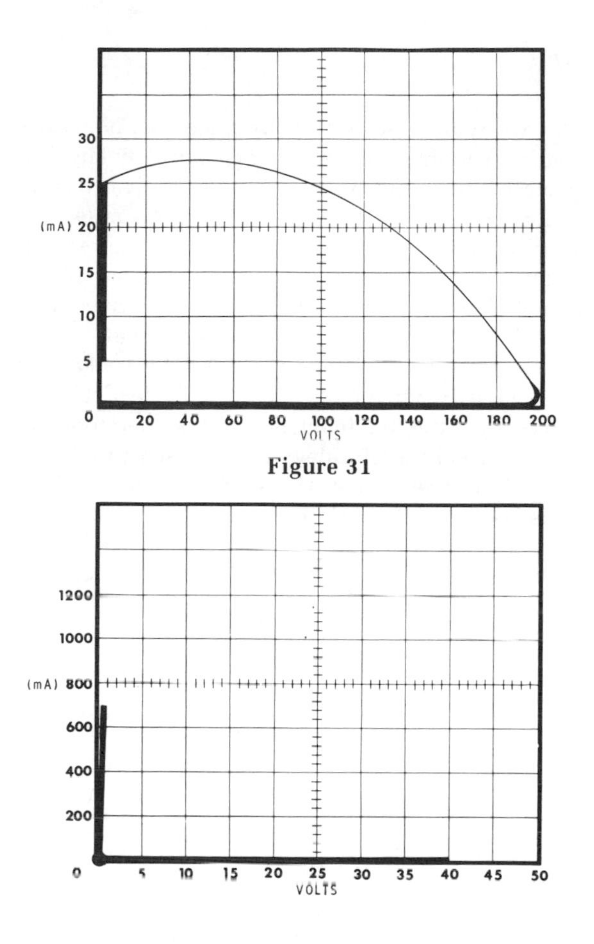

Figure 31

Figure 32

In addition to the tests just mentioned, the SCR can be tested for reverse breakdown voltage and leakage current. For this test, the gate is left open, and the test is conducted exactly as it was for the standard PN junction.

TRIAC

The triac is nothing more than dual SCR's arranged such that they conduct in different directions. The tests are the same except that each test must be performed twice, once in each direction, and no reverse breakdown test or leakage test can be performed.

Summary

You have seen how the transistor curve tracer can be used to perform dynamic tests on a number of components. Virtually any device can be tested, and curve tester applications are limited only by the imagination of the operator. Everything from resistors to integrated circuits will provide a display; however, it is up to you as the operator to interpret the display. Since most curve tracers provide two inputs, it is usually practical to compare a known good device against the suspect device.

A drawback of the curve tracer and the transistor tester is that neither checks the device at high frequency. The sweep rate of the curve tracer is usually derived from the line frequency and is, therefore, only 120 Hz. Thus, the only sure check of a device to be used in a high frequency circuit is to try it in that circuit. However, most of the time when the device checks good, it will work in the circuit.

SPECTRUM ANALYZER

Introduction

There are two ways of observing an electrical waveform. The waveform may be observed in either the time domain or the frequency domain. Different information is obtained from each observation.

If a pulse or square wave is analyzed in the time domain as shown in Figure 33, information can be obtained about rise time, pulse width, pulse recurrence time, or any other information that can be derived from a plot of current or voltage versus time. Information of this type is essential when designing or working with logic circuits.

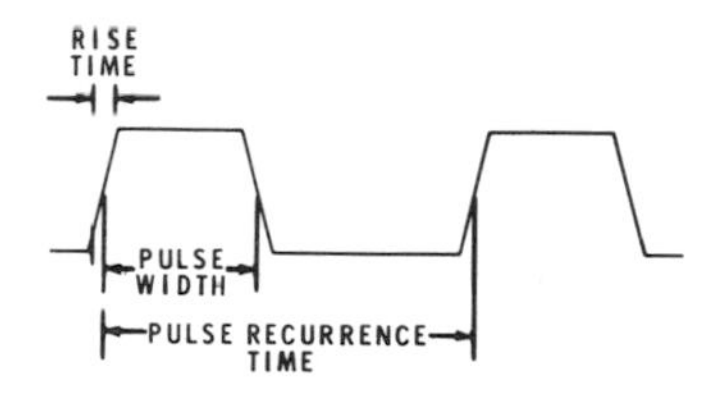

Figure 33

A

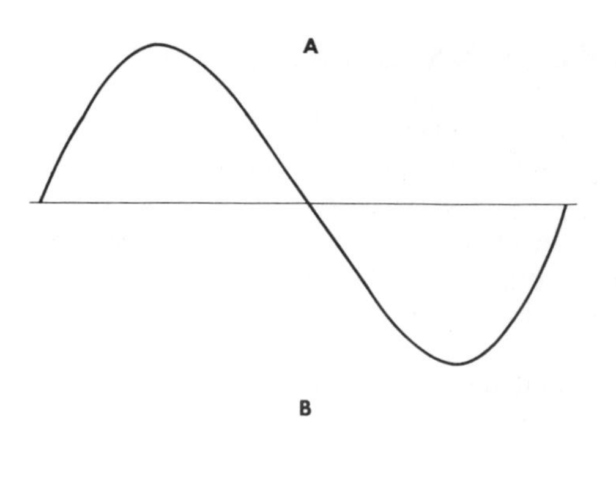

B

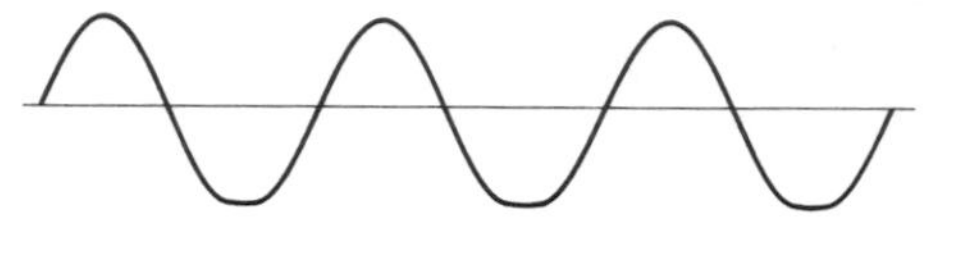

C

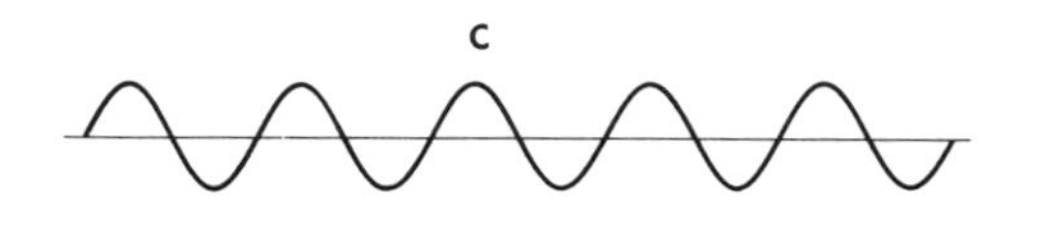

D

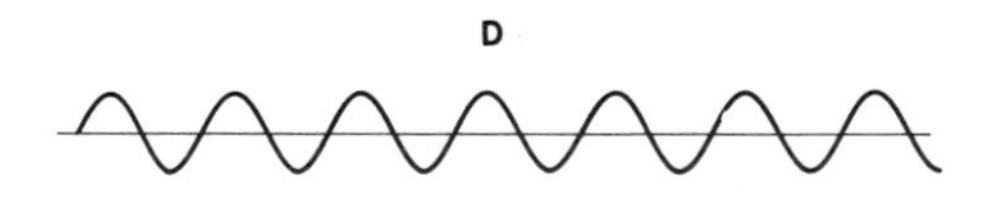

E

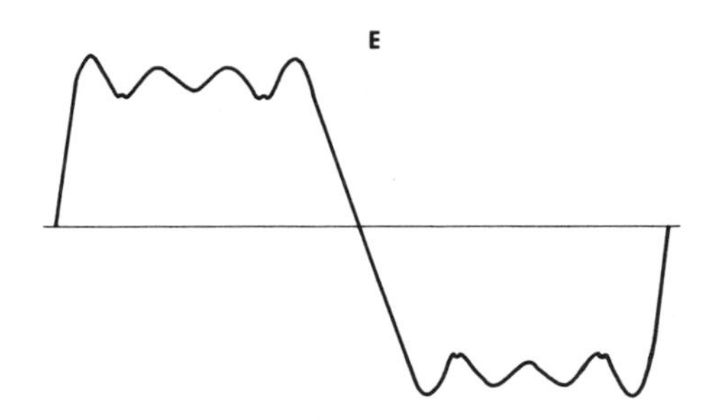

Figure 34

It is also possible to analyze a square wave using the frequency domain. A square wave can be considered to contain a fundamental sine wave plus an infinite number of odd harmonics added in phase. Figure 34A shows one cycle of the fundamental waveform. Figure 34B shows the third order harmonic, which is three times the frequency and one third the amplitude of the fundamental. Figure 34C is the fifth harmonic and Figure 34D is the seventh harmonic. If all the instantaneous values are added, the resultant waveform appears as shown in Figure 34E. It can readily be seen that this waveform is taking on the shape of a square wave. The more harmonics that are added, the more nearly perfect the square wave becomes.

If you observe a square wave displayed on an oscilloscope, you will see the result of all the odd harmonics added in phase. Therefore, oscilloscopes will not show the individual frequencies or amplitudes. To separate the various harmonics, the square wave must be studied in the frequency domain.

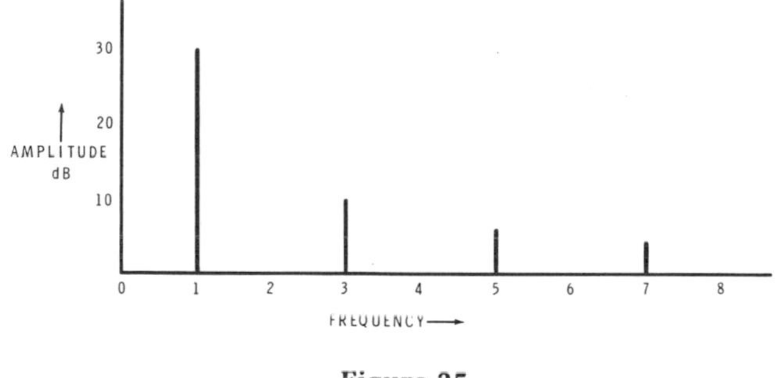

Figure 35

Figure 35 shows a frequency domain representation of the square wave we have been observing. The amplitudes and frequency relationships shown here are the same as those in Figure 34. The fundamental frequency, or first harmonic, is 30 units in amplitude. It could be any frequency, but for purposes of explanation, assume 1 MHz.

If the fundamental is 1 MHz, the third harmonic is 3 MHz and so on through 7 MHz. Thus, while it is usually possible to detect the presence of harmonics using time domain measurements, only frequency domain measurements allow the frequency and amplitude to be measured.

Square waves are not the only waves that contain harmonics. In fact, all waveforms, excepting pure sine waves, contain some harmonics or other "distortion" factors. In radio transmitters, harmonics are present along with modulation sidebands. To design and maintain this equipment, it is essential that the engineer or technician be able to accurately detect and measure these factors. The spectrum analyzer was designed for this type of measurement. It indicates the voltage or power at each frequency.

Types

There are two basic types of spectrum analyzers, swept-tuned and real-time. The swept-tuned analyzer starts at one end of the band and "sweeps" to the other end. Thus, each frequency in the band is sampled in turn. Because it looks at only one frequency at a time, the swept-tuned analyzer may not detect transient signals.

On the other hand, the real-time analyzer looks at all frequencies within the band at the same time. Therefore, real-time analyzers are capable of displaying transient signals.

REAL-TIME

A block diagram of the real-time analyzer is shown in Figure 36A. This analyzer consists of a large number of channels, each containing a band pass filter and a detector. Each channel overlaps the adjacent channel giving continuous coverage over the entire bandwidth of the analyzer. Thus, the bandwidth, or range, of the analyzer depends on the number of filters, while the resolution depends upon the selectivity of each individual filter as shown in Figure 36B.

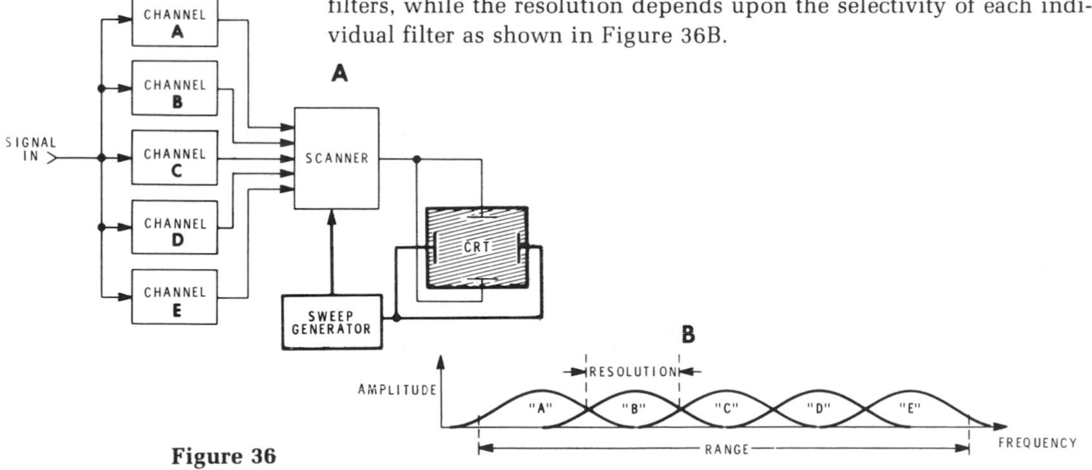

Figure 36

The output from each channel is fed to an electronic scanner that switches between the channels at a very high rate. The output from the scanner is then fed to the vertical plates of a CRT. The scanner is driven by a sweep generator that also drives the horizontal plates of the CRT. Thus, the channel outputs are synchronized to the sweep on the CRT.

For the real-time analyzer to cover a wide range of frequencies, a large number of channels is required. This makes it very expensive. This type of analyzer also lacks flexibility as a fixed frequency range is covered. It finds its major application in the audio and low frequency ranges.

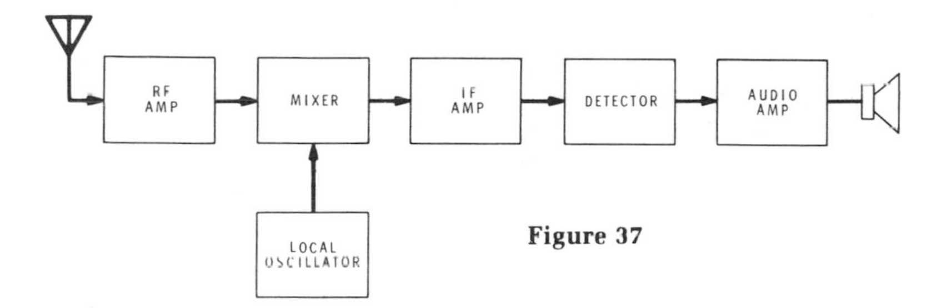

Figure 37

SWEPT — TUNED

To explain the operation of the swept-tuned spectrum analyzer, we will review the operation of a simple superheterodyne receiver as shown in Figure 37. A number of signals may be present at the antenna. These are amplified by the RF amp and fed to the mixer.

A signal from the local oscillator is also fed to the mixer. The local oscillator signal is variable in frequency. When the RF and local oscillator signals mix in the mixer, the result will be the two original frequencies, the sum, and the difference.

The IF amplifier is selective and passes only a relatively narrow band of frequencies. When the results of the mixing action fall within the bandwidth of the IF amplifier, a signal is passed to the detector.

The detector removes the audio from the IF. This audio is then amplified and sent to the speaker.

As the local oscillator is tuned, different RF inputs will cause an output at the speaker. The output from the speaker will depend upon the frequency and power of the received signal.

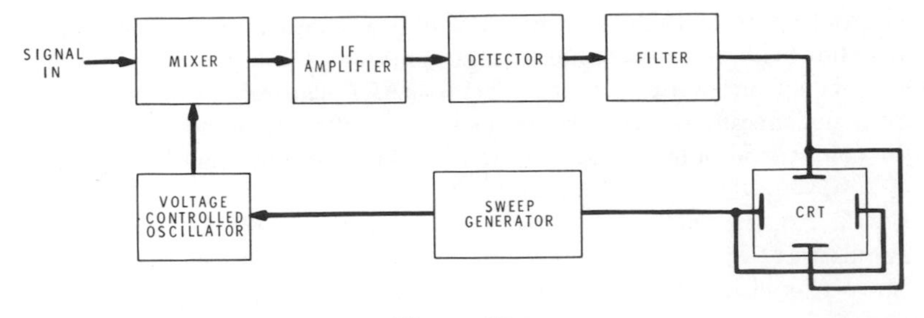

Figure 38

The superheterodyne spectrum analyzer is very similar as shown in Figure 38. The unknown frequency is connected to the mixer input. When the frequency difference between the input and the local oscillator is equal to the IF, a response will be seen on the CRT. If the local oscillator is at a higher frequency than the input, then

$$f_{in} = f_{lo} - f_{IF}$$

where

f_{in} = input frequency

f_{lo} = local oscillator frequency

f_{IF} = intermediate frequency

For example, if the IF is 50 MHz, and the local oscillator can be tuned from 50 to 100 MHz, the analyzer has a range of 0 to 50 MHz. A 20 MHz signal at the input will mix with a 70 MHz signal from the local oscillator, and a difference frequency of 50 MHz will result. Thus, an indication will be seen on the CRT.

In the spectrum analyzer, the local oscillator is controlled by a sweep generator. The same sweep generator produces a sweep on the CRT. Thus, when the sweep is at the extreme left side of the CRT, the voltage controlled oscillator (VCO) is at its lowest frequency. As the sweep develops, the frequency from the VCO increases at the same rate that the trace moves across the CRT. Therefore, the position of the vertical deflection is proportional to the frequency of the input. Obviously, the amplitude of the deflection represents the amplitude or power of the input. There are other types of swept-tuned spectrum analyzers other than the superheterodyne; however, the superheterodyne is much more flexible and is, by far, the most common type in use today.

Requirements

If a spectrum analyzer is to be useful, it must be capable of accurate frequency and amplitude measurements. To have this capability, the analyzer must meet the following requirements:

1. Flat response across the entire frequency band.
2. Adequate tuning range.
3. Adequate resolution.
4. High sensitivity.
5. Amplitude and frequency calibration.

Resolution

The resolution of the analyzer is determined by the bandwidth of the IF filters. This bandwidth is usually considered to be the 3dB bandwidth of the filters.

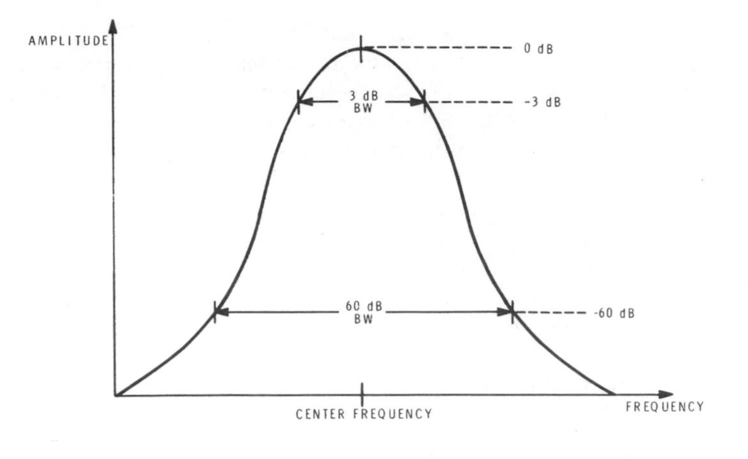

Figure 39

Figure 39 shows the shape of a typical filter response. All filters have a **shape factor**. The shape factor is usually a ratio of the 60 dB bandwidth to the 3 dB bandwidth; the smaller the number, the more narrow the bandwidth. For example, if the IF bandwidth is 1 kHz, at the 3 dB points, then the analyzer can resolve two signals of equal amplitude that are 1 kHz or more apart. However, if the signals are not of equal amplitude, they must differ in frequency by more than the IF bandwidth. How much they must differ depends on the shape factor. If the curve shown in Figure 39 represents a filter with a shape factor of 10:1, the two signals with amplitudes differing by 60 dB must have a frequency difference of 5 times the IF bandwidth before they can be resolved. The filter shape factor may also be specified as the ratio of the 60 dB bandwidth to the 6 dB bandwidth, or the ratio of the 40 dB bandwidth to the 3 dB bandwidth. This makes it somewhat difficult to compare filters. For example, a filter might have a 20:1 shape factor when the ratio of 60 dB to 3 dB is used. The same filter would have a shape factor of approximately 10:1 if the 60 dB to 6 dB ratio were used. Therefore, shape factor is meaningful only when the dB bandwidths are the same for the filters being compared.

Since the resolution of the analyzer depends on the IF bandwidth, it would appear that by decreasing bandwidth, resolution would be increased. This is true up to a point. However, the actual resolution and bandwidth is limited by the stability of the analyzer.

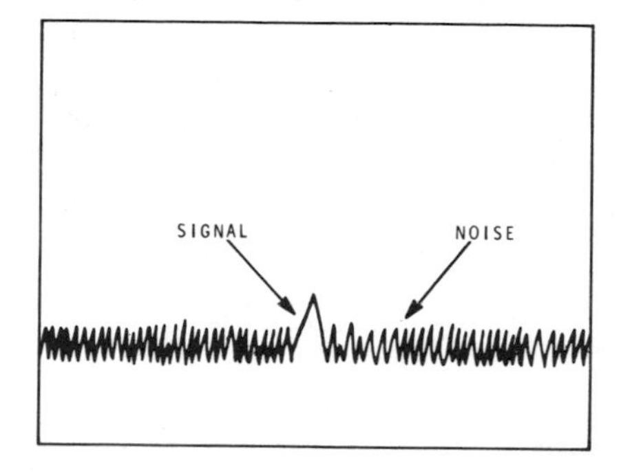

Figure 41

Sensitivity

The sensitivity of the analyzer is a measure of the minimum signal that the analyzer can detect. The main limiting factor on sensitivity is the internal generated noise of the analyzer. The analyzer always displays the signal plus the noise as shown in Figure 41. Here, the signal level is equal to the noise level and is superimposed on the noise. Therefore, the signal will appear 3dB above the noise. This 3dB difference is the minimum that is consistently discernable. Therefore, the sensitivity could be expressed as the point where:

$$\frac{S + N}{N} = 2$$

WHERE:

S = signal power

N = average noise

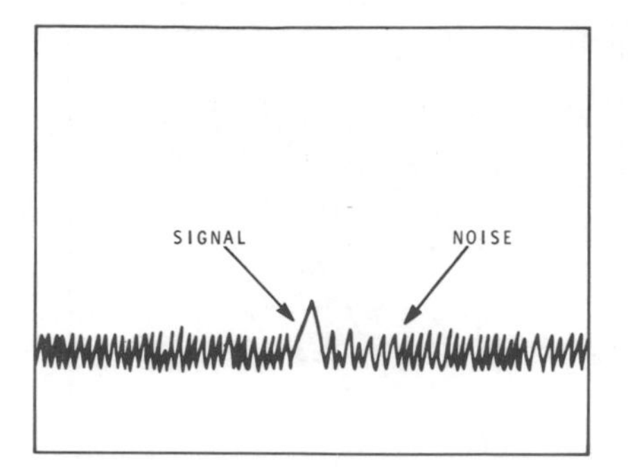

Figure 41

Sensitivity

The sensitivity of the analyzer is a measure of the minimum signal that the analyzer can detect. The main limiting factor on sensitivity is the internal generated noise of the analyzer. The analyzer always displays the signal plus the noise as shown in Figure 41. Here, the signal level is equal to the noise level and is superimposed on the noise. Therefore, the signal will appear 3dB above the noise. This 3dB difference is the minimum that is consistently discernable. Therefore, the sensitivity could be expressed as the point where:

$$\frac{S + N}{N} = 2$$

WHERE:

S = signal power

N = average noise

Frequency Response

Since the major purpose of the spectrum analyzer is to display signal levels at different frequencies, the response should be as flat as possible across the entire frequency range. The necessary flatness can be very difficult to achieve, since it must apply to very large as well as very small signals. One of the major causes of nonlinearity is conversion loss in the mixer. If the local oscillator voltage is too large compared to the input signal, this conversion loss will vary with frequency. It is possible, therefore, that an analyzer could have a flat response with large input signals but not with weak inputs.

Applications

The spectrum analyzer is an excellent instrument for any application that requires the display of power or voltage versus frequency. One such application is measuring modulation.

There are several types of modulation. However, they all involve varying the amplitude, frequency or phase of the carrier. This variation generates additional frequencies called sidebands. The spectrum analyzer can measure these frequency components easily.

AMPLITUDE MODULATION

In amplitude modulation, the carrrier amplitude is varied by the modulating signal. During the modulation process, two additional frequency components are generated. These are the sum and difference of the carrier and the modulation frequencies. The sum of the carrier and the modulation frequency is called the **upper sideband** (USB) and the difference between the carrier and the modulation frequency is called the **lower sideband** (LSB).

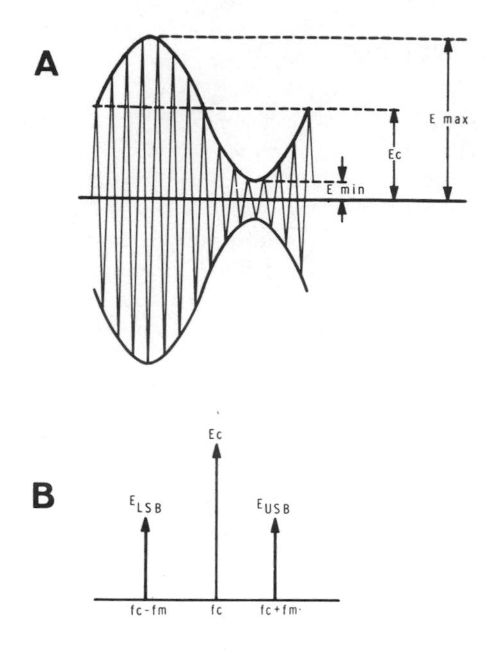

Figure 42

The amount of power in sidebands is determined by the amount or percentage of modulation (m). It is possible to determine the percentage of modulation using either the time domain or the frequency domain. Figure 42A shows how a modulated waveform appears when viewed in the time domain, i.e. on an oscilloscope. The percentage of modulation can be calculated from the time domain presentation by:

$$m = \frac{E_{max} - E_{min}}{E_{max} + E_{min}}$$

When m = 1, or 100% modulation, each sideband amplitude will equal one half of the carrier amplitude. Thus, it is relatively easy to calculate carrier power and sideband power from the time domain. However, it is difficult to determine the various frequency components. Viewing the signal in the frequency domain overcomes these problems. The spectrum analyzer shows both the power and frequency of the sideband components as shown in Figure 42B.

The sideband frequencies can be determined by:

$$USB = f_c + f_m$$

$$LSB = f_c - f_m$$

Where

USB = upper sideband

LSB = lower sideband

f_c = carrier frequency

f_m = modulation frequency

Both carrier and sideband amplitude can be read directly on the spectrum analyzer, and percentage of modulation can be calculated by:

$$m = \frac{2\,E_{sb}}{E_c}$$

The spectrum analyzer can also display levels of modulation and distortion much lower than the oscilloscope.

FREQUENCY MODULATION

Frequency modulation or FM is somewhat different from amplitude modulation. In AM, the amplitude of the carrier is varied proportional to the amplitude of the modulating signal; however, in FM, the frequency of the carrier is varied proportional to the frequency of the modulating signal. The amplitude of the modulated signal always stays constant, regardless of modulation frequency and amplitude.

Figure 43A shows a time domain representation of a frequency modulated signal. Notice that the period, and thus the frequency, of the FM signal changes with the amplitude of the modulating signal. Side bands are developed with FM and will appear on the spectrum analyzer much as they do with AM. However, with FM, power is transferred from the carrier to the sidebands. The distribution of power depends on the modulation index (m). Figure 43B shows an FM spectrum with a modulation index of 0.2. Here, only two small sidebands are generated, and most of the power is in the carrier.

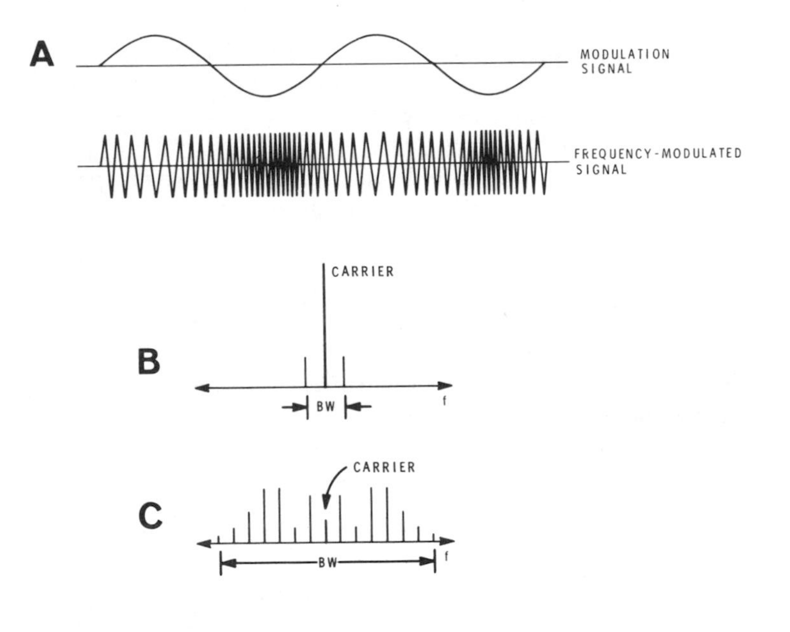

Figure 43

In Figure 43C, the modulation index is 5. Notice that the carrier power has decreased significantly, with more power shifted to the sidebands. Modulation index can be calculated by:

$$m = \frac{\Delta \text{ f peak}}{f_m}$$

where:

Δ f peak	=	peak frequency deviation of the carrier
m	=	modulation index
f_m	=	modulation frequency

In theory, FM generates an infinite number of sidebands; however, these sidebands become insignificant beyond a certain frequency range from the carrier. Any sideband with a voltage that is less than 1 percent of the unmodulated carrier can be considered insignificant. Thus, by counting the number of sidebands, we can determine the required bandwidth.

PULSE MODULATION

The original purpose of the spectrum analyzer was to view the output of a pulsed radar trasmitter. Many properties of the radar pulse, such as pulse width, bandwidth, duty cycle, peak power, and average power can be accurately measured.

In Figure 34, you saw that a square wave was formed when the odd harmonics were added in phase. The fundamental frequency determined the square wave rate, while the harmonics determine the wave shape. By selecting different harmonics and changing the phase, it is possible to describe any waveform in this manner. The spectrum analyzer takes the composite wave and breaks it down into its individual frequency components.

Keep in mind that, up to now, we have been discussing the square wave or pulse as it stands alone. However, in a practical application, the pulse is used to amplitude modulate an RF carrier.

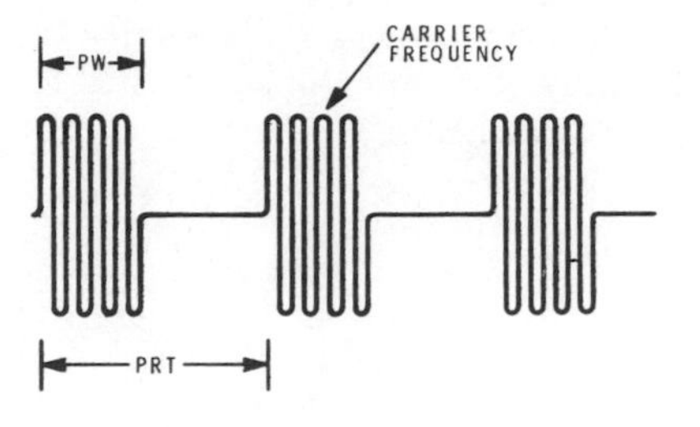

Figure 44A

Figure 44A shows a pulse modulation waveform in the time domain.

Notice that this pulse is somewhat different from the pulses you have studied in the past. This pulse consists of a short "burst" of RF followed by a rest period then another pulse. The pulse width (PW) is the period of time that the RF is present while the pulse recurrence time (PRT) is the time from the start of one pulse to the start of the next pulse. Once PRT is known, it is easy to find pulse recurrence frequency (PRF). We know that frequency is the reciprocal of time; therefore, $PRF = \frac{1}{PRT}$. Conversly, $PRT = \frac{1}{PRF}$. When using an oscilloscope, you will have to convert from PRT to PRF, and when using a spectrum analyzer, you will have to convert from PRF to PRT.

The carrier frequency (f_c) is the frequency of the RF that comprises the pulse.

You saw in Figure 42 that when a carrier is modulated, sidebands were formed above and below the carrier frequency. With pulse modulation, a large number of sidebands are produced. In fact, a pair of sidebands is produced for each harmonic contained in the modulating pulse. These sidebands are referred to as **spectral lines** on the analyzer.

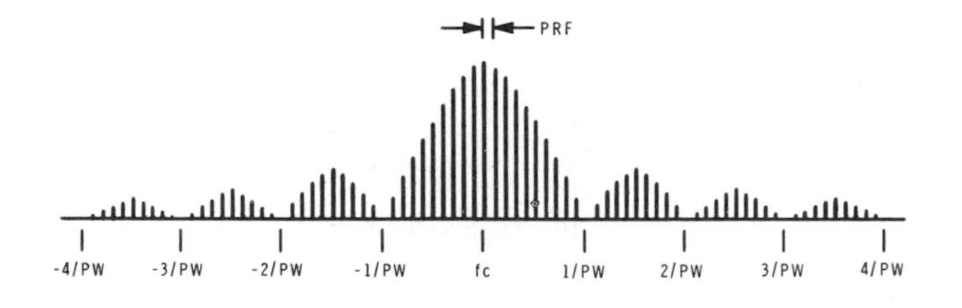

Figure 44B

Figure 44B shows how the spectral plot of a pulse might appear. The center line, labeled f_C, represents the carrier frequency. Each of the other lines represents a sideband. A sideband is the carrier frequency plus or minus the PRF, or one of the harmonics. Thus, the separation between the spectral lines is determined by the PRF. A high PRF results in widely spaced lines while a low PRF produces closely spaced lines.

The individual spectral lines draw a pattern on the CRT that is in the form of a "main lobe" in the center and "side lobes" on each side. The width of the lobes, or the point where they go through zero is determined by the pulse width.

A narrow pulse will contain higher frequency harmonics than a wide pulse. Since the main lobe contains the carrier and the upper and lower sidebands, it will be twice as wide as the side lobes, or equal to 2/pw in width. Thus a narrow pulse causes wide spectrum lobes, and a wide pulse causes narrow spectrum lobes.

As mentioned earlier, for a perfect pulse, the number of side lobes is infinite. However, the perfect pulse does not exist; therefore, the number of side lobes is limited. The number of side lobes is determined by the pulse purity or shape and is not affected by PW or PRF. Thus, with the spectrum analyzer, it is possible to determine not only PW and PRF, but pulse shape as well.

LOGIC TESTERS

Principles

Before starting the study of logic testers, we will briefly review some of the measurements that are required in digital circuits. Figure 47 shows a typical digital pulse train that might occur in a TTL circuit. If this waveform is observed on an oscilloscope, all of the characteristics, such as peak amplitude, pulse width, pulse recurrence time, etc., can be determined. However, most of this is probably unneeded information.

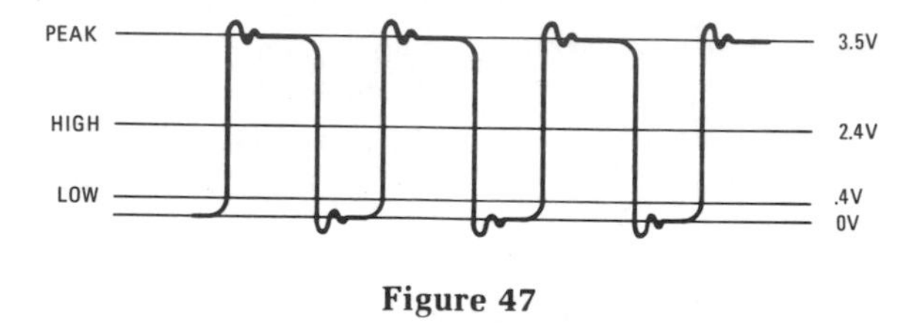

Figure 47

In TTL logic, any time the signal voltage goes "above" approximately +2.4 V, the circuits see a "high" or a logic "1". Any time the signal voltage drops below approximately +0.4 V, the circuits see a "low" or a logic "0". Therefore, all the technician needs to know is if the signal goes "high" or "low".

While the actual high or low level will vary with logic types, it is possible to build a simple tester that will work with most types of logic. Figure 48 shows such a simple logic tester. If a high is applied to the signal input, it will be inverted by A1, and a low will be felt at the input of A2. At the same time, this low will be felt at LED1, the 0 indicator, causing it to be off. The low is then inverted by A2, causing a high at LED2. Thus, LED2 conducts, indicating a 1 or a high at the signal input. A low signal in has the opposite effect, lighting LED1 and turning off LED2. Thus, a 0 is indicated.

If the input alternates between 1 and 0 more than a few times per second, the eye will be unable to follow the change, and both LED's will be lit.

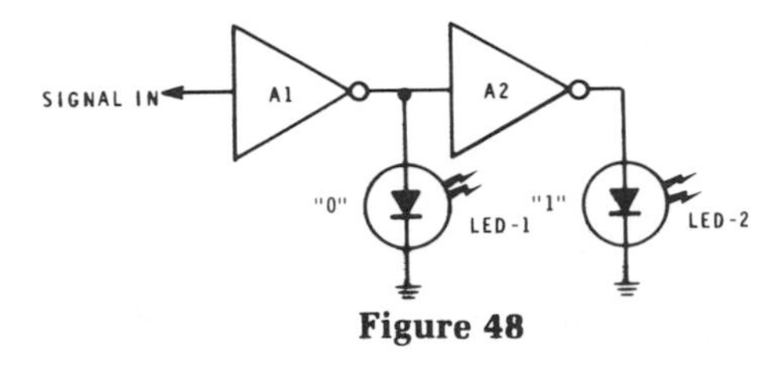

Figure 48

Another drawback of this simple circuit is in the input circuit. If the input is open, not connected to either a high or a low, the circuit will usually interpret it as a high and indicate a 1. Also, there is no way to detect a short transient pulse. Even with these limitations, many useful measurements can be made with such a simple device.

Logic Probe

A common piece of test equipment that uses the principles shown in Figure 48 with added circuits to overcome some of the drawbacks is the Logic Probe shown in Figure 49. This probe indicates whether the logic level is high or low. The probe also has a memory that indicates if there has been a change in logic level. A memory light stays on until the reset button is pressed. This allows the probe to detect and display transient pulses as short as 10 nsec in duration. Unlike the simple test circuit in Figure 48, this probe will not indicate a high when the circuit is open. Instead, it will indicate a "bad" level. A number of methods could be used to overcome the drawbacks of the simple logic probe shown in Figure 48.

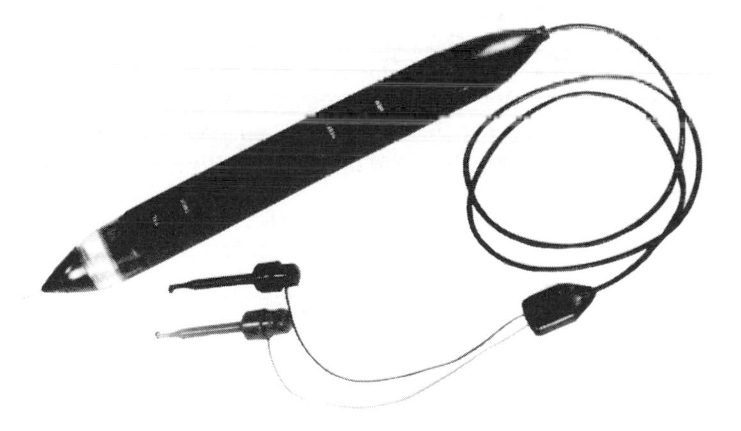

Figure 49

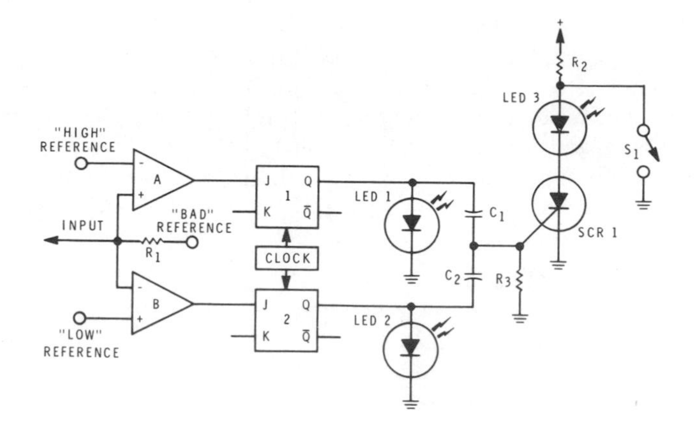

Figure 50

Perhaps, the major disadvantage is the inablility to detect a bad level of input signal. Fortunately, this is easily overcome by the addition of two comparators in the input circuit as shown in Figure 50.

Comparator **A** has the high reference voltage applied to the inverting input. For instance, if any voltage greater than 2.4 V is considered a high, then 2.4 V is used as a high reference. The output from comparator **A** will be low until the input signal exceeds the reference, at which time, the output will go high.

Comparator **B** has the low reference potential applied to the non-inverting input. For instance, if the low level is 0.5 V or less, then the output will be low for any level higher than that. Thus, the output from both comparators will be low for any input voltage between the high and low levels. If the input goes high, the output of comparator A will be high while the output of B stays low. If the signal in goes low, the output of B will be high, but A will remain low.

However, there remains one small problem. If the input is open, 0 V will be applied to both A and B. B will interpret this as a low and the output will be high. Obviously, we don't want to indicate a low with nothing in. This can be cured by connecting the input to a bad reference voltage through a very large resistance. This bad reference, a voltage somewhere between the low and the high references, will hold both outputs at a low level with the input open. The large value of R_1 will allow any input, either high or low, to overcome the bad reference and be felt on the inputs producing the appropriate output.

The outputs from the comparators are connected to the "J" inputs of a pair of J-K flip-flops. If the K input to the flip-flop is left open, it will appear as a high of all times. Thus, anytime J is high, the flip-flop will toggle with each clock pulse. Each time Q goes high, the corresponding LED will light. For instance, if the input is connected to ground, the inverting input of comparator B goes low, causing the output to go high. The high of J and K causes the flip flop to toggle with each clock pulse. LED2 will flash at the clock rate. A similar action occurs in the high channel when the input is connected to a high. Therefore, LED1 will flash for a steady high, LED2 will flash for steady low, while both LED's will flash with a pulsed input.

Transient pulses can be detected by the addition of a simple memory circuit as shown in Figure 50. The Q outputs of the flip-flop are connected to the gate of a silicon controlled rectifier. Anytime the gate goes positive, the SCR will conduct and continue to conduct until the positive voltage is removed from the anode. Thus, anytime a pulse, either high or low, is felt at the input, LED3 will light and stay lit. S_1 is a reset switch that removes the anode voltage and cuts off the SCR.

While the circuit shown here will perform as a logic probe, some additional refinements would be nice. For instance, input filtering and protection would extend the usefulness of the probe as would dual reference voltages which would allow the probe to be used on MOS and TTL circuits. Latching circuits and a reset pulse from the clock would allow shorter transients to be read, and an internal voltage regulator would allow the use of different supply voltages. Most quality probes have these refinements.

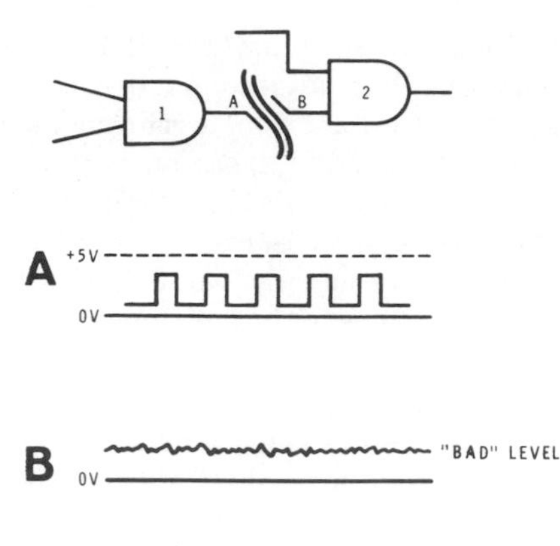

Figure 51

Using the Logic Probe

The logic probe is normally used to "signal trace" in digital circuits. For example, when troubleshooting the circuit shown in Figure 51, the pulsed waveform will be detected at point A. However, point B will float at a bad level which will cause no indication on the digital probe. TTL logic will interpret this as a high. An open of this type can be either internal or external to the chip. In either case, the logic probe is a big help in locating it. Other common failures are internal opens or shorts that cause the output to be held at either a high or a low level. These too, are easily detected by the probe.

Logic Pulser

A companion to the logic probe is the logic pulser. The pulser is nothing more than a pulse generator with relatively high current capability. Pulse current from the probe must be capable of pulsing the circuit either high or low. That is, if the test point is high, the pulser must drive it low, and if it is low, the pulser must drive it high. The pulser must do this unless the circuit is shorted to either VCC or ground.

By using these two instruments, it is possible to trace rapidly through a circuit, pulsing the circuit and observing the response. Together, the logic probe and the pulser make a valuable troubleshooting tool.